Autodesk Revit 2020 Architecture Basics

From the Ground Up

Elise Moss

SDC

PUBLICATIONS

SDC Publications
P.O. Box 1334
Mission, KS 66222
913-262-2664
www.SDCpublications.com
Publisher: Stephen Schroff

Examination Copies
Books received as examination copies are for review purposes only and may not be made available for student use. Resale of examination copies is prohibited.

Electronic Files
Any electronic files associated with this book are licensed to the original user only. These files may not be transferred to any other party.

Trademarks
The following are registered trademarks of Autodesk, Inc.: AutoCAD, AutoCAD Architecture, Revit, Autodesk, AutoCAD Design Center, Autodesk Device Interface, VizRender, and HEIDI.
Microsoft, Windows, Word, and Excel are either registered trademarks or trademarks of Microsoft Corporation.
All other trademarks are trademarks of their respective holders.

The authors and publisher of this book have used their best efforts in preparing this book. These efforts include the development, research, and testing of material presented. The author and publisher shall not be held liable in any event for incidental or consequential damages with, or arising out of, the furnishing, performance, or use of the material herein.

ISBN-13: 978-1-63057-263-1
ISBN-10: 1-63057-263-2

Printed and bound in the United States of America.

Preface

Revit is a parametric 3D modeling software used primarily for architectural work. Traditionally, architects have been very happy working in 2D, first on paper, and then in 2D CAD, usually in AutoCAD.

The advantages of working in 3D are not initially apparent to most architectural users. The benefits come when you start creating your documentation and you realize that your views are automatically defined for you with your 3D model. Your schedules and views automatically update when you change features. You can explore your conceptual designs faster and in more depth.

Revit will not make you a better architect. However, it will allow you to communicate your ideas and designs faster, easier, and more beautifully. I wrote the first edition of this text more than ten years ago. Since that first edition, Revit has become the primary 3D CAD software used in the AEC industry. Revit knowledge has become a valuable skill in today's market and will continue to be in demand for at least another decade.

The book is geared towards users who have no experience in 3D modeling and very little or no experience with AutoCAD. Some experience with a computer and using the Internet is assumed.

I have endeavored to make this text as easy to understand and as error-free as possible…however, errors may be present. Please feel free to email me if you have any problems with any of the exercises or questions about Revit in general.

Acknowledgements

A special thanks to Gerry Ramsey, Scott Davis, James Balding, Rob Starz, and all the other Revit users out there who provided me with valuable insights into the way they use Revit.

Thanks to Stephen Schroff, Zach Werner and Karla Werner who work tirelessly to bring these manuscripts to you, the user, and provide the important moral support authors need.

My eternal gratitude to my life partner, Ari, my biggest cheerleader throughout our years together.

Elise Moss
Elise_moss@mossdesigns.com

TABLE OF CONTENTS

Class Files

To download the files that are required for this book, type the following in the address bar of your web browser:

SDCpublications.com/downloads/978-1-63057-263-1

Notes:

Lesson 1
The Revit Interface

Go to Start→Programs→Autodesk → Revit 2020.

You can also type Revit in the search field under Programs and it will be listed.

When you first start Revit, you will see this screen:

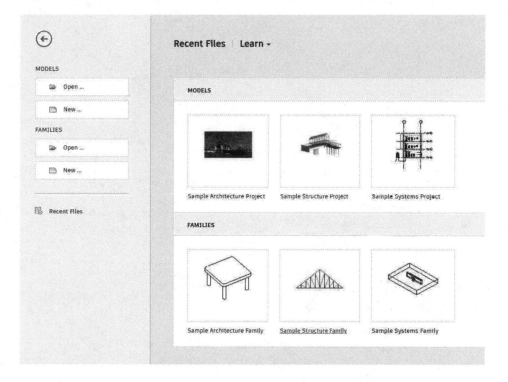

It is divided into three sections:

The Left section has two panels. The top panel is to Open or Start a new project. Revit calls the 3D building model a project. Some students find this confusing.

The bottom panel in the left section is used to open, create, or manage Revit families. Revit buildings are created using Revit families. Doors, windows, walls, floors, etc., are all families.

Recent Files show recent files which have been opened or modified as well as sample files.

There is a tab for Learning.

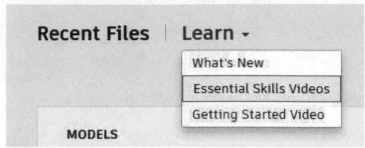

The Learn flyout has three sections.

What's New launches a browser tab listing the new features in this release.

Essential Skills Videos launches a browser tab listing several on-line videos new users can watch and follow along to learn the software.

You should be able to identify the different areas of the user interface in order to easily navigate around the software.

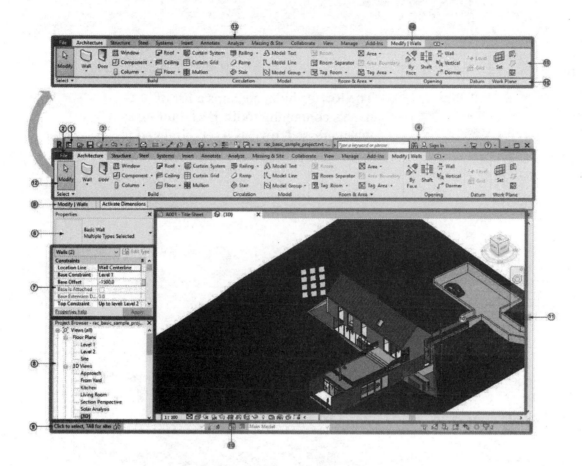

1	Revit Home	9	Status Bar
2	File tab	10	View Control Bar
3	Quick Access Toolbar	11	Drawing Area/Window
4	Info Center	12	Ribbon
5	Options Bar	13	Tabs on the ribbon
6	Type Selector	14	Contextual Ribbon – appears based on current selection
7	Properties Palette	15	Tools on the current tab of the ribbon
8	Project Browser	16	Ribbon Panels – used to organize Ribbon Tools

The Revit Ribbon

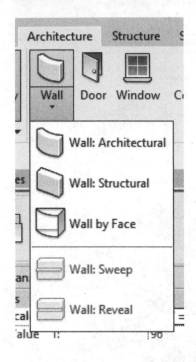

The Revit ribbon contains a list of panels containing tools. Each button, when pressed, reveals a set of related commands.

The Quick Access Toolbar (QAT)

Most Windows users are familiar with the standard tools: New, Open, Save, Undo, and Redo.

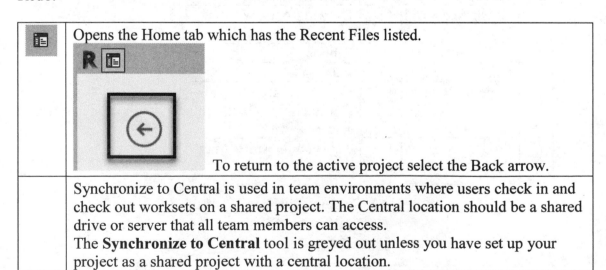

	Opens the Home tab which has the Recent Files listed.
	To return to the active project select the Back arrow.
	Synchronize to Central is used in team environments where users check in and check out worksets on a shared project. The Central location should be a shared drive or server that all team members can access.
	The **Synchronize to Central** tool is greyed out unless you have set up your project as a shared project with a central location.

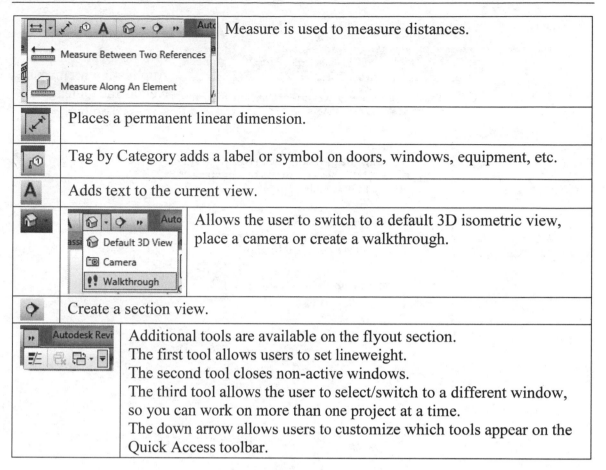	Measure is used to measure distances.
	Places a permanent linear dimension.
	Tag by Category adds a label or symbol on doors, windows, equipment, etc.
	Adds text to the current view.
	Allows the user to switch to a default 3D isometric view, place a camera or create a walkthrough.
	Create a section view.
	Additional tools are available on the flyout section. The first tool allows users to set lineweight. The second tool closes non-active windows. The third tool allows the user to select/switch to a different window, so you can work on more than one project at a time. The down arrow allows users to customize which tools appear on the Quick Access toolbar.

Printing

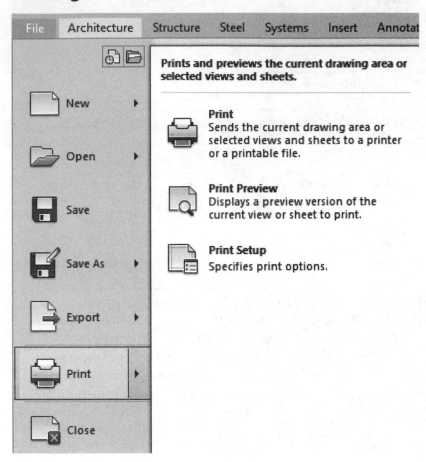

Print is located in the Application Menu as well as the Quick Access Toolbar.

The Print dialog is fairly straightforward.
Select the desired printer from the drop-down list, which shows installed printers.

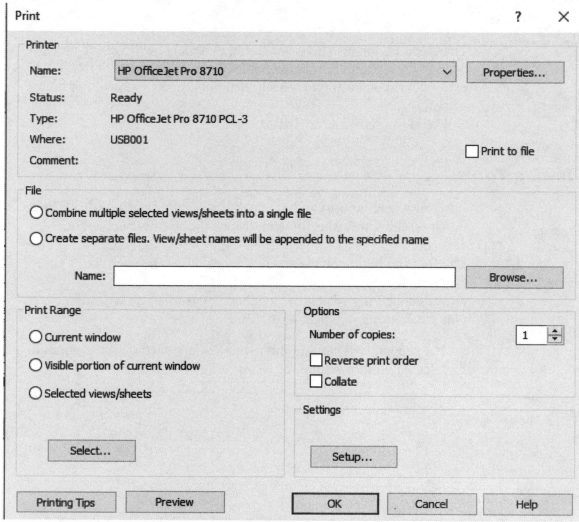

You can set to 'Print to File' by enabling the check box next to Print to File.
The Print Range area of the dialog allows you to print the current window, a zoomed in portion of the window, and selected views/sheets.

Undo

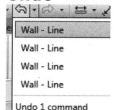

The Undo tool allows the user to select multiple actions to undo.
To do this, use the drop down arrow next to the Undo button; you can select which recent action you want to undo. You cannot skip over actions (for example, you cannot undo 'Note' without undoing the two walls on top of it).
Ctrl-Z also acts as UNDO.

Redo

The Redo button also gives you a list of actions, which have recently been undone. Redo is only available immediately after an UNDO. For example, if you perform UNDO, then WALL, REDO will not be active.
Ctrl-Y is the shortcut for REDO.

Viewing Tools

A scroll wheel mouse can replace the use of the steering wheel. Press down on the scroll wheel to pan. Rotate the scroll wheel to zoom in and out.
The Rewind button on the steering wheel takes the user back to the previous view.
Different steering wheels are available depending on whether or not you are working in a Plan or 3D view.

The second tool has a flyout menu that allows the user to zoom to a selected window/region or zoom to fit (extents).

✓ Zoom in Region
Zoom Out(2x)
Zoom to Fit
Zoom All to Fit
Zoom Sheet Size
Previous Pan/Zoom
Next Pan/Zoom

➢ Orient to a view allows the user to render an elevation straight on without perspective. Orient to a plane allows the user to create sweeps along non-orthogonal paths.

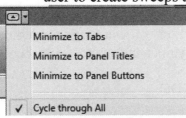

Minimize to Tabs

Minimize to Panel Titles

Minimize to Panel Buttons

✓ Cycle through All

➢ If you right click on the Revit Ribbon, you can minimize the ribbon to gain a larger display window.

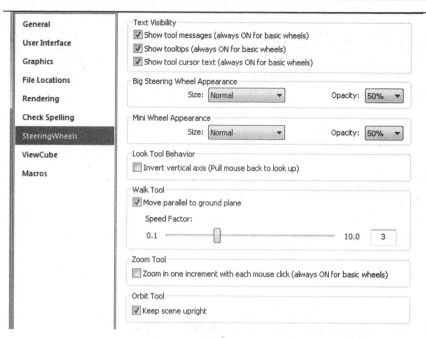

You can control the appearance of the steering wheels by right clicking on the steering wheel and selecting Options.

Some users have a 3D Connexion device – this is a mouse that is used by the left hand to zoom/pan/orbit while the right hand selects and edits.

Revit 2020 will detect if a 3D Connexion device is installed and add an interface to support the device.

It takes some practice to get used to using a 3D mouse, but it does boost your speed.

For more information on the 3D mouse, go to 3dconnexion.com.

Exercise 1-1:
Using the Steering Wheel & ViewCube

Drawing Name: *basic_project.rvt*
Estimated Time: 30 minutes

This exercise reinforces the following skills:

- ❑ ViewCube
- ❑ 2D Steering Wheel
- ❑ 3D Steering Wheel
- ❑ Project Browser
- ❑ Shortcut Menus
- ❑ Mouse

1.    Select Home from the QAT or press **Ctl+D**.

2. The Home screen launches.

Sample Architecture Project

3. Select the Sample Architectural Project file. The file name is
    rac_basic_sample_project.rvt.

    File name: basic_project

    If you do not see the Basic Sample Project listed, you can use the file basic_project included with the Class Files available for download from the publisher's website. To download the Class Files, type the following in the address bar of your web browser:
    SDCpublications.com/downloads/978-1-63057-263-1

4.
    Site
    3D Views
    — Approach
    — From Yard
    — Kitchen
    — Living Room
    — Section Perspective
    — Solar Analysis
    — {3D}

    In the Project Browser:
    Expand the 3D Views.
    Double left click on the {3D} view.
    This activates the view.
    The active view is already in BOLD.

5.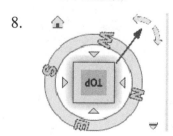

    If you have a mouse with a scroll wheel, experiment with the following motions:
    If you roll the wheel up and down, you can zoom in and out.
    Hold down the SHIFT key and press down the scroll wheel at the same time. This will rotate or orbit the model.
    Release the SHIFT key. Press down the scroll wheel. This will pan the model.

6. When you are in a 3D view, a tool called the ViewCube is visible in the upper right corner of the screen.

7. Click on the top of the cube as shown.

8. The display changes to a plan view.
    Use the rotate arrows to rotate the view.

9.  Click on the little house (home/default view tool) to return to the default view.

The little house disappears and reappears when your mouse approaches the cube.

10.  Select the Steering Wheel tool located on the View Control toolbar.

11. A steering wheel pops up.

 Notice that as you mouse over sections of the steering wheel they highlight.
Mouse over the Zoom section and hold down the left mouse button. The display should zoom in and out.
Mouse over the Orbit section and hold down the left mouse button. The display should orbit.
Mouse over the Pan section and hold down the left mouse button. The display should pan.

12.

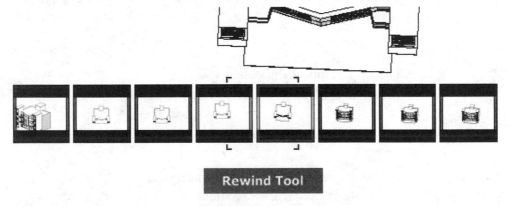

Select the Rewind tool and hold down the left mouse button.
A selection of previous views is displayed.
You no longer have to back through previous views. You can skip to the previous view you want.
Select a previous view to activate.
Close the steering wheel by selecting the X in the top right corner.

13.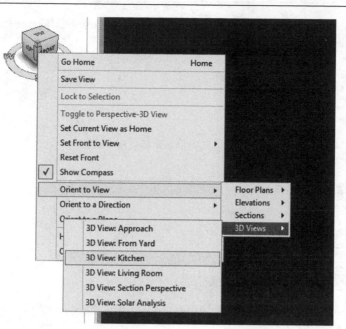

Place the mouse cursor over the View Cube. Right click and a shortcut menu appears.
Select **Orient to View→ 3d Views→3D View: Kitchen**.

14.

Double click on **Level 1** in the Project Browser. This will open the Level 1 floor plan view.

Be careful not to select the ceiling plan instead of the floor plan.

15.

This is an elevation marker. It defines an elevation view.

16.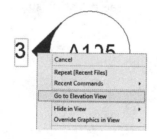

If you right click on the triangle portion of the elevation view, you can select **Go to Elevation View** to see the view associated with the elevation marker.

17. 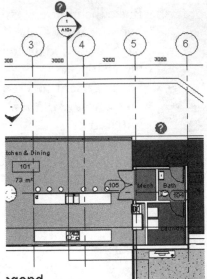 Locate the section line on Level 1 labeled A104.

18. The question mark symbols in the drawing link to help html pages. Click on one of the question marks.

19. 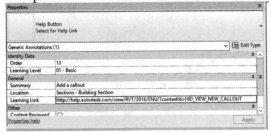 Look in the Properties palette.
In the Learning Links field, a link to a webpage is shown. Click in that field to launch the webpage.

20. 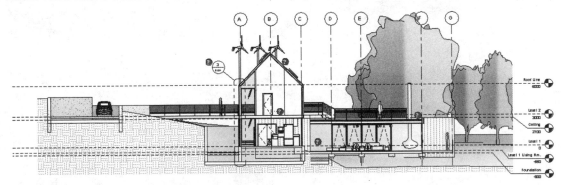 A browser will open to the page pertaining to the question mark symbol.

21. Double left click on the arrow portion of the section line.
22. This view has a callout.

Double left click on the callout bubble.

The bubble is located on the left.

23.

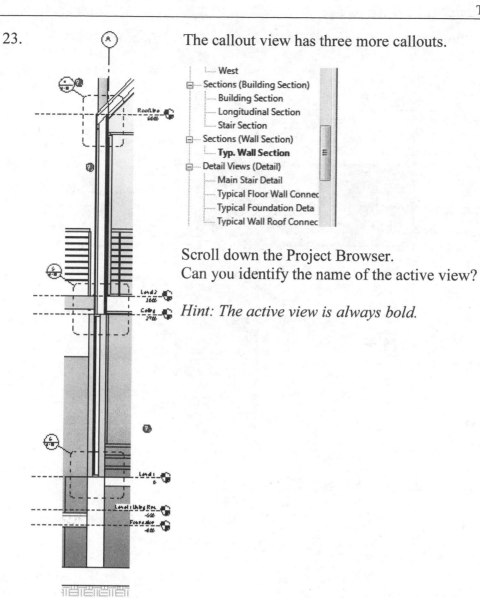

The callout view has three more callouts.

> ····· West
> ⊟···· Sections (Building Section)
> ····· Building Section
> ····· Longitudinal Section
> ····· Stair Section
> ⊟···· Sections (Wall Section)
> ····· **Typ. Wall Section**
> ⊟···· Detail Views (Detail)
> ····· Main Stair Detail
> ····· Typical Floor Wall Connec
> ····· Typical Foundation Deta
> ····· Typical Wall Roof Connec

Scroll down the Project Browser.
Can you identify the name of the active view?

Hint: The active view is always bold.

24.

To the right of the view, there are levels.
Some of the levels are blue and some of
them are black.
The blue levels are story levels; they have
views associated with them.
The black levels are reference levels; they
do not have views associated with them.
Double left click on the Level 1 bubble.

Level 1
0

Level 1 Living Rm.
-550

Foundation
-800

25. The Level 1 floor plan is opened. *We are back where we started!*

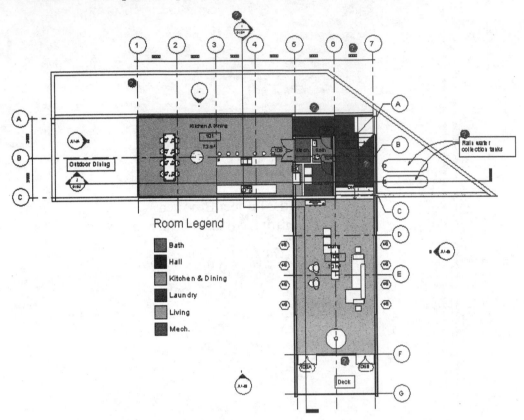

Level 1 is also in BOLD in the project browser.

26.

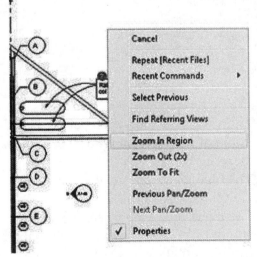

Right click in the window.
On the shortcut menu:
Select **Zoom In Region**.

This is the same as the Zoom Window tool in AutoCAD.

27.
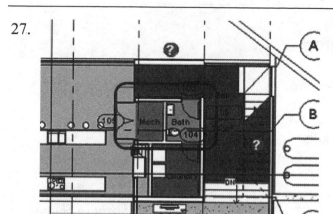
Place a rectangle with your mouse to zoom into the bathroom area.

28.

Right click in the window.
On the shortcut menu:
Select **Zoom To Fit**.
This is the same as the Zoom Extents tool in AutoCAD.
You can also double click on the scroll wheel to zoom to fit.

29. Close the file without saving.
Close by pressing **Ctl+W** or using **File→Close**.

Exercise 1-2:
Changing the View Background

Drawing Name: *basic_project.rvt*
Estimated Time: 5 minutes

This exercise reinforces the following skills:
- ❑ Graphics mode
- ❑ Options

Many of my students come from an AutoCAD background and they want to change their background display to black.

1.   Go to **File→Open→Project**.

2.   Locate the file called *basic_project.rvt*.
       This is included with the Class Files you downloaded from the publisher's website.

3.   Select the drop-down on the Application Menu.
       Select **Options**.

4.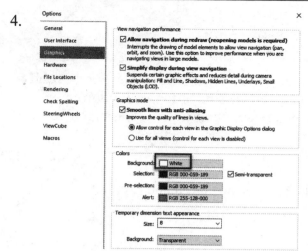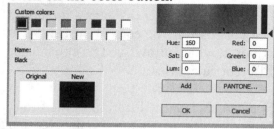

Select **Graphics**.
Locate the color tab next to Background.
Click on the color button.

Select the Black color.
Press **OK** twice.

5. Close the file without saving.

Revit's Project Browser

The Project Browser displays a hierarchy for all views, schedules, sheets, families and elements used in the current project. It has an interface similar to the Windows file explorer, which allows you to expand and collapse each branch as well as organize the different categories.

You can change the location of the Project Browser by dragging the title bar to the desired location. Changes to the size and the location are saved and restored when Revit is launched.

You can search for entries in the Project Browser by right-clicking inside the browser and selecting Search from the right-click menu to open a dialog.

The Project Browser is used to navigate around your project by opening different views. You can also drag and drop views onto sheets or families into views.

Revit's Properties Palette

The Properties Palette is a contextually based dialog. If nothing is selected, it will display the properties of the active view. If something is selected, the properties of the selected item(s) will be displayed.

By default, the Properties Palette is docked on the left side of the drawing area.

The Type Selector is a tool that identifies the selected family and provides a drop-down from which you can select a different type.

To make the Type Selector available when the Properties palette is closed, right-click within the Type Selector, and click Add to Quick Access Toolbar. To make the Type Selector available on the Modify tab, right-click within the Properties palette, and click Add to Ribbon Modify Tab. Each time you select an element, it will be reflected on the Modify tab.

Immediately below the Type Selector is the Properties filter; this may be used when more than one element is selected. When different types of elements are selected, such as walls and doors, only the instance properties common to the elements selected will display on the palette.

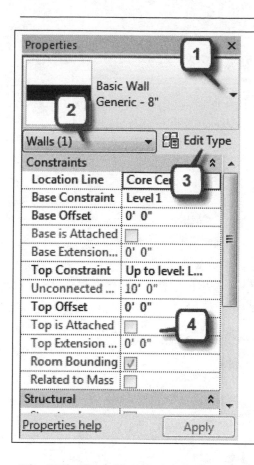

1. Type Selector

2. Properties filter

3. Edit Type button

4. Instance properties

The Edit Type button is used to modify or duplicate the element family type. Revit uses three different family classes to create a building model: system, loadable, and in-place masses. System families are walls, floors, roofs, and ceilings. Loadable families are doors, windows, columns, and furniture. In-place masses are basic shapes which can be used to define volumetric objects and are project-specific.

When you place an element into a building project, such as a wall, it has two types of properties: Type Properties and Instance Properties. The Type Properties are the structural materials defined for that wall and they do not change regardless of where the wall is placed. The Instance Properties are unique to that element, such as the length of the wall, the volume (which changes depending on how long the wall is), the location of the wall, and the wall height.

Exercise 1-3:
Closing and Opening the Project Browser and Properties Palette

Drawing Name: *basic_project.rvt*
Estimated Time: 5 minutes

This exercise reinforces the following skills:
- ❑ User Interface
- ❑ Ribbon
- ❑ Project Browser
- ❑ Properties panel

Many of my students will accidentally close the project browser and/or the properties palette and want to bring them back. Other students prefer to keep these items closed most of the time so they have a larger working area.

1. 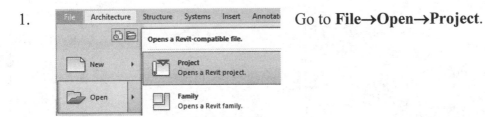 Go to **File→Open→Project**.

2. File name: basic_project   Locate the file called *basic_project.rvt*.
    This is included with the Class Files you downloaded from the publisher's website.

3. Properties   Close the Properties palette by clicking on the x in the corner.

4. basic_project - Project Browser   Close the Project Browser by clicking on the x in the corner.

5. 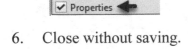 Activate the View ribbon.
    Go to the User Interface dropdown at the far right.
    Place a check next to the Project Browser and Properties to make them visible.

6. Close without saving.

Revit's System Browser

The System Browser is useful in complex projects where you want to coordinate different building components, such as an HVAC system, structural framing, or plumbing.

The System Browser opens a separate window that displays a hierarchical list of all the components in each discipline in a project, either by the system type or by ones.

Customizing the View of the System Browser

The options in the View bar allow you to sort and customize the display of systems in the System Browser.

- **Systems**: displays components by major and minor systems created for each discipline.

- **Zones**: displays zones and spaces. Expand each zone to display the spaces assigned to the zone.

- **All Disciplines**: displays components in separate folders for each discipline (mechanical, piping, and electrical). Piping includes plumbing and fire protection.

- **Mechanical**: displays only components for the Mechanical discipline.

- **Piping**: displays only components for the Piping disciplines (Piping, Plumbing, and Fire Protection).

- **Electrical**: displays only components for the Electrical discipline.

- **AutoFit All Columns**: adjusts the width of all columns to fit the text in the headings.

  Note: You can also double-click a column heading to automatically adjust the width of a column.

- **Column Settings**: opens the Column Settings dialog where you specify the columnar information displayed for each discipline. Expand individual categories (General, Mechanical, Piping, Electrical) as desired, and select the properties that you want to appear as column headings. You can also select columns, and click Hide or Show to select column headings that display in the table.

Exercise 1-4:
Using the System Browser

Drawing Name: *basic_project.rvt*
Estimated Time: 10 minutes

This exercise reinforces the following skills:
- ❑ User Interface
- ❑ Ribbon
- ❑ System Browser

This browser allows the user to locate and identify different elements used in lighting, mechanical, and plumbing inside the project.

1. 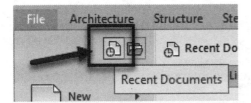 Go to **File**.

   Select the **Recent Documents** icon.

2. 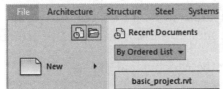 Locate the file called *basic_project.rvt.* *This is included with the Class Files you downloaded from the publisher's website.*

3. Activate the **View** ribbon.

4. Go to the User Interface drop-down list.
   Place a check next to **System Browser**.

5. 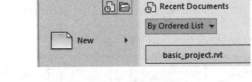 Expand the **Unassigned** category.
   Expand the **Piping** category.
   Expand the **Domestic-Cold Water**.
   Locate the **Lavatory–TOTO** in the Master Bath.

6. 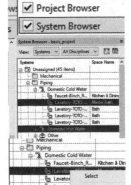 Right click on the **Lavatory –TOTO** in the Master Bath.
   Select **Show**.

7. Press **OK**.

8. Select the **Show** button.

9.   Press the **Show** button until you see this view.

Press **Close**.

10.   Note that the active view is the **Level 2** floor plan. *This is bold in the Project Browser.*

11.   Zoom out to see the Master Bath.

To close the System Browser, select the x button in the upper right corner of the dialog.

12.   Close the file without saving.

Revit's Ribbon

The ribbon displays when you create or open a project file. It provides all the tools necessary to create a project or family.

An arrow next to a panel title indicates that you can expand the panel to display related tools and controls.

By default, an expanded panel closes automatically when you click outside the panel. To keep a panel expanded while its ribbon tab is displayed, click the push pin icon in the bottom-left corner of the expanded panel.

Exercise 1-5:
Changing the Ribbon Display

Drawing Name: *basic_project.rvt*
Estimated Time: 15 minutes

This exercise reinforces the following skills:
- User Interface
- Ribbon

Many of my students will accidentally collapse the ribbon and want to bring it back. Other students prefer to keep the ribbon collapsed most of the time so they have a larger working area.

1. 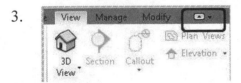   Select **basic_project** from the Recent Files window.

2.    Activate **Level 1** in the Project Browser.

3. 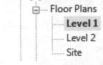   On the ribbon: Locate the two small up and down arrows.

4.   Left click on the white button.

5.   The ribbon collapses to tabs.

6. 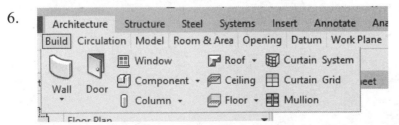 Left click on the word **Architecture**.
Hover the mouse over the word **Build** and the Build tools panel will appear.
The build tools will be grayed out in a 3D view.

7.    Click on the white button until the ribbon is restored.

8.    The ribbon is full-size again.

9.    Click on the black arrow.

10. 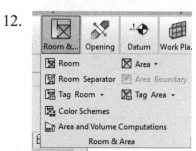   Click on **Minimize to Panel Buttons**.

11.   The ribbon changes to panel buttons.

12.   Left click on a panel button and the tools for the button will display.
      Some buttons are grayed out depending on what view is active.

13.   Click on the white button to display the full ribbon.

14. Close without saving.

The Modify Ribbon

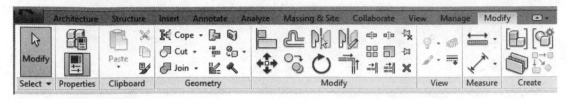

When you select an entity, you automatically switch to Modify mode. A Modify ribbon will appear with different options.

Select a wall in the view. Note how the ribbon changes.

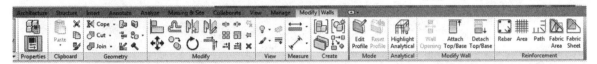

Revit uses three types of dimensions: Listening, Temporary, and Permanent. Dimensions are placed using the relative positions of elements to each other. Permanent dimensions are driven by the value of the temporary dimension. Listening dimensions are displayed as you draw or place an element.

Exercise 1-6
Temporary, Permanent and Listening Dimensions

Drawing Name: dimensions.rvt
Estimated Time: 30 minutes

This exercise reinforces the following skills:
- ☐ Project Browser
- ☐ Scale
- ☐ Graphical Scale
- ☐ Numerical Scale
- ☐ Temporary Dimensions
- ☐ Permanent Dimensions
- ☐ Listening Dimensions
- ☐ Type Selector

1. Browse to *dimensions.rvt* in the Class Files you downloaded from the publisher's website.
    Save the file to a folder.
    Open the file.

    The file has four walls.
    The horizontal walls are 80′ in length.
    The vertical walls are 52′ in length.

    We want to change the vertical walls so that they are 60′ in length.

2. 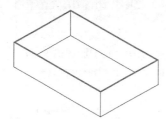 Select the House icon on the Quick Access toolbar.

3. *This switches the display to a default 3D view.*

   You see that the displayed elements are walls.

4. 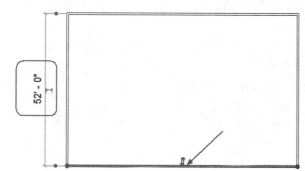 Double left click on Level 1 under Floor Plans in the Project Browser.

   The view display changes to the Level 1 floor plan.

5. Select the bottom horizontal wall.

   A temporary dimension appears showing the vertical distance between the selected wall and the wall above it.

6. A small dimension icon is next to the temporary dimension.

   Left click on this icon to place a permanent dimension.

   Left click anywhere in the window to release the selection.

7. Select the permanent dimension extension line, not the text.

   A lock appears. If you left click on the lock, it would prevent you from modifying the dimension.

   If you select the permanent dimension, you cannot edit the dimension value, only the dimension style.

8. 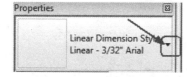 The Properties palette shows the dimension style for the dimension selected.

   Select the small down arrow.
   This is the Type Selector.

The Type Selector shows the dimension styles available in the project.
If you do not see the drop-down arrow on the right, expand the properties palette and it should become visible.

9.

Left click on the Linear - 1/4″ Arial dimension style to select this dimension style.

Left click anywhere in the drawing area to release the selection and change the dimension style.

10. The dimension updates to the new dimension style.

11. Select the right vertical wall.

12. A temporary dimension appears showing the distance between the selected wall and the closest vertical wall.

Left click on the permanent dimension icon to place a permanent dimension.

13. A permanent dimension is placed.

Select the horizontal permanent dimension extension line, not the text.

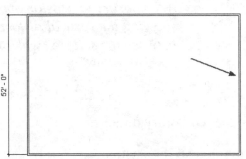

14.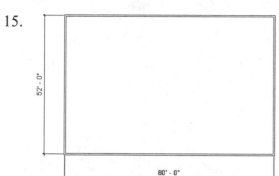

Use the Type Selector to change the dimension style of the horizontal dimension to the Linear - 1/4″ Arial dimension style.

Left click anywhere in the drawing area to release the selection and change the dimension style.

15.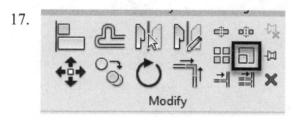

You have placed two *permanent* dimensions and assigned them a new dimension type.

16. Hold down the Control key and select both **horizontal** (top and bottom side) walls so that they highlight.

    NOTE: *We select the horizontal walls to change the vertical wall length, and we select the vertical walls to change the horizontal wall length.*

17.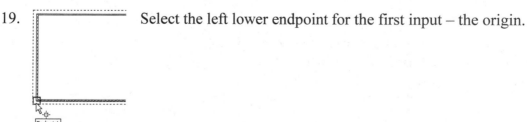

Select the **Scale** tool located on the Modify Walls ribbon.

In the Options bar, you may select Graphical or Numerical methods for resizing the selected objects.

18. | Modify | Walls | ◉ Graphical ○ Numerical Scale: 2 |

    Enable **Graphical**.

    The Graphical option requires three inputs.

    Input 1: Origin or Base Point
    Input 2: Original or Reference Length
    Input 3: Desired Length

19.

    Select the left lower endpoint for the first input – the origin.

20.
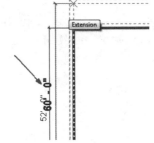
Select the left upper endpoint for the second input – the reference length.

21.
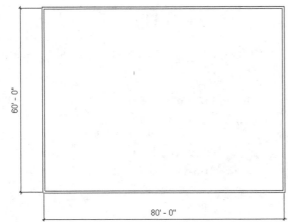
Extend your cursor until you see a dimension of 60'.

The dimension you see as you move the cursor is a *listening dimension.*

You can also type 60' and press ENTER.

22.
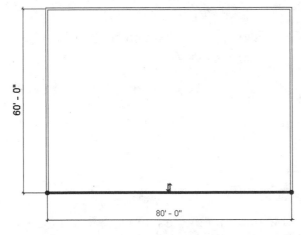
Left click for the third input – the desired length.
Left click anywhere in the window to release the selection and exit the scale command.
Note that the permanent dimension updates.

23.

Select the bottom horizontal wall to display the temporary dimension.

24.

Left click on the vertical temporary dimension.
An edit box will open. Type **50 0**.
Revit does not require you to enter units.
Left click to enter and release the selection.

25.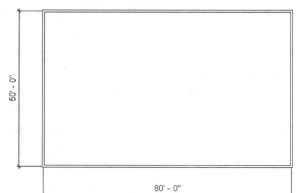

The permanent dimension updates.

Note if you select the permanent dimension, you cannot edit the dimension value, only the dimension style.

26. Now, we will change the horizontal walls using the Numerical option of the Resize tool.
Hold down the Control key and select the left and right vertical walls.

27.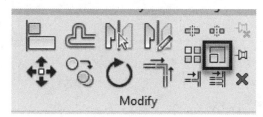

Select the **Scale** tool.

28. | Modify | Walls | ○ Graphical ● Numerical Scale: 0.5 |

Enable **Numerical**.
Set the Scale to **0.5**.
This will change the wall length from **80'** to **40'**.
The Numerical option requires only one input for the Origin or Base Point.

29.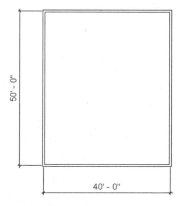

Select the left lower endpoint for the first input – the origin.

30. The walls immediately adjust. Left click in the drawing area to release the selection.
Note that the permanent dimension automatically updates.

31. Close without saving.

The Group tool works in a similar way as the AutoCAD GROUP. Basically, you are creating a selection set of similar or dissimilar objects, which can then be selected as a single unit. Once selected, you can move, copy, rotate, mirror, or delete them. Once you create a Group, you can add or remove members of that group. Existing groups are listed in your browser panel and can be dragged and dropped into views as required. Typical details, office layouts, bathroom layouts, etc. can be grouped, and then saved out of the project for use on other projects.

Tips & Tricks: If you hold down the Control key as you drag a selected object or group, Revit will automatically create a copy of the selected object or group.

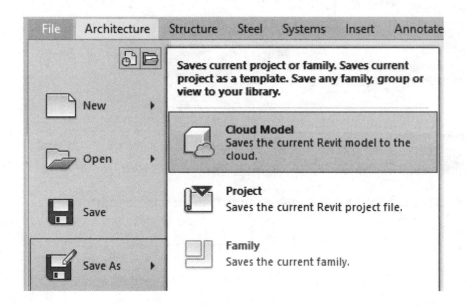

Revit has an option to save projects to the Cloud. In order to save to the cloud, the user needs to have a subscription for their software.

This option allows you to store your projects to a secure Autodesk server, so you can access your files regardless of your location. Users can share their project with each other and even view and comment on projects without using Revit. Instead you use a browser interface.

Autodesk hopes that if you like this interface you will opt for their more expensive option – BIM360, which is used by many AEC firms to collaborate on large projects.

The Collaborate Ribbon

Collaborate tools are used if more than one user is working on the same project. The tools can also be used to check data sets coming in from other sources.

➢ If you plan to export to .dwg or import existing .dwg details, it is important to set your import/export setting. This allows you to control the layers that Revit will export to and the appearance of imported .dwg files.

➢ Pressing the ESC Key twice will always take you to the Modify command.

➢ As you get more proficient with Revit, you may want to create your own templates based on your favorite settings. A custom template based on your office standards (including line styles, rendering materials, common details, door, window and room schedules, commonly used wall types, door types and window types and typical sheets) will increase project production.

Revit's Menu

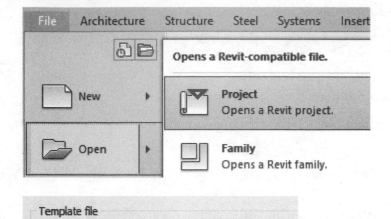

When you want to start a new project/building, you go to File→New→Project or use the hot key by pressing Control and 'N' at the same time.

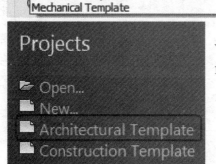

When you start a new project, you use a default template (default.rte). This template creates two Levels (default floor heights) and sets the View Window to the Floor Plan view of Level 1.

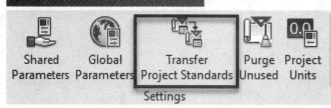

You can also select the desired template to start a new project from the *Recent Files* window.

You can transfer the Project settings of an old project to a new one by opening both projects in one session of Revit, then with your new project active, select **Manage→Settings→Transfer Project Standards**.
Check the items you wish to transfer, and then click 'OK'.

Transfer Project Settings is useful if you have created a custom system family, such as a wall, floor, or ceiling, and want to use it in a different project.

The View Ribbon

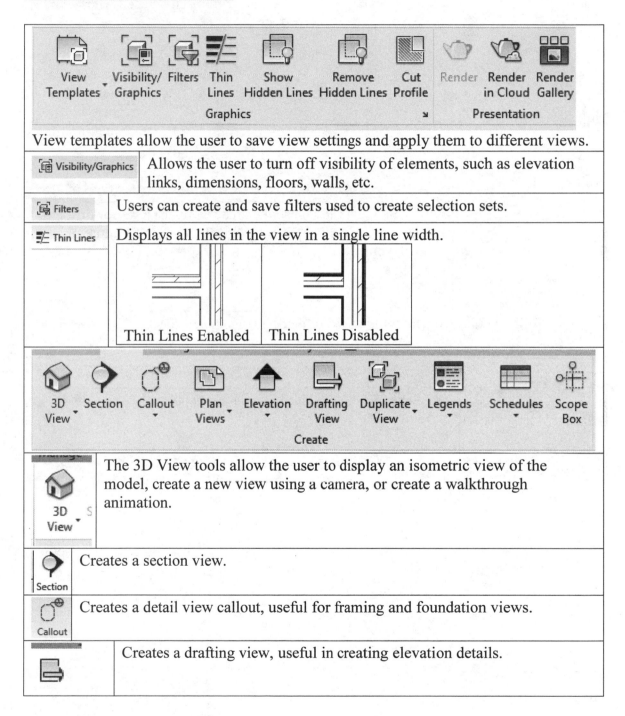

View templates allow the user to save view settings and apply them to different views.

Visibility/Graphics	Allows the user to turn off visibility of elements, such as elevation links, dimensions, floors, walls, etc.
Filters	Users can create and save filters used to create selection sets.
Thin Lines	Displays all lines in the view in a single line width.

Thin Lines Enabled · Thin Lines Disabled

3D View	The 3D View tools allow the user to display an isometric view of the model, create a new view using a camera, or create a walkthrough animation.
Section	Creates a section view.
Callout	Creates a detail view callout, useful for framing and foundation views.
	Creates a drafting view, useful in creating elevation details.

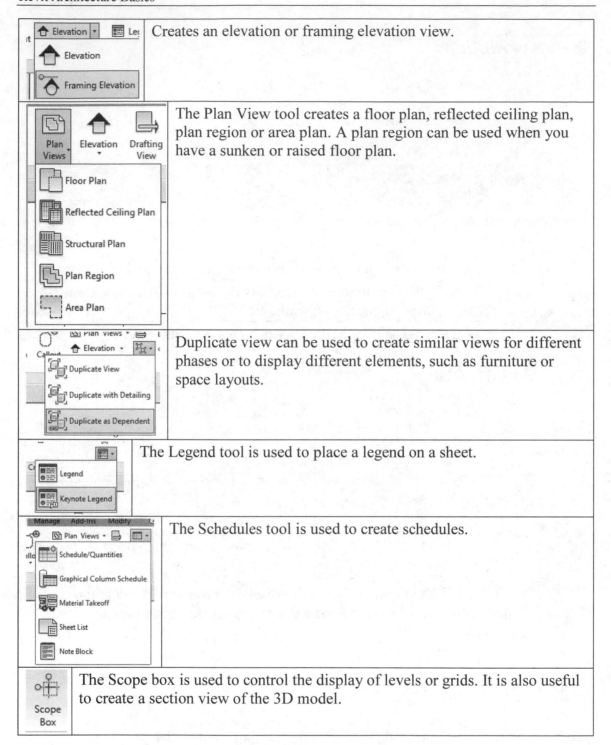

	Creates an elevation or framing elevation view.
	The Plan View tool creates a floor plan, reflected ceiling plan, plan region or area plan. A plan region can be used when you have a sunken or raised floor plan.
	Duplicate view can be used to create similar views for different phases or to display different elements, such as furniture or space layouts.
	The Legend tool is used to place a legend on a sheet.
	The Schedules tool is used to create schedules.
	The Scope box is used to control the display of levels or grids. It is also useful to create a section view of the 3D model.

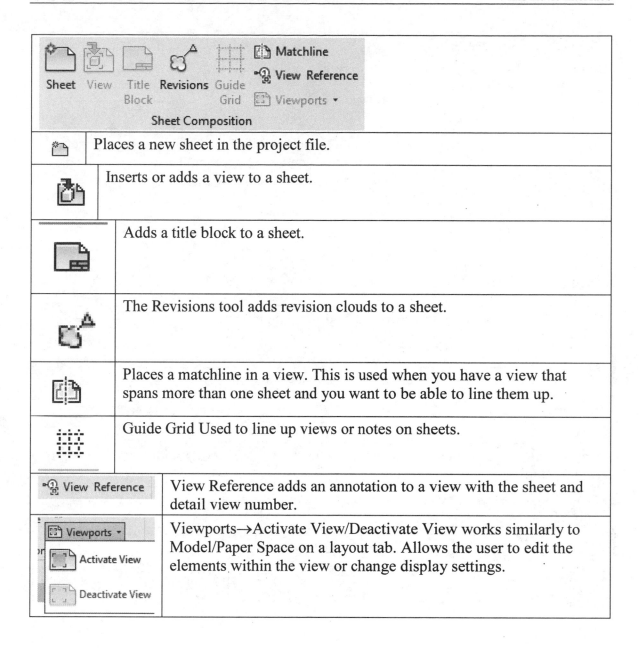

	Places a new sheet in the project file.
	Inserts or adds a view to a sheet.
	Adds a title block to a sheet.
	The Revisions tool adds revision clouds to a sheet.
	Places a matchline in a view. This is used when you have a view that spans more than one sheet and you want to be able to line them up.
	Guide Grid Used to line up views or notes on sheets.
View Reference	View Reference adds an annotation to a view with the sheet and detail view number.
Viewports · / Activate View / Deactivate View	Viewports→Activate View/Deactivate View works similarly to Model/Paper Space on a layout tab. Allows the user to edit the elements within the view or change display settings.

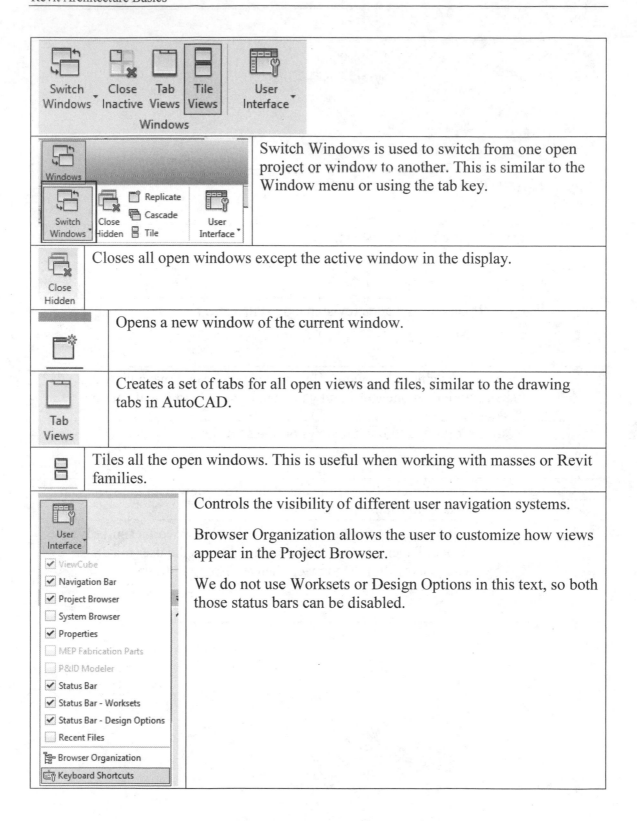

	Switch Windows is used to switch from one open project or window to another. This is similar to the Window menu or using the tab key.
	Closes all open windows except the active window in the display.
	Opens a new window of the current window.
	Creates a set of tabs for all open views and files, similar to the drawing tabs in AutoCAD.
	Tiles all the open windows. This is useful when working with masses or Revit families.
	Controls the visibility of different user navigation systems. Browser Organization allows the user to customize how views appear in the Project Browser. We do not use Worksets or Design Options in this text, so both those status bars can be disabled.

Creating standard view templates (exterior elevations, enlarged floor plans, etc.) in your project template or at the beginning of a project will save lots of time down the road.

The Manage Ribbon

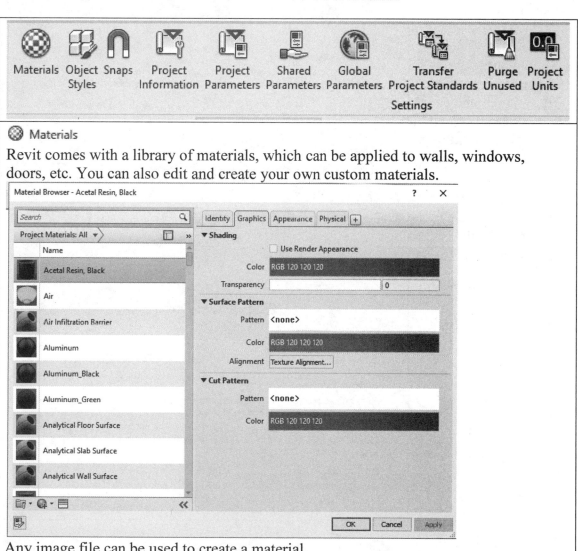

Any image file can be used to create a material.
My favorite sources for materials are fabric and paint websites.

Object Styles is used to control the line color, line weight, and line style applied to different Revit elements. It works similarly to layers in AutoCAD, except it ensures that all the similar elements, like doors, look the same.

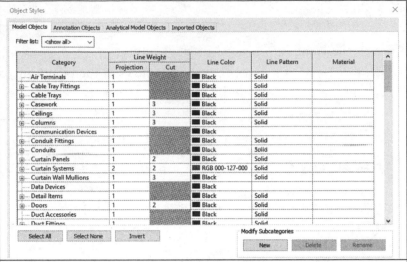

Snaps in Revit are similar to object snaps in AutoCAD. You can set how your length and dimensions snap as well as which object types your cursor will snap to when making a selection.

Like AutoCAD, it is best not to make your snaps all-encompassing so that you snap to everything as that defeats the purpose and usefulness of snaps.

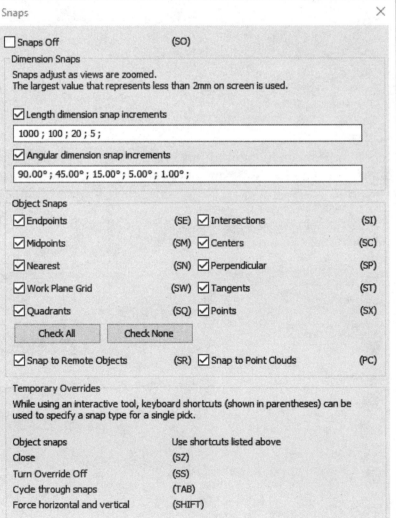

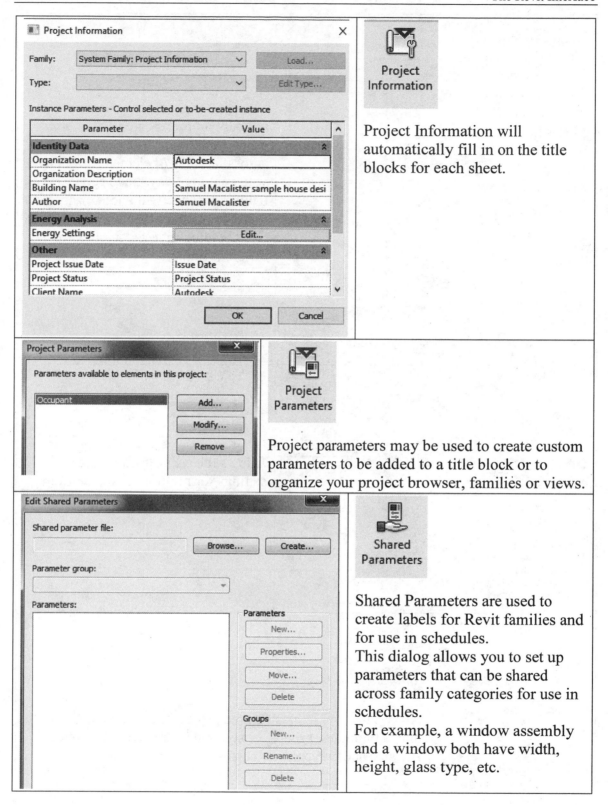

Project Information will automatically fill in on the title blocks for each sheet.

Project parameters may be used to create custom parameters to be added to a title block or to organize your project browser, families or views.

Shared Parameters are used to create labels for Revit families and for use in schedules.
This dialog allows you to set up parameters that can be shared across family categories for use in schedules.
For example, a window assembly and a window both have width, height, glass type, etc.

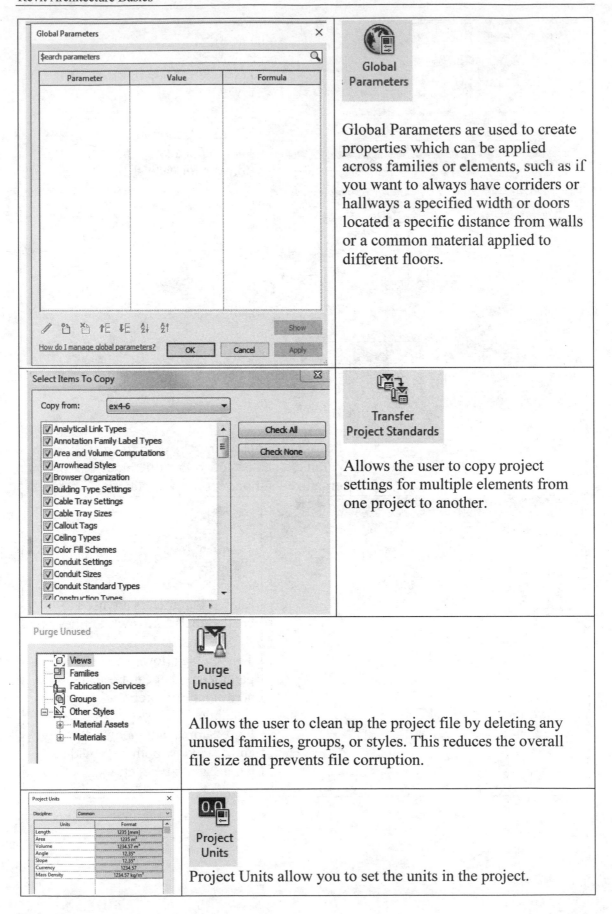

Global Parameters

Global Parameters are used to create properties which can be applied across families or elements, such as if you want to always have corriders or hallways a specified width or doors located a specific distance from walls or a common material applied to different floors.

Transfer Project Standards

Allows the user to copy project settings for multiple elements from one project to another.

Purge Unused

Allows the user to clean up the project file by deleting any unused families, groups, or styles. This reduces the overall file size and prevents file corruption.

Project Units

Project Units allow you to set the units in the project.

	Unfortunately, any dimensions that are placed do not automatically update to the new units. If you change the units after dimension annotations have been placed, you have to place new dimension annotations.
	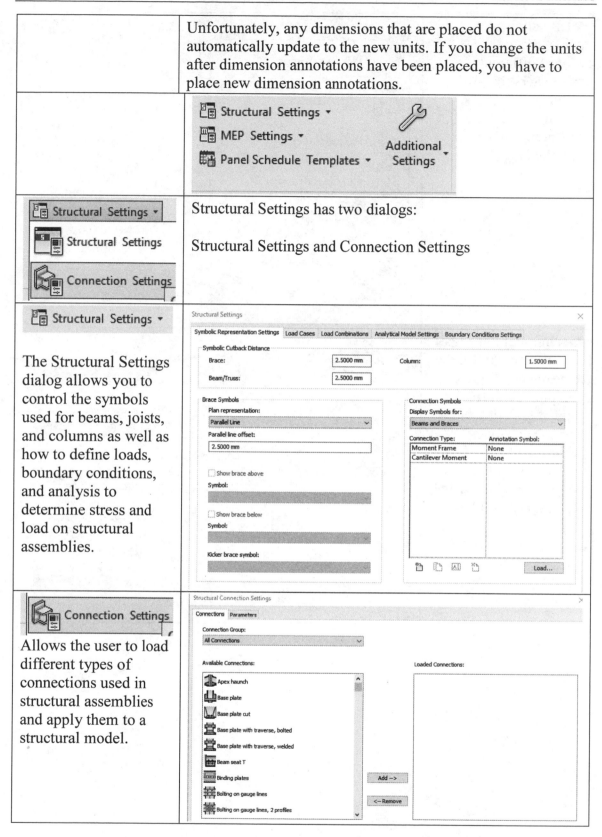
Structural Settings ▾ Structural Settings Connection Settings	Structural Settings has two dialogs: Structural Settings and Connection Settings
Structural Settings ▾ The Structural Settings dialog allows you to control the symbols used for beams, joists, and columns as well as how to define loads, boundary conditions, and analysis to determine stress and load on structural assemblies.	
Connection Settings Allows the user to load different types of connections used in structural assemblies and apply them to a structural model.	

1-45

MEP Settings Allow the user to control different parameters for MEP documentation, including the appearance of the specified elements.	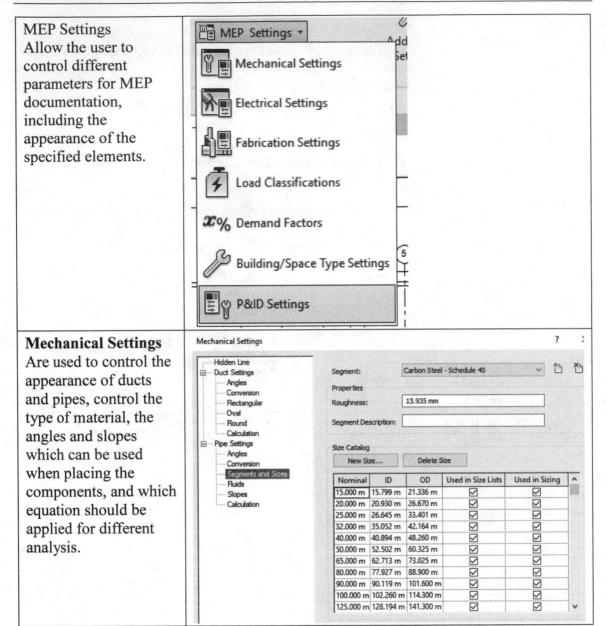
Mechanical Settings Are used to control the appearance of ducts and pipes, control the type of material, the angles and slopes which can be used when placing the components, and which equation should be applied for different analysis.	

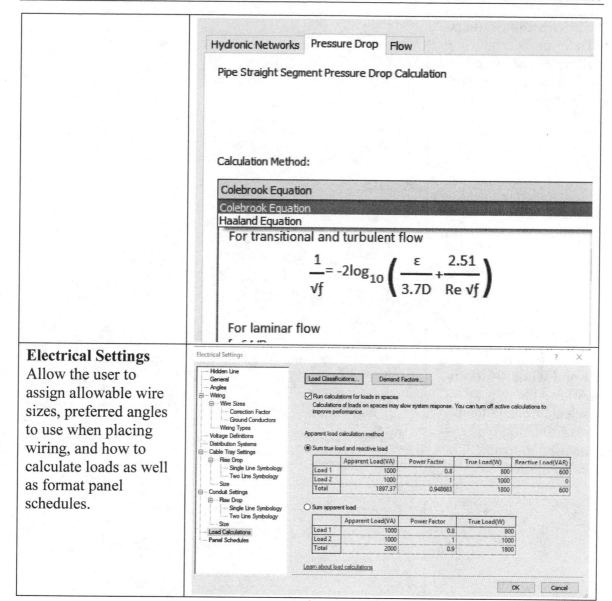

For transitional and turbulent flow

$$\frac{1}{\sqrt{f}} = -2\log_{10}\left(\frac{\varepsilon}{3.7D} + \frac{2.51}{Re\sqrt{f}}\right)$$

For laminar flow

Electrical Settings
Allow the user to assign allowable wire sizes, preferred angles to use when placing wiring, and how to calculate loads as well as format panel schedules.

Fabrication Settings
Determines which libraries are loaded and available for ductwork, electrical, and piping.

Load Classifications
Allows the user to select or create load classification types to be used in load calculations.

Demand Factors
Allows the user to select or create demand factor types to be used in load calculations.

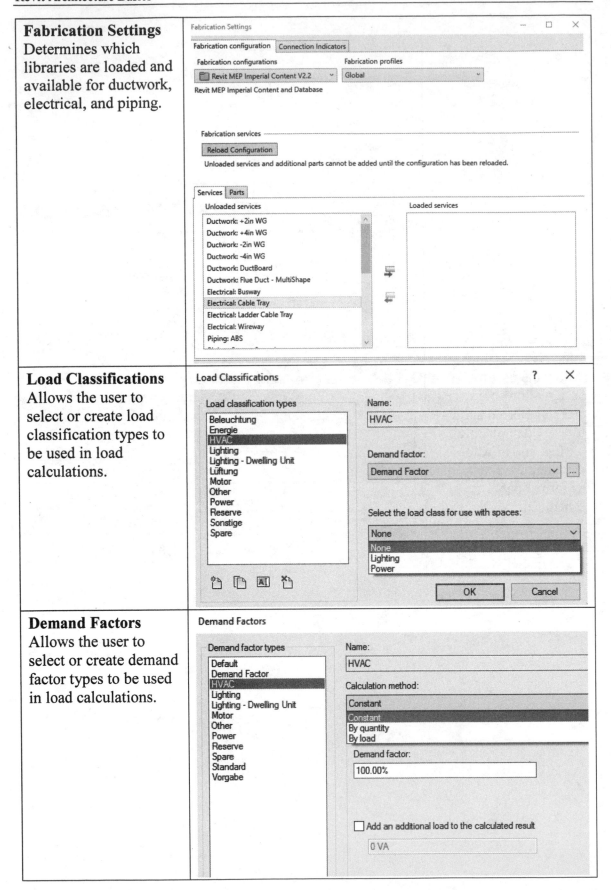

Building/Space Type Settings

Are used for energy analysis. Users can assign different values to different building or space types or create their own custom types.

To facilitate modeling an AutoCAD P&ID drawing in Revit, you'll need to map P&ID components and schematic lines to real-world 3D model components, such as a P&ID valve symbol with a Revit family, such as a ball valve.

Manage Project is similar to the External Reference Manager in AutoCAD. This is used when you have inserted external CAD files, such as topo files to be used in the model.

Starting View sets the view you want to open automatically whenever you open a file.

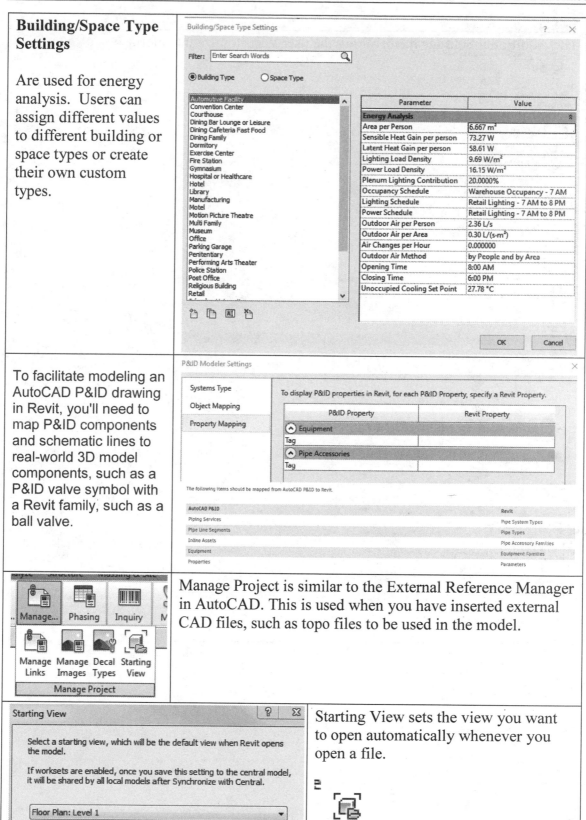

The Additional Settings menu allows the user to customize the interface.

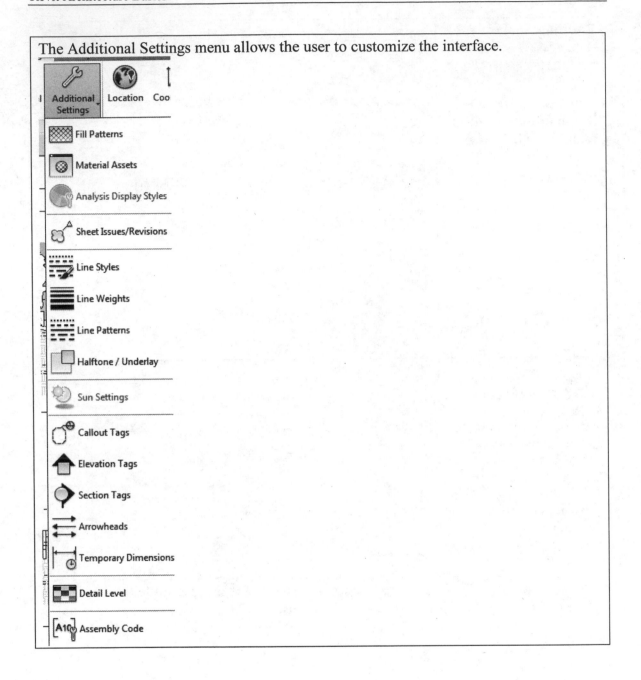

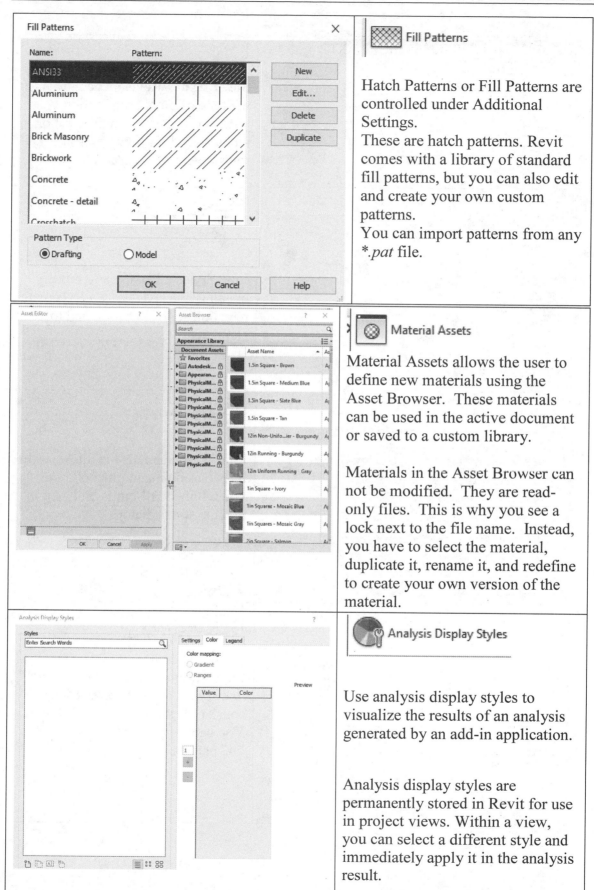

Fill Patterns

Hatch Patterns or Fill Patterns are controlled under Additional Settings.
These are hatch patterns. Revit comes with a library of standard fill patterns, but you can also edit and create your own custom patterns.
You can import patterns from any *.pat* file.

Material Assets

Material Assets allows the user to define new materials using the Asset Browser. These materials can be used in the active document or saved to a custom library.

Materials in the Asset Browser can not be modified. They are read-only files. This is why you see a lock next to the file name. Instead, you have to select the material, duplicate it, rename it, and redefine to create your own version of the material.

Analysis Display Styles

Use analysis display styles to visualize the results of an analysis generated by an add-in application.

Analysis display styles are permanently stored in Revit for use in project views. Within a view, you can select a different style and immediately apply it in the analysis result.

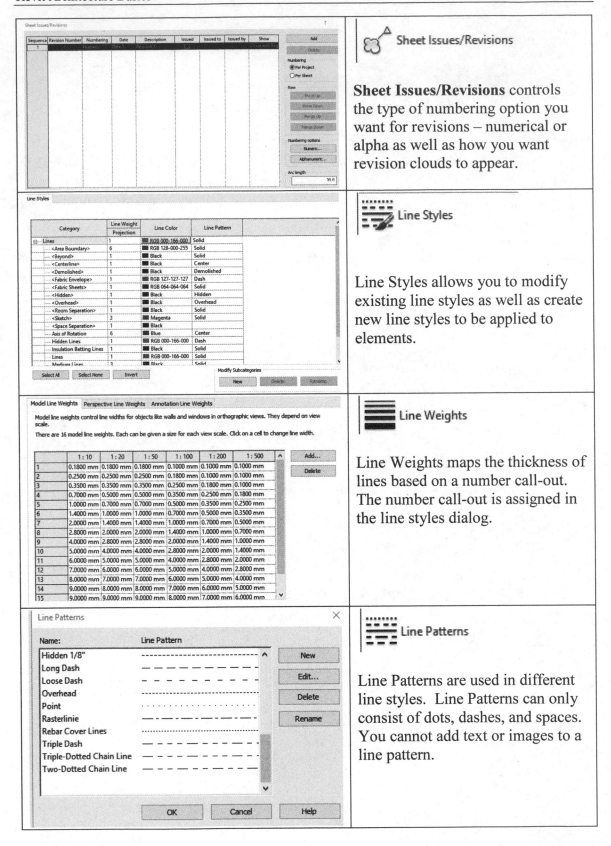

Sheet Issues/Revisions

Sheet Issues/Revisions controls the type of numbering option you want for revisions – numerical or alpha as well as how you want revision clouds to appear.

Line Styles

Line Styles allows you to modify existing line styles as well as create new line styles to be applied to elements.

Line Weights

Line Weights maps the thickness of lines based on a number call-out. The number call-out is assigned in the line styles dialog.

Line Patterns

Line Patterns are used in different line styles. Line Patterns can only consist of dots, dashes, and spaces. You cannot add text or images to a line pattern.

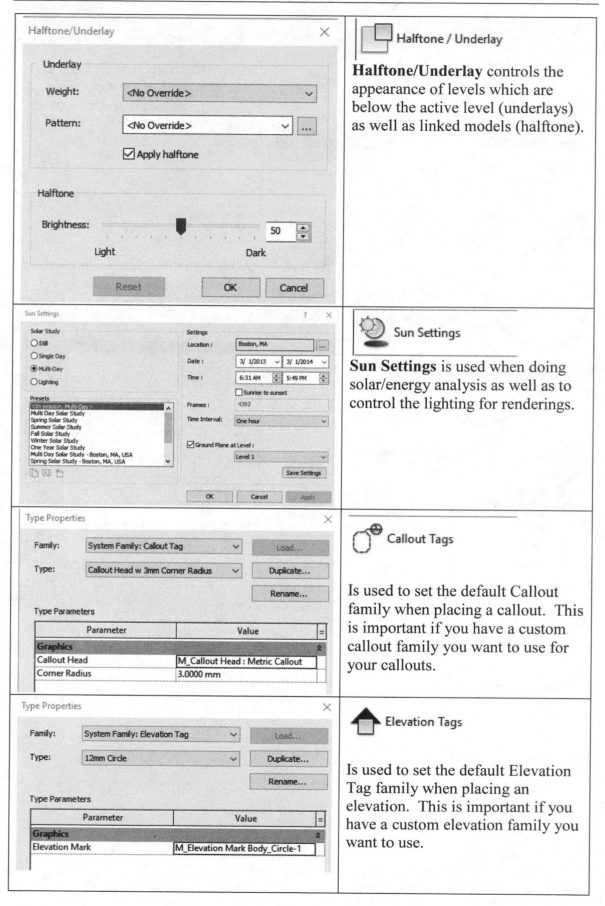

Halftone / Underlay

Halftone/Underlay controls the appearance of levels which are below the active level (underlays) as well as linked models (halftone).

Sun Settings

Sun Settings is used when doing solar/energy analysis as well as to control the lighting for renderings.

Callout Tags

Is used to set the default Callout family when placing a callout. This is important if you have a custom callout family you want to use for your callouts.

Elevation Tags

Is used to set the default Elevation Tag family when placing an elevation. This is important if you have a custom elevation family you want to use.

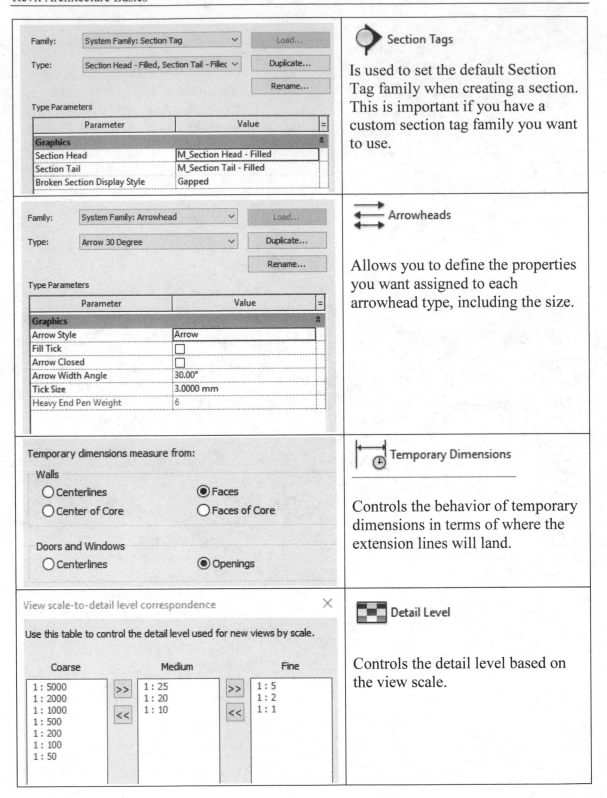

Section Tags

Is used to set the default Section Tag family when creating a section. This is important if you have a custom section tag family you want to use.

Arrowheads

Allows you to define the properties you want assigned to each arrowhead type, including the size.

Temporary Dimensions

Controls the behavior of temporary dimensions in terms of where the extension lines will land.

Detail Level

Controls the detail level based on the view scale.

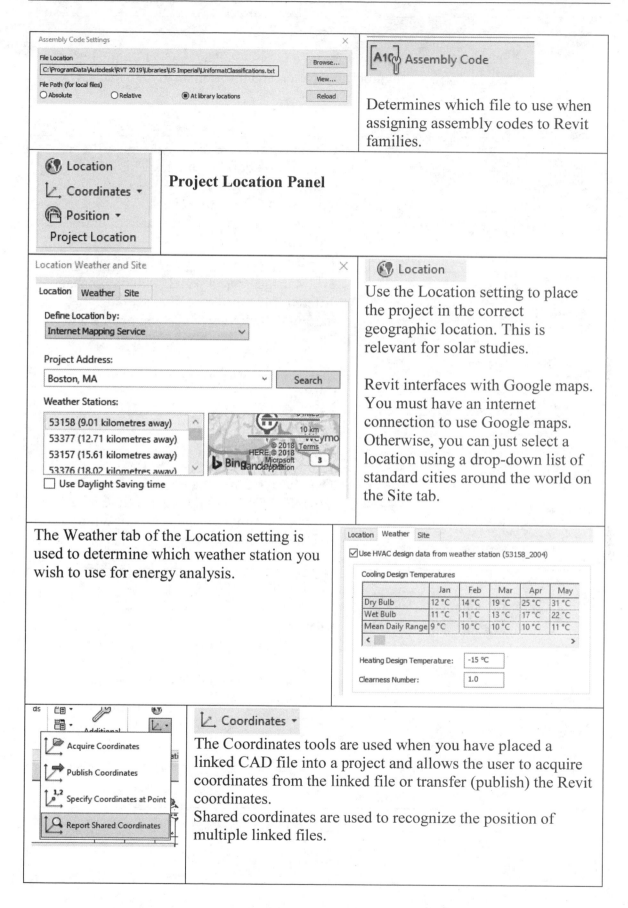

Assembly Code Settings

File Location
C:\ProgramData\Autodesk\RVT 2019\Libraries\US Imperial\UniformatClassifications.txt   Browse...
                                                                                      View...
File Path (for local files)
○ Absolute   ○ Relative   ● At library locations   Reload

Assembly Code

Determines which file to use when assigning assembly codes to Revit families.

Location
Coordinates ▾
Position ▾
Project Location

Project Location Panel

Location Weather and Site

Location | Weather | Site

Define Location by:
Internet Mapping Service ▾

Project Address:
Boston, MA ▾   Search

Weather Stations:
53158 (9.01 kilometres away)
53377 (12.71 kilometres away)
53157 (15.61 kilometres away)
53376 (18.02 kilometres away)
☐ Use Daylight Saving time

10 km
© 2018 Terms
HERE, © 2018
Microsoft
Bing Corporation

Location

Use the Location setting to place the project in the correct geographic location. This is relevant for solar studies.

Revit interfaces with Google maps. You must have an internet connection to use Google maps. Otherwise, you can just select a location using a drop-down list of standard cities around the world on the Site tab.

The Weather tab of the Location setting is used to determine which weather station you wish to use for energy analysis.

Location | Weather | Site

☑ Use HVAC design data from weather station (53158_2004)

Cooling Design Temperatures

	Jan	Feb	Mar	Apr	May
Dry Bulb	12 °C	14 °C	19 °C	25 °C	31 °C
Wet Bulb	11 °C	11 °C	13 °C	17 °C	22 °C
Mean Daily Range	9 °C	10 °C	10 °C	10 °C	11 °C

Heating Design Temperature: -15 °C
Clearness Number: 1.0

Acquire Coordinates
Publish Coordinates
Specify Coordinates at Point
Report Shared Coordinates

Coordinates ▾

The Coordinates tools are used when you have placed a linked CAD file into a project and allows the user to acquire coordinates from the linked file or transfer (publish) the Revit coordinates.
Shared coordinates are used to recognize the position of multiple linked files.

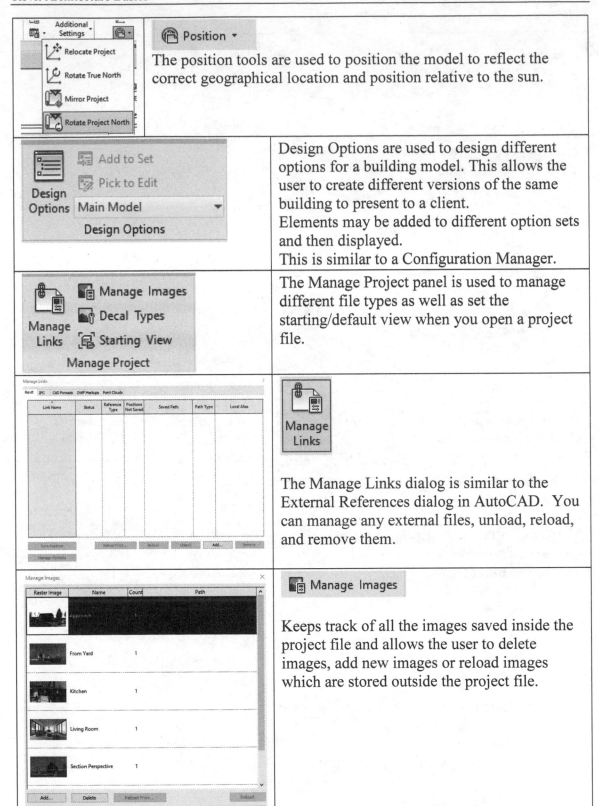

Position ▾	The position tools are used to position the model to reflect the correct geographical location and position relative to the sun.
	Design Options are used to design different options for a building model. This allows the user to create different versions of the same building to present to a client. Elements may be added to different option sets and then displayed. This is similar to a Configuration Manager.
	The Manage Project panel is used to manage different file types as well as set the starting/default view when you open a project file.
	The Manage Links dialog is similar to the External References dialog in AutoCAD. You can manage any external files, unload, reload, and remove them.
	Manage Images Keeps track of all the images saved inside the project file and allows the user to delete images, add new images or reload images which are stored outside the project file.

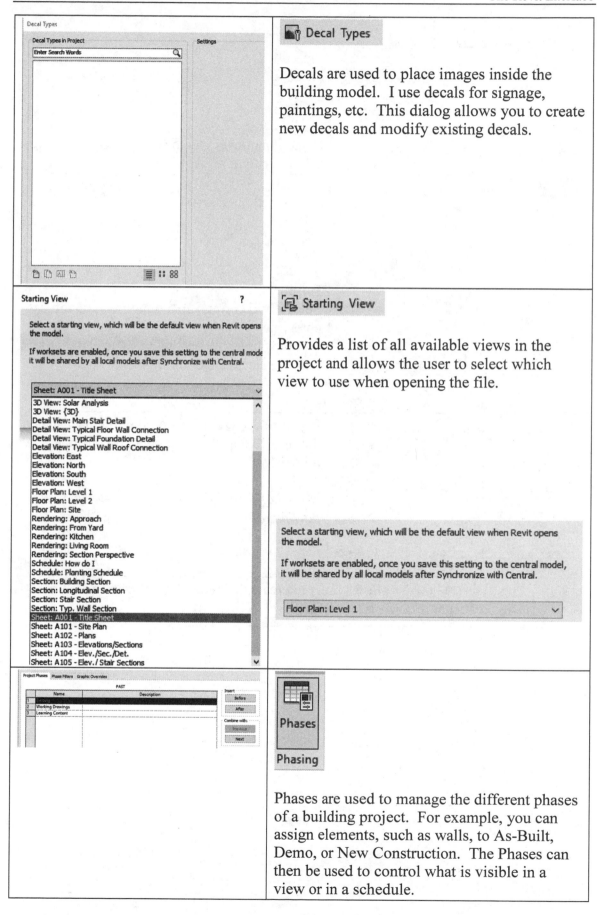

Decal Types

Decals are used to place images inside the building model. I use decals for signage, paintings, etc. This dialog allows you to create new decals and modify existing decals.

Starting View

Provides a list of all available views in the project and allows the user to select which view to use when opening the file.

Select a starting view, which will be the default view when Revit opens the model.

If worksets are enabled, once you save this setting to the central model, it will be shared by all local models after Synchronize with Central.

Floor Plan: Level 1

Phases

Phasing

Phases are used to manage the different phases of a building project. For example, you can assign elements, such as walls, to As-Built, Demo, or New Construction. The Phases can then be used to control what is visible in a view or in a schedule.

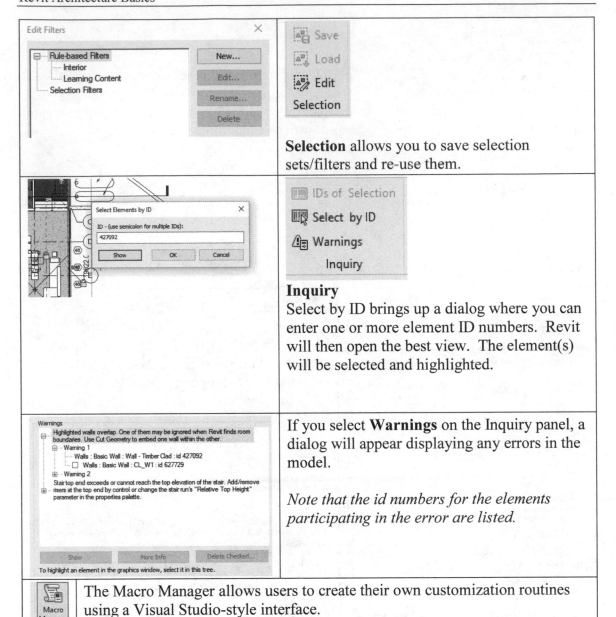

Selection allows you to save selection sets/filters and re-use them.

Inquiry

Select by ID brings up a dialog where you can enter one or more element ID numbers. Revit will then open the best view. The element(s) will be selected and highlighted.

If you select **Warnings** on the Inquiry panel, a dialog will appear displaying any errors in the model.

Note that the id numbers for the elements participating in the error are listed.

The Macro Manager allows users to create their own customization routines using a Visual Studio-style interface.

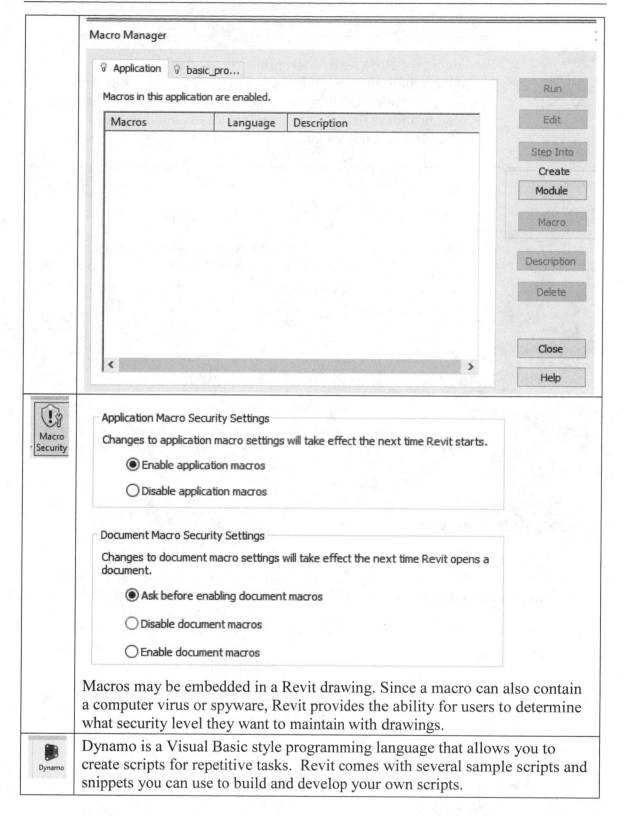

Macros may be embedded in a Revit drawing. Since a macro can also contain a computer virus or spyware, Revit provides the ability for users to determine what security level they want to maintain with drawings.

Dynamo is a Visual Basic style programming language that allows you to create scripts for repetitive tasks. Revit comes with several sample scripts and snippets you can use to build and develop your own scripts.

> ➢ By default, the Browser lists all sheets and all views.
> ➢ If you have multiple users of Revit in your office, place your family libraries and rendering libraries on a server and designate the path under the File Locations tab under Options; this allows multiple users to access the same libraries and materials.

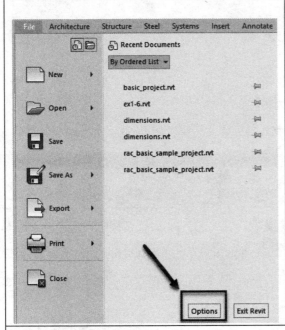

To access the System Options:

Go to the Revit menu and select the Options button located at the bottom of the dialog.

Save Reminder interval:

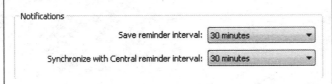

Notifications

You can select any option from the drop-down. This is not an auto-save feature. Your file will not be saved unless you select the 'Save' button. All this does is bring up a reminder dialog at the selected interval.

User Name	This indicates the user name that checks in and checks out documents. You can only change the user name when there are no open projects. The username can also be used as a property for sheets.
Username Elise	**Username** smoss@peralta.edu You are currently signed in. Your Autodesk ID is used as the username. If you need to change your username, you will need to sign out. Sign Out If you use Autodesk 360 and are signed in, then the Autodesk ID is used as the username. Signing into Autodesk 360 allows you to save a backup of your project on the cloud as well as render in the cloud – using Autodesk's servers.

Journal File Cleanup	Determines how many past versions of a project's transcriptions can be stored. You can set the number of project versions to be saved and delete any number exceeding that value after a set number of days. Transcripts are used to recover a project file if it gets damaged or corrupted.
Journal File Cleanup When number of journals exceeds: 3 then Delete journals older than (days): 10 There is currently no option to place the journal file anywhere but the default location.	The transcripts give you the ability to clean up previously created journals, which are located in "C:\Program Files\<Revit product name and version>\Journals." They can be opened with a WordPad, NotePad, or any other text-based program.

Worksharing Update Frequency Less Frequent ——————— More Frequent Every 5 seconds	If you are in a worksharing environment, where you and other team members are working on the same project, you can set how often the common project is updated.
View Options Default view discipline: Architectural Architectural Structural Mechanical Electrical Plumbing Coordination OK	The View options set the display style for views depending on discipline.

The User Interface Options

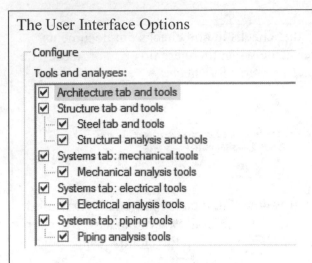

Enabling the Tools and analyses controls the visibility of the ribbons. You can hide the ribbons for those tools which you don't need or use by unchecking those ribbons.

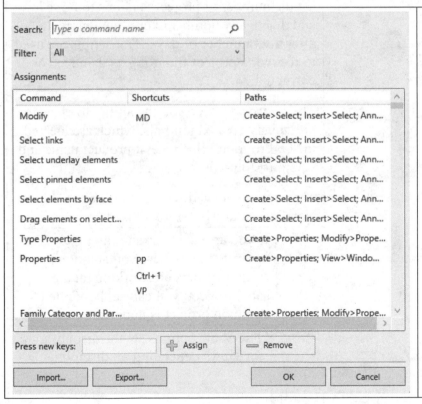

Keyboard Shortcuts (also located under User Interface on the View ribbon) allow you to assign keyboard shortcuts to your favorite commands.

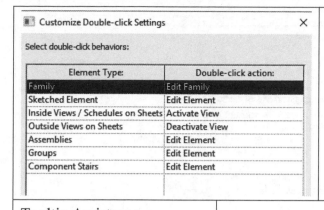

Double-click Options allow you to determine what happens when you double-left mouse click on different types of elements.

Tooltip Assistance

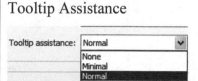

This setting controls the number of help messages you will see as you work.

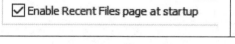

You can enable/disable whether the Recent Files page is visible when you launch Revit.

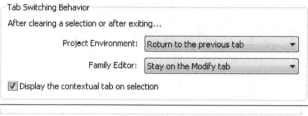

Users can also control what happens when they tab from one window to another.

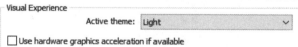

Active Theme specifies whether the user interface uses light or dark colors.

If you enable the hardware graphics acceleration, you will have improved performance. This only works if you have a graphics card which has been tested and certified to work with Autodesk software.

The Graphics tab

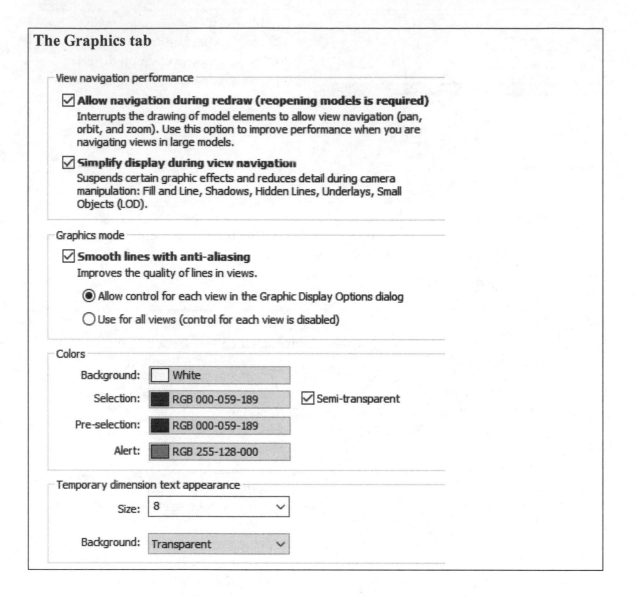

Background	Allows you to set the background color of the display window.
Selection Color	Set the color to be used when an object(s) is selected. You can also enable to be semi-transparent.
Pre-selection Color	Set the color to be used when the cursor hovers over an object.
Alert Color	Sets the color for elements that are selected when an error occurs.
Temporary Dimension Text	Users can set the font size of the temporary dimension text and change the background of the text box.

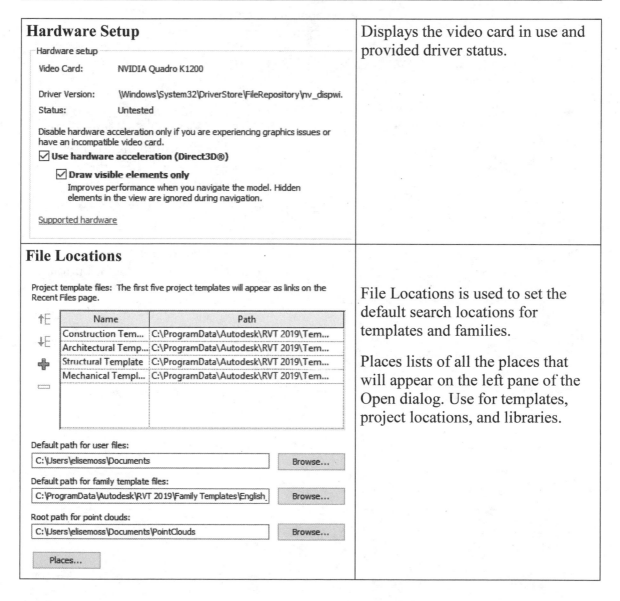

Hardware Setup	Displays the video card in use and provided driver status.
File Locations	File Locations is used to set the default search locations for templates and families. Places lists of all the places that will appear on the left pane of the Open dialog. Use for templates, project locations, and libraries.

Rendering

Rendering controls where you are storing your AccuRender files and directories where you are storing your materials.

This allows you to set your paths so Revit can locate materials and files easily.

If you press the Get More RPC button, your browser will launch to the Archvision website. You must download and install a plug-in to manage your RPC downloads. There is a free exchange service for content, but you have to create a login account.

Check Spelling

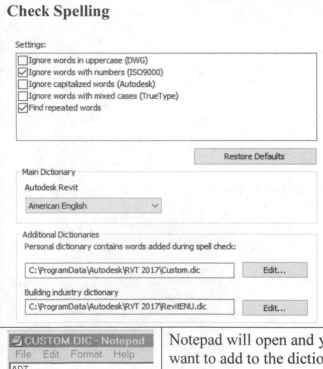

Check Spelling allows you to use your Microsoft Office dictionary as well as any custom dictionaries you may have set up. These dictionaries are helpful if you have a lot of notes and specifications on your drawings.

To add a word to the Custom Dictionary, press the **Edit** button.

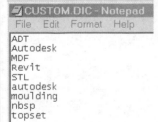

Notepad will open and you can just type in any words you want to add to the dictionary.

Save the file.

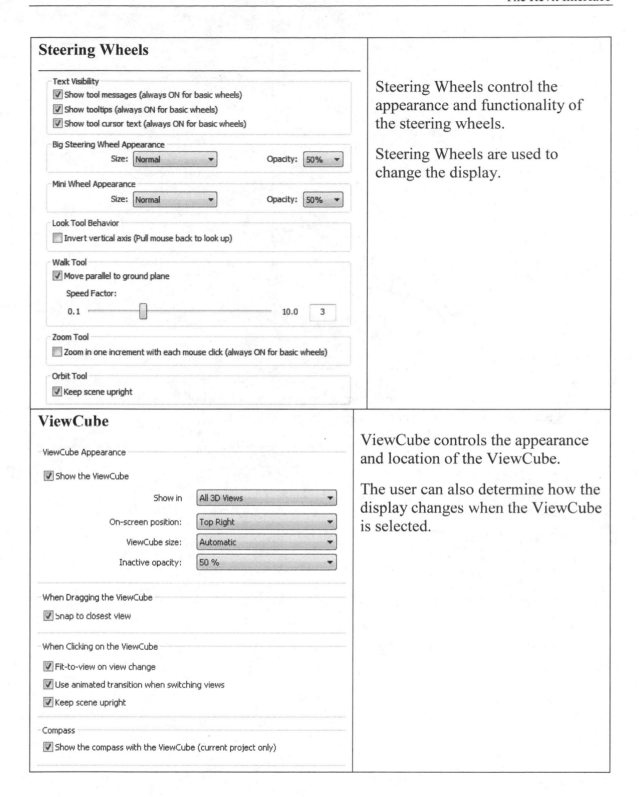

Steering Wheels

Steering Wheels control the appearance and functionality of the steering wheels.

Steering Wheels are used to change the display.

ViewCube

ViewCube controls the appearance and location of the ViewCube.

The user can also determine how the display changes when the ViewCube is selected.

Macros

Application Macro Security Settings

Changes to application macro settings will take effect the next time Revit starts.

- ◉ Enable application macros
- ○ Disable application macros

Document Macro Security Settings

Changes to document macro settings will take effect the next time Revit opens a document.

- ◉ Ask before enabling document macros
- ○ Disable document macros
- ○ Enable document macros

Macros is where the user sets the security level for Macros.

To boost productivity, store all your custom templates on the network for easy access and set the pointer to the correct folder.

The Help Menu

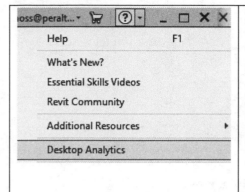

Revit Help (also reached by function key F1) brings up the Help dialog.

What's New allows veteran users to quickly come up to speed on the latest release.

Essential Skills Videos is a small set of videos to help get you started. They are worth watching, especially if your computer skills are rusty.

Revit Community is an internet-based website with customer forums where you can search for solutions to your questions or post your questions. Most of the forums are monitored by Autodesk employees who are friendly and knowledgeable.

Additional Resources launches your browser and opens to a link on Autodesk's site. Autodesk Building Solutions take you to a Youtube channel where you can watch video tutorials.

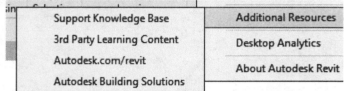

Desktop Analytics are used by Autodesk to track your usage of their software remotely using your internet connection. You can opt out if you do not wish to participate.

About Revit Architecture 2020 reaches a splash screen with information about the version and build of the copy of Revit you are working with. If you are unsure about which Service Pack you have installed, this is where you will find that information.

Setting File Locations

Drawing Name: Close all open files
Estimated Time: 5 minutes

This exercise reinforces the following skills:
- ❑ Options
- ❑ File Locations

1. Close all open files or projects.

2. 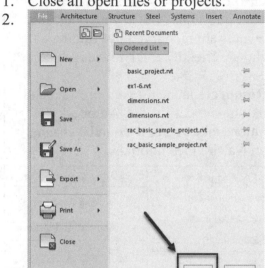 Go to the **File Menu**.

   Select the **Options** button at the bottom of the window.

3.   ⌐ File Locations   Select the **File Locations** tab.

4. Default path for user files:

   C:\Users\Elise\Documents   Browse...   In the **Default path for users** section, pick the **Browse** button.

5. Default path for user files:

   Navigate to the local or network folder where you will save your files. When the correct folder is highlighted, pick **Open**. Your instructor or CAD manager can provide you with this file information.

I recommend to my students to bring a flash drive to class and back up each day's work onto the flash drive. That way you will never lose your valuable work. Some students forget their flash drive. For those students, make a habit of uploading your file to Google Drive, Dropbox, Autodesk 360, or email your file to yourself.

Exercise 1-8
Adding the Default Template to Recent Files

Drawing Name: Close all open files
Estimated Time: 5 minutes

This exercise reinforces the following skills:
- Options
- File Locations
- Project Templates
- Recent Files

1. 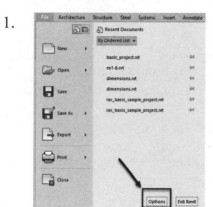 Select **Options** on the File Menu.

2. 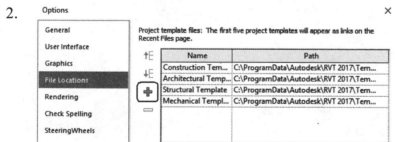 Activate File Locations.

    Press the **Plus** icon to add a project template.

3. 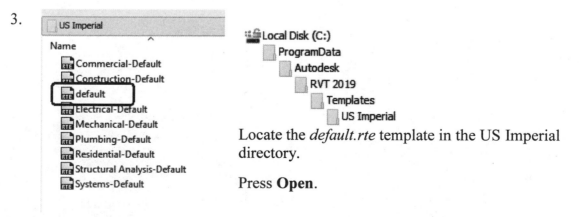 Locate the *default.rte* template in the US Imperial directory.

    Press **Open**.

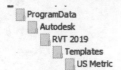

If you prefer to work in metric, select the US Metric folder and select *m_default.rte*.

4.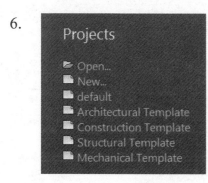

Note the file is now listed.
Select the Up arrow to move it to be ahead of the Construction template.

5.

It is now listed above the Construction template.
Organize the remaining templates in your preferred order.
Press **OK**.

6.

Projects

Open...
New...
default
Architectural Template
Construction Template
Structural Template
Mechanical Template

The default template is now available on the Recent Files page.

You can list up to five different templates for use on the Recent Files page.

Exercise 1-9
Turning Off the Visibility of Ribbons

Drawing Name: Close all open files
Estimated Time: 5 minutes

This exercise reinforces the following skills:
- Options
- User Interface
- Ribbon Tools

1.	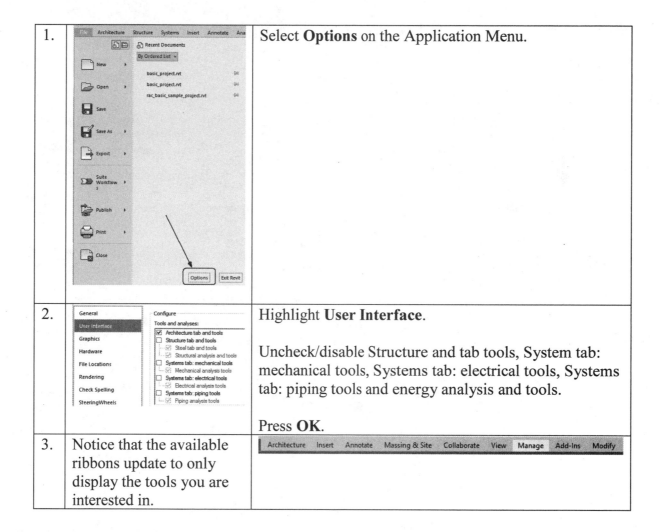	Select **Options** on the Application Menu.
2.		Highlight **User Interface**. Uncheck/disable Structure and tab tools, System tab: mechanical tools, Systems tab: electrical tools, Systems tab: piping tools and energy analysis and tools. Press **OK**.
3.	Notice that the available ribbons update to only display the tools you are interested in.	

Notes:

Lesson 1 Quiz

True or False

1. Revit Warnings do not interrupt commands, so they can be ignored and later reviewed.
2. Deactivate View on the View menu is used to remove Views from Sheets.
3. The Visibility\Graphics dialog is used to control what elements are visible in a view.

Multiple Choice [Select the Best Answer]

4. Temporary dimensions:

    A. drive the values that appear in the permanent dimensions.
    B. display the relative position of the selected element.
    C. display the size of the selected element.
    D. display only when an element is selected.

5. When you select an object, you can access the element properties by:

    A. Right clicking and selecting Element Properties.
    B. Selecting Element Properties→Instance Properties on the Ribbon Bar.
    C. Going to File→Properties.
    D. Using the Properties pane located on the left of the screen.

6. The interface item that changes appearance constantly depending on entity selections and command status is the _____.

    A. Ribbon
    B. Project Browser
    C. Menu Bar
    D. Status Bar

7. Display Options available in Revit are _____.

    A. Wireframe
    B. Hidden Line
    C. Shaded
    D. Consistent Colors
    E. All of the Above

8. The shortcut key to bring up the Help menu is ________.

   A. F2
   B. F3
   C. VV
   D. F1

9. To add elements to a selection set, hold down the ____ key.

   A. CONTROL
   B. SHIFT
   C. ESCAPE
   D. TAB

10. The color blue on a level bubble indicates that the level:

   A. has an associated view.
   B. is a reference level.
   C. is a non-story level.
   D. was created using the line tool.

ANSWERS:

1) T; 2) F; 3) T; 4) A, B, & D 5) D; 6) A; 7) E; 8) D; 9) A; 10) A

Lesson 2
Mass Elements

Mass Elements are used to give you a conceptual idea of the space and shape of a building without having to take the time to put in a lot of detail. It allows you to create alternative designs quickly and easily and get approval before you put in a lot of effort.

Massing Tools

Show Mass	Show Mass controls the visibility of mass entities.
In-Place Mass	Creates a solid shape.
Place Mass	Inserts a mass group into the active project.
Curtain System / Roof / Wall / Floor — Model by Face	Model by Face: Converts a face into a Roof, Curtain Wall System, Wall, or Floor.

When creating a conceptual mass to be used in a project, follow these steps:

1. Create a sketch of the desired shape(s).
2. Create levels to control the height of the shapes.
3. Create reference planes to control the width and depth of the shapes.
4. Draw a sketch of the profile of the shape.
5. Use the Massing tools to create the shape.

Masses can be used to create a component that will be used in a project, such as a column, casework, or lighting fixture, or they can be used to create a conceptual building.

Exercise 2-1
Shapes

Drawing Name: shapes.rfa
Estimated Time: 5 minutes

This exercise reinforces the following skills:

- ❑ Creating basic shapes using massing tools
- ❑ Create an extrude
- ❑ Modify the extrude height
- ❑ Create a revolve
- ❑ Create a sweep
- ❑ Create a blend
- ❑ Modify a blend

1.

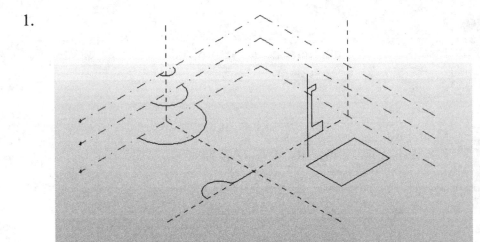

There are several sketches in the file.

Each set of sketches will be used to create a specific type of mass form.

Revit knows what type of shape you want to create based on what you select. It's all in the ingredients!

2.

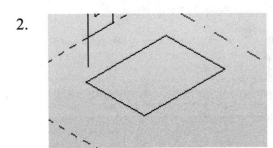

The most basic mass form is an extrude. This mass form requires a single closed polygonal sketch.

The sketch should have no gaps or self-intersecting lines.

Select the rectangle so it highlights.

3.   When the sketch is selected, you will see grips activated at the vertices.

4.   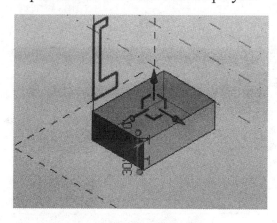 Select **Create Form→Solid Form** to create the extrude.

You won't see the Create Form tool on the ribbon unless a sketch is selected.

5.   A preview of the extrude is displayed.

You can use the triad on the top of the extrude to extend the shape in any of the three directions.

There are two dimensions:

The bottom dimension controls the height of the extrude.
The top dimension controls the distance from top face of the extrude and the next level.

6.   Activate the **East Elevation**.

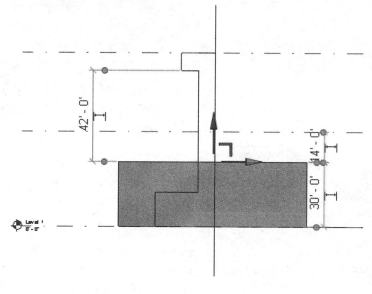

You can see how the dimensions are indicating the relative distances between the top face and the upper level, the bottom face, and another sketch.

7.

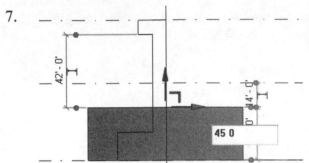

Change the 30'-0" dimension to **45'-0"**.
To change the dimension, left click on the 30' 0" dimension text.

Press *Enter*.

8.

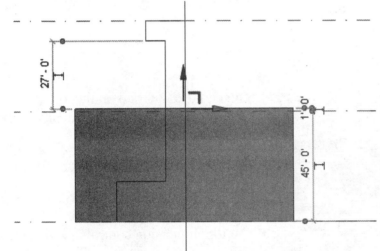

Note how the other relative dimensions update.

Left Click anywhere in the window to exit defining the mass.

If you press Enter, you will create a second mass.

9.

Switch to a 3D view.

10.

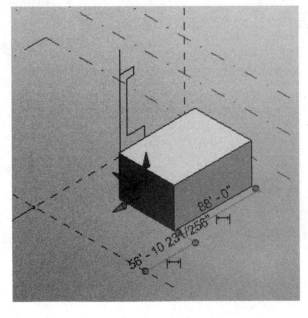

Select the Back/North face of the block. *Use the Viewcube to verify the orientation.*

To select the face, place the mouse over the face and press the tab key until the entire face highlights.

This activates the triad and also displays the temporary dimensions.

11.

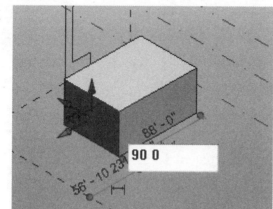

You will see two temporary dimensions. One dimension indicates the distance of the face to the opposite face (the length of the block). One dimension indicates the distance of the face to the closest work plane.

Change the 88' 0" dimension to **90'-0"**.

Left click in the window to release the selection and complete the change.

If you press Enter, you will create a second mass.

12.

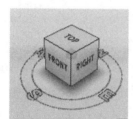

Use the View Cube to reorient the view so you can clearly see the next sketch.

This sketch will be used to create a Revolve.

A Revolve requires a closed polygonal shape PLUS a reference line which can be used as the axis of revolution.

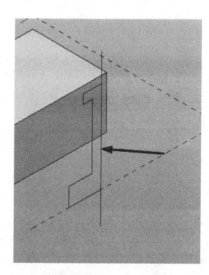

13.

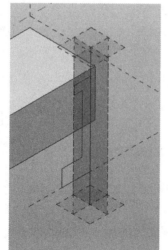

Hold down the CONTROL key and select the Axis Line (this is a reference line) and the sketch.

Note that a reference line automatically defines four reference planes. These reference planes can be used to place sketches.

14. 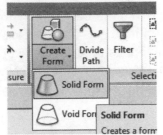 Select **Create Form→Solid Form**.

15. 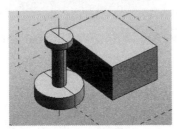 A Revolve will be created.

Our next mass form will be a SWEEP.

A sweep requires two sketches. One sketch must be a closed polygonal shape. This sketch is called the profile. The second sketch can be open or closed and is called the path. The profile travels along the path to create the sweep.

16. 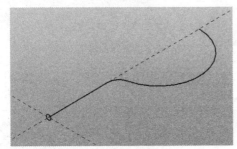 Hold down the CONTROL key and select the small circle and the sketch that looks like a question mark.

The two sketches will highlight.

17. 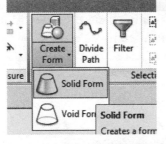 Select **Create Form→Solid Form**.

18. 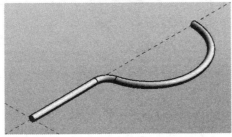 The sweep will be created.

The most common error when creating a sweep is to make the profile too big for the path. If the profile self-intersects as it travels along the path, you will get an error message. Try making the profile smaller to create a successful sweep. Sweeps are useful for creating gutters, railings, piping, and lighting fixtures.

19.

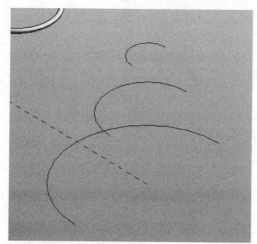

Our final shape will be a BLEND or LOFT. A blend is created using two or more open or closed sketches. Each sketch must be placed on a different reference plane or level.

Hold down the CONTROL key and select the three arcs.

20.

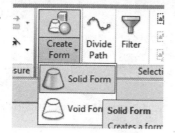

Select **Create Form→Solid Form**.

21.

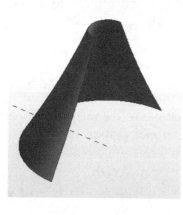

A blend shape is created.

Blends are used for complex profiles and shapes.

Close without saving.

Challenge Task:

Can you create the four basic shapes from scratch?

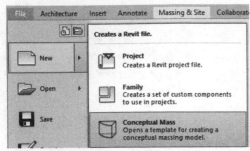

Create an extrude, revolve, sweep and blend using New→Conceptual Mass from the Applications menu.

Steps to make an Extrude:

1. Set your active plane using the Set Work Plane tool.
2. Create a single polygonal sketch with no gaps or self-intersecting lines.
3. Select the sketch.
4. Create Form→Solid Form.
5. Green Check to finish the mass.

Steps to make a Revolve:

1. Set your active plane using the Set Work Plane tool.
2. Switch to a 3D view.
3. Draw a reference line to be used as the axis of revolution.
4. Pick one of the planes defined by the reference line as the active plane to place your sketch.
5. Create a single polygonal sketch – no gaps, overlapping or self-intersecting lines – for the revolve shape.
6. Hold down the CONTROL key and select BOTH the sketch and reference line. If the reference line is not selected, you will get an extrude.
7. Create Form→Solid Form.
8. Green Check to finish the mass.

Steps to make a Sweep:

1. Activate Level 1 floor plan view. You need to select one reference plane for the profile and one reference plane for the path. The reference planes must be perpendicular to each other. Set Level 1 for the path's reference plane.
2. Draw a path on Level 1. The path can be a closed or open sketch.
3. Create a reference plane to use for the profile. Draw a reference plane on Level 1 – name it profile plane. Make this the active plane for the profile sketch.
4. Switch to a 3D view. Create a single polygonal sketch – no gaps, overlapping or self-intersecting lines. The profile sketch should be close to the path or intersect it so it can follow the path easily. If it is too far away from the path, it will not sweep properly or you will get an error.
5. Hold down the CONTROL key and select BOTH the path and the profile. If only one object is selected, you will get an extrude.
6. Create Form→Solid Form.
7. Green Check to finish the mass.

Steps to make a Blend:

1. Blends require two or more sketches. Each sketch should be on a parallel reference plane. You can add levels or reference planes for each sketch. If you want your blend to be vertical, use levels. If you want your blend to be horizontal, use reference planes.
   a. To add levels, switch to an elevation view and select the Level tool.
   b. To add reference planes, switch to a floor plan view and select the Reference Plane tool. Name the reference planes to make them easy to select.
2. Set the active reference plane using the Option Bar or the Set Reference Plane tool.
3. Create a single polygonal sketch – no gaps, overlapping or self-intersecting lines.
4. Select at least one more reference plane to create a second sketch. Make this the active plane.
5. Create a single polygonal sketch – no gaps, overlapping or self-intersecting lines.
6. Hold down the CONTROL key and select all the sketches created. If only one object is selected, you will get an extrude.
7. Create Form→Solid Form.
8. Green Check to finish the mass.

Create a Conceptual Model

Drawing Name: default.rte [m_default.rte]
Estimated Time: 5 minutes

This exercise reinforces the following skills:

- ❑ Switching Elevation Views
- ❑ Setting Project Units
- ❑ Add a Level

This tutorial uses metric or Imperial units. Metric units will be designated in brackets.

Revit uses a level to define another floor or story in a building.

1. 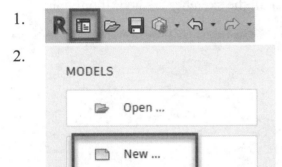 Select the **Recent Files** icon on the Quick Access Toolbar.

2. Select **New…**

3. Select the *default* template in the drop-down list.

   This template was added to the list in Exercise 1-8.

4.  Under the Template file, select **Browse**.

5.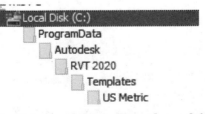

   Locate the *Imperial Templates* folder under *ProgramData/Autodesk/RVT 2020*.

   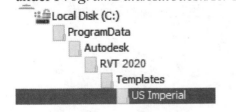

   Locate the *Metric Templates* folder under *ProgramData/Autodesk/RVT 2020*.

6.

US Metric
Name
Construction-DefaultMetric
Construction-DefaultUS-Canada
DefaultMetric
DefaultUS-Canada
Electrical-Default_Metric
Mechanical-Default_Metric
Plumbing-Default_Metric
Structural Analysis-DefaultMetric
Structural Analysis-DefaultMetric-Up
Structural Analysis-DefaultUS-Canada
Systems-Default_Metric

Notice the types of templates available in each of these folders.

The number of templates available was determined when the software was installed. You can add more templates by modifying the installation using the Control Panel or download additional templates from Autodesk's content library.

7.

Template file

default.rte ▾ Browse...

Select the *default.rte [DefaultMetric.rte]* template.

Press **OK**.

Brackets indicate metric, which can be selected as an alternative.

If you accidentally picked Metric when you wanted Imperial or vice versa, you can change the units at any time.

To change Project Units, go to the **Manage** Ribbon.

Select **Settings→Project Units**.

Project Units

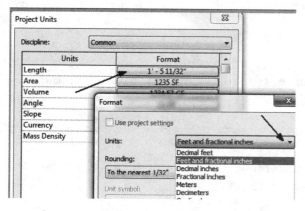

Left click the Length button, then select the desired units from the drop-down list.

8.

Elevations (Building Elevation)
- East
- North
- South
- West

Double click **East** under Elevations.

This activates the East view orientation.

9. Architecture   Select the **Architecture** ribbon.

10. Level   Select the **Level** tool under Datum. (This adds a floor elevation.)

11. 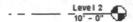 Move your mouse to set an elevation of **12′ [3650 mm]**.
Pick to start the elevation line.

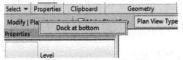

12.

In the Options bar, enable **Make Plan View**.
The Options bar can be located below the ribbon or at the bottom of the screen.

 If the Options bar is located below the ribbon and you
would prefer it on the bottom of the screen, right click and select **Dock at Bottom**.

If the Options bar is located at the bottom of the screen
and you would prefer it to be below the ribbon, right click on the Options bar and
select **Dock at Top**.

Make Plan View should be enabled if you want Revit to automatically create a floor
plan view of this level. If you forget to check this box, you can create the floor plan
view later using the **View** Ribbon.

Double click on the blue elevation symbol to automatically switch to the
floor plan view for that elevation.

---

13. 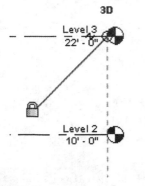 Pick to place the end point to position the level indicator
above the other indicators.

14. Basically, you place a new level by picking two points at the desired height.

Right click and select **Cancel** to exit the Level command.

*Revit is always looking for references even among annotations; you will notice that
your level tags snap and lock together so when you move one to the right or left, all
those in line with it will follow.*

The jogged line allows the user to create a jog if desired.

If you need to adjust the position of the tag, just click on the line; 3 blue grips will appear. These can be clicked and dragged as needed. You can also right click on a level tag and select 'Hide annotation in view' and the tag and level line will disappear in that view only.

Hide Annotation in View is only enabled if an object is selected first.

15. Save the file as a project as *ex2-2.rvt*.

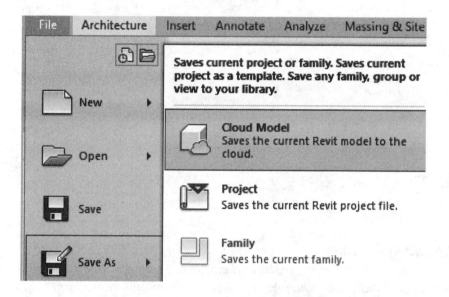

Revit has an option where you can save the current file, whether it is a project file or a family, to a cloud server. In order for this option to be available, you have to pay a subscription fee to Autodesk for the cloud space.

Exercise 2-3
Adding an In-Place Mass

Drawing Name: ex2-2.rvt
Estimated Time: 10 minutes

This exercise reinforces the following skills:

- Switching Elevation Views
- Add Mass

1. 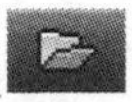 Open or continue working in the file *ex2-2.rvt*.

2. 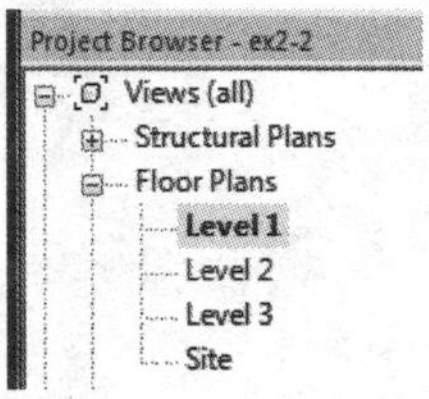

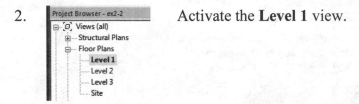

Activate the **Level 1** view.

3. 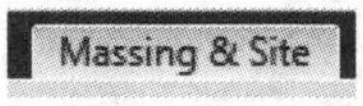

Select the **Massing & Site** ribbon.

4. 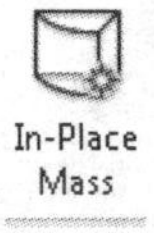

Select the **In-Place Mass** tool.

Revit uses three different family categories. System families are families which are defined inside the project, such as floors, walls, and ceilings. Loadable families are external files which can be loaded into the project, such as doors, windows, and furniture. The third family category is in-place masses, which are families created on-the-fly. In-place masses are only available inside the project where they are created and are usually unique to the project since they aren't loadable. However, you can use the Copy and Paste function to copy in-place masses from one project to another, if you decide to re-use it. That can be a hassle because you need to remember which project is storing the desired in-place mass.

5. 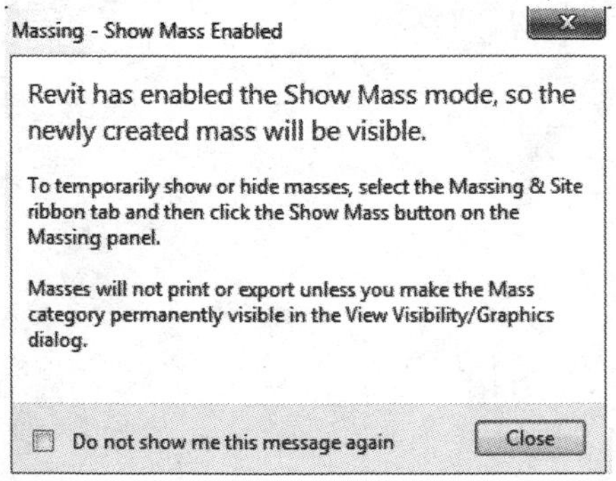

Masses, by default, are invisible. However, in order to create and edit masses you need to see what you are doing. Revit brings up a dialog to let you know that the software is switching the visibility of masses to ON, so you can work.

Press **Close**.

If you don't want to be bugged by this dialog, enable the ***Don't show me this message again*** *option.*

6.  Enter **Mass 1** in the Name field.

   Press **OK**.

Next, we create the boundary sketch to define our mass. This is the footprint for the conceptual building.

7. Select the **Line** tool located under the Draw panel.

8. Enable **Chain** in the Options bar.

   This allows you to draw lines without always having to pick the start point.
   If you hold down the SHIFT key while you draw, this puts you in orthogonal mode.

9. Create the shape shown.
   The left figure shows the units in Imperial units (feet and inches).
   The right figure shows the units in millimeters.

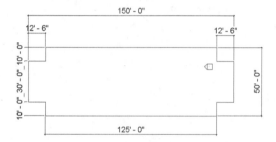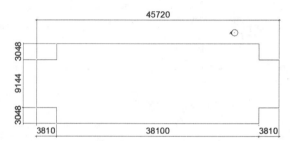

   You can draw using listening dimensions or enter the dimension values as you draw.

   For feet and inches, you do not need to enter the ' or " symbols. Just place a space between the numbers.

   Revit doesn't have a CLOSE command for the LINE tool unlike AutoCAD, so you do have to draw that last line.

10. Exit out of drawing mode by right clicking and selecting Cancel twice, selecting ESC on the keyboard or by selecting the Modify button on the ribbon.

11. Switch to a 3D view.
    Activate the **View** ribbon and select **3D View**.

    You can also switch to a 3D view from the Quick Access toolbar by selecting the house icon.

12. Window around the entire sketch so it is highlighted.

13. 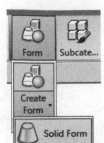 Select **Form→Create Form→Solid Form**.

You must select the sketch to create the form.

If the sketch has any gaps, overlapping lines, or self-intersecting lines, you will get an error. Exit the command and inspect the sketch to make sure it is a cleanly closed polygon.

14. 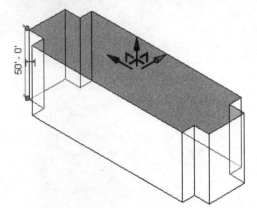   An extrusion distance is displayed. This can be edited, if desired.

15.   Select the green check box to **Finish Mass**.

The Mass is created.

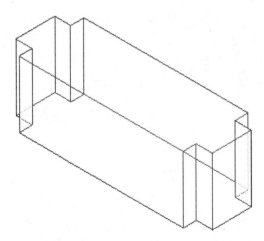

16. Save the file as *ex2-3.rvt*.

  Object tracking will only work if the sketch objects are active and available in the current sketch. You can use Pick to copy entities into the current sketch.

Exercise 2-4
Modifying Mass Elements

Drawing Name: ex2-3.rvt
Estimated Time: 30 minutes

This exercise reinforces the following skills:

- ❑ Show Mass
- ❑ Align
- ❑ Modify Mass
- ❑ Mirror
- ❑ Create Form
- ❑ Save View

1.   Open *ex2-3.rvt*.

2.   If you don't see the mass, **Show Mass Form and Floors** on the Massing & Site ribbon to turn mass visibility ON.

Some students may experience this issue if they close the file and then re-open it for a later class.

3.   Activate the **East** Elevation.

4.   We see that top of the building does not align with Level 3.

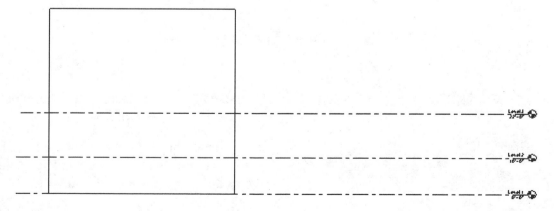

To adjust the horizontal position of the level lines, simply select the line and use the grip to extend or shorten it.

5. Select the **Modify** Ribbon.

6. Select the **Align** tool.

When using Align, the first element selected acts as the source, and the second element selected shifts position to align with the first element.

7.
Select the top level line (Level 3) then select the top of the extrusion.

Right click and select **Cancel** twice to exit the Align command.

8. The top of the extrusion now aligns to Level 3.

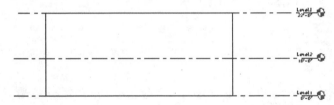

The lock would constrain or lock the top of the extrusion to the level. If the level elevation changes, then the extrusion will automatically update.

9.
⊟ Floor Plans
     Level 1
     Level 2
     Level 3
     Site

Activate **Level 2** under Floor Plans.

10. Select **In-Place Mass** from the Massing & Site ribbon.

In-Place
Mass

11. Name: Tower

Name the new mass **Tower**.

Press **OK**.

12.
You can use object tracking to locate the intersection between the two corners.

To activate object tracking, enable the **Pick Lines** tool located under Draw. Then select the two lines you want to align with.

13. 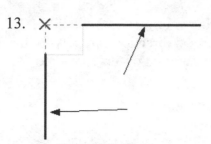 Select the two lines indicated to be used for object tracking to locate the center of the circle.

14. 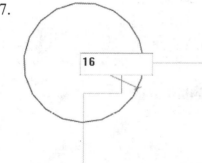 Select the **Circle** tool under Draw.

15. Placement Plane: Level : Level 2 ▼ | 📖 | ☐ Make surface from closed loops | ☑ Chain Offset: 0' 0" | ☐ Radius: 1' 0"

Uncheck Make surface from closed loops on the Options bar. Enable Chain.

16. When you see the large X and the tooltip says Intersection, you will have located the intersection.

Pick to locate the center of the circle at the intersection.

17. Enter a radius of **16'-0"** [**4880**].

Cancel out of the command.

When you used the Pick Line tool, you copied those lines into the current sketch. Once the lines were part of the current sketch, they could be used for object tracking.

18. Select the circle sketch so it is highlighted.

19. Select the **Draw Mirror Axis** tool on the Modify panel.

20.  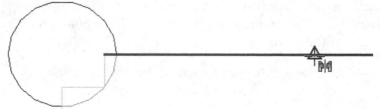 Locate the midpoint of the small horizontal line and pick.

21.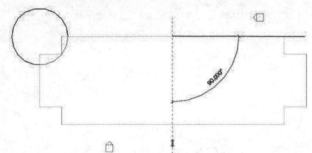

Bring your mouse down in the Vertical direction and pick for the second point of the mirror axis.

22.

The circle sketch is mirrored.

Left click to release the selection.

23. 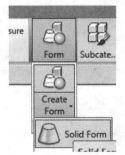 Switch to a 3D view using the Project Browser.

24.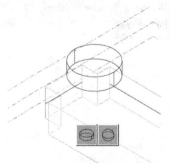

Select **one** of the circles so it is highlighted.
Remember you can only extrude one closed polygon at a time.

Select **Form→Create Form→Solid Form**.

25.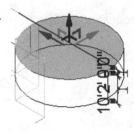

A small toolbar will appear with two options for extruding the circle.

Select the option that looks like a cylinder.

26.

A preview of the extrusion will appear with the temporary dimension. You can edit the temporary dimension to modify the height of the extrusion.

Press ENTER or left click to accept the default height.

27.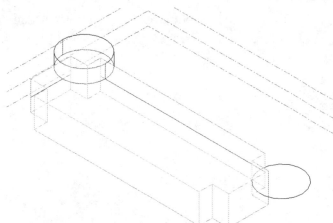

If you press ENTER more than once, additional cylinders will be placed.

The circle is extruded.

28.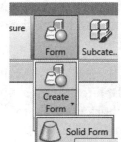

Select the remaining circle so it is highlighted.

Select **Form→Create Form→Solid Form**.

29.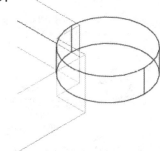

A small toolbar will appear with two options for extruding the circle.

Select the option that looks like a cylinder.

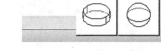

30.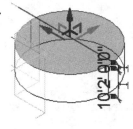

A preview of the extrusion will appear with the temporary dimension. You can edit the temporary dimension to modify the height of the extrusion.

Press ENTER or left click to accept the default height.

If you press ENTER more than once, you will keep adding cylinders.

Press ESC or right click and select CANCEL to exit the command.

31.

Both circles are now extruded.

32.

Select the two lines used to locate the circle sketch.

Right click and select **Delete** from the shortcut menu to delete the lines.

Cancel

Repeat [Solid Form]
Recent Commands   >

Create Similar
Edit Family
Select Previous
Select All Instances   >
Delete

Reset Crop Boundary to Model
Reset Crop Boundary to Screen

You can also press the Delete key on the keyboard or use the Delete tool on the Modify panel.

33.   Finish Cancel   Select **Finish Mass**.
       Mass Mass
       In-Place Editor   *If you do not delete the lines before you finish the mass, you will get an error message.*

34.   Elevations (Building Elevation)   Activate the **South** Elevation.
           East
           North
           South
           West

35. Select each level line.

    Maximize 3D Extents   Right click and select **Maximize 3D Extents**.
                           This will extend each level line so it covers the entire model.

36.   Activate the Modify ribbon.

           Select the **Align** tool from the Modify Panel.

37.   ☑ Multiple Alignment   Prefer: Wall faces   On the Options bar, enable **Multiple Alignment**.

38. 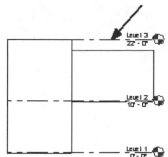 Select the Level 3 line as the source object.

39. Select the top of the two towers as the edges to be shifted.
Right click and select CANCEL twice to exit the command or press ESC.

40. 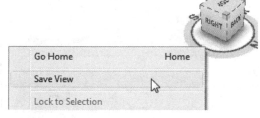 Switch to a 3D view using the Project Browser.

41. Use the ViewCube located in the upper right of the screen to orbit the model.

42. To save the new orientation, right click on the ViewCube and select **Save View**.

Go Home	Home
Save View	
Lock to Selection	

43. Name: 3D Ortho   Enter **3D Ortho** for the name of the view.

Press **OK**.

44. The **Saved** view is now listed in the Project browser under 3D Views.

45. Save the file as *ex2-4.rvt*.

➢ Pick a mass element to activate the element's grips. You can use the grips to change the element's shape, size, and location.
➢ You can only use the **View→Orient** menu to activate 3D views when you are already in 3D view mode.

Exercise 2-5
Create Wall by Face

Drawing Name: ex2-4.rvt
Estimated Time: 15 minutes

This exercise reinforces the following skills:

- ❑ Wall by Face
- ❑ Trim
- ❑ Show Mass

You can add doors and windows to your conceptual model to make it easier to visualize.

1. Open *ex2-4.rvt*.

2. Activate the **3D Ortho** view under 3D Views.

3. Activate the **Massing & Site** ribbon.

4. Select **Model by Face→Wall**.

5. 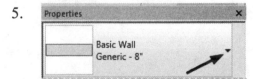 Note the wall type currently enabled in the Properties pane. A different wall type can be selected from the drop-down list available using the small down arrow.

   Imperial:
   Set the Default Wall Type to:
   Basic Wall: Generic- 8″.

   Metric:
   Set the Default Wall Type to:
   Basic Wall: Generic- 200 mm.

6. 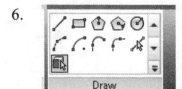 Enable **Pick Faces** from the Draw Panel on the ribbon.

7.

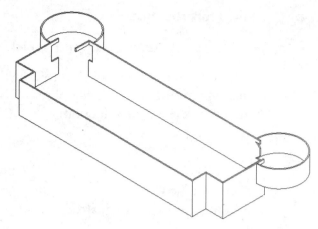

Select each wall and cylinder. The cylinder will be halved by the walls, so you will have to select each half.

You will have to do some cleanup work on the corners where the towers are.

Right click and select CANCEL to exit the command.

Some students will accidentally pick the same face more than once. You will see an error message that you have overlapping/duplicate walls. Simply delete the extra walls.

8.

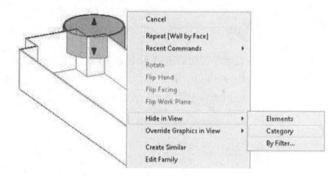

Select any visible mass.
Right click and select **Hide in View→ Category**.

This will turn off the visibility of masses.

9.

Activate **Level 1** floor plan.

10.

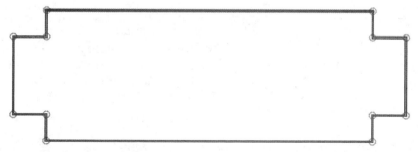

Window around all the walls to select.

11.

Select the **Filter** tool from the ribbon.

Filter

12.

Uncheck all the boxes EXCEPT walls.
Press **OK**.

There are some duplicate walls in this selection.

13.

In the Properties pane:

Set the Top Constraint to **up to Level 3**.

Right click and select Cancel to release the selection or left click in the display window to release the selection.

14.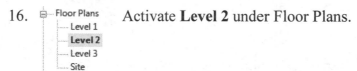

Hold down the Ctrl Key.
Select the four walls indicated.

15.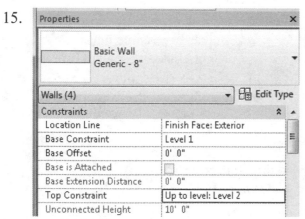

In the Properties pane:

Set the Top Constraint to **Up to Level 2**.

Right click and select Cancel to release the selection or left click in the display window to release the selection.

16. Floor Plans
    Level 1
    Level 2
    Level 3
    Site

Activate **Level 2** under Floor Plans.

17.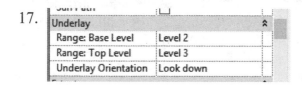

In the Properties Pane:
Go to the **Underlay** category.

Set the Range Base Level to **Level 2**.
Set the Range Top Level to **Level 3**.
Set the Underlay orientation to **Look down**.

This will turn off the visibility of all entities located below Level 2.

Each view has its own settings. We turned off the visibility of masses on Level 1, but we also need to turn off the visibility of masses on the Level 2 view.

18.

Show Mass
by View Settings

On the ribbon: Toggle the Show Mass by View Settings tool to turn the visibility of masses OFF.

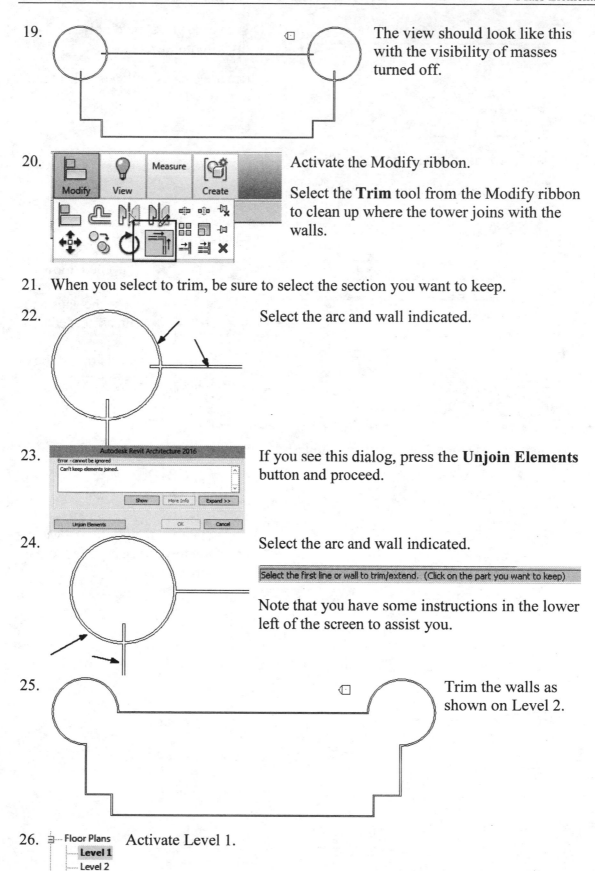

19. The view should look like this with the visibility of masses turned off.

20. Activate the Modify ribbon.

    Select the **Trim** tool from the Modify ribbon to clean up where the tower joins with the walls.

21. When you select to trim, be sure to select the section you want to keep.

22. Select the arc and wall indicated.

23. If you see this dialog, press the **Unjoin Elements** button and proceed.

24. Select the arc and wall indicated.

    Select the first line or wall to trim/extend. (Click on the part you want to keep)

    Note that you have some instructions in the lower left of the screen to assist you.

25. Trim the walls as shown on Level 2.

26. Floor Plans   Activate Level 1.
    Level 1
    Level 2

27.

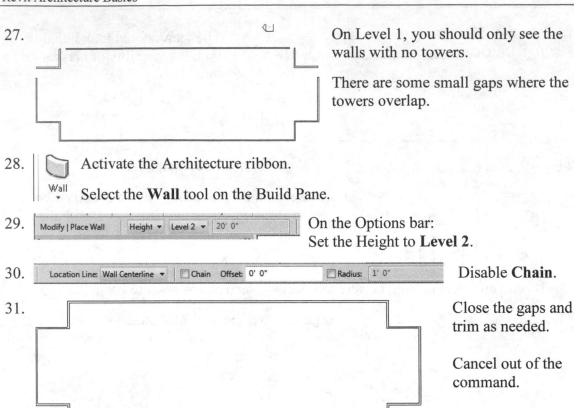

On Level 1, you should only see the walls with no towers.

There are some small gaps where the towers overlap.

28. Activate the Architecture ribbon.

Wall Select the **Wall** tool on the Build Pane.

29. | Modify | Place Wall | Height ▾ | Level 2 ▾ | 20' 0" |

On the Options bar:
Set the Height to **Level 2**.

30. | Location Line: Wall Centerline ▾ | ☐ Chain Offset: 0' 0" | ☐ Radius: 1' 0" |

Disable **Chain**.

31.

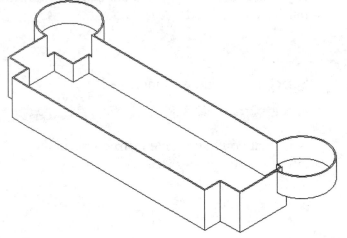

Close the gaps and trim as needed.

Cancel out of the command.

32. Switch to a 3D view and orbit your model.

Check to make sure the walls and towers are adjusted to the correct levels.

33. Save as *ex2-5.rvt*.

Exercise 2-6
Adding Doors and Windows

Drawing Name: ex2-5.rvt
Estimated Time: 30 minutes

This exercise reinforces the following skills:

- ❑ Basics
- ❑ Door
- ❑ Load from Library
- ❑ Window
- ❑ Array
- ❑ Mirror
- ❑ Shading

You can add doors and windows to your conceptual model to make it easier to visualize.

1. Open *ex2-5.rvt*.

2. Activate **Level 1** under Floor Plans.

3. Level 1 should appear like this.

4. Activate the **Architecture ribbon**.

5. Select the **Door** tool under the Build panel.

6. Select **Load Family** under the Mode panel.

    Doors are loadable families.

7.

Browse to the **Doors** folder under the Imperial or Metric library – use Imperial if you are using Imperial units or use Metric if you are using Metric units.

As you highlight each file in the folder, you can see a preview of the family.

Note that the files are in alphabetical order.

8. For Imperial Units: Go to the Commercial folder. Locate the *Door-Exterior-Double.rfa* file.

For Metric Units:
Locate the *M_Door-Exterior-Double.rfa* file.

Press **Open**.

9.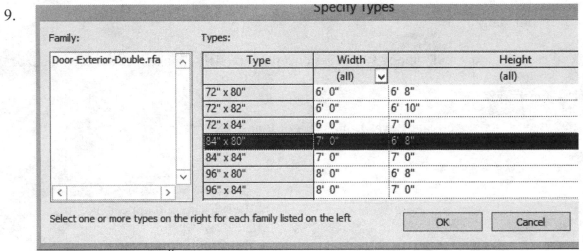

For Imperial Units: Highlight the **84" x 80"** type and press **OK**.

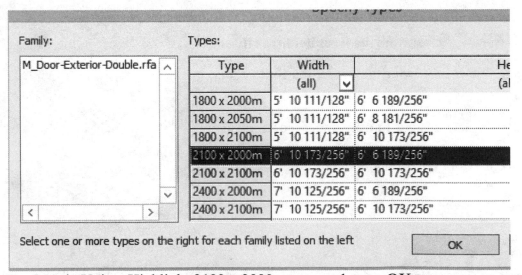

For Metric Units: Highlight **2100 x 2000m** type and press **OK**.

10.

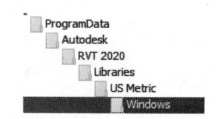

62' - 2"   62' - 2"

Place the door so it is centered on the wall as shown.

Doors are wall-hosted, so you will only see a door preview when you place your cursor over a wall.

11.

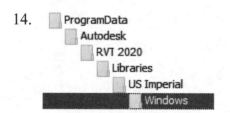

If you press the space bar before you pick to place, you can control the orientation of the door.

After you have placed the door, you can flip the door by picking on it then pick on the vertical or horizontal arrows.

12. ▦   Pick the **Window** tool from the Build panel.

Window *Windows are model families.*

13. ⬇   Select **Load Family** from the Mode panel.

Load
Family

14.

```
ProgramData                        ProgramData
  Autodesk                           Autodesk
    RVT 2020                           RVT 2020
      Libraries                          Libraries
        US Imperial                        US Metric
          Windows                            Windows
```

Browse to the **Windows** folder under the Imperial or Metric library – use Imperial if you are using Imperial units or use Metric if you are using Metric units.

15.

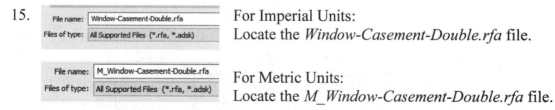

For Imperial Units:
Locate the *Window-Casement-Double.rfa* file.

For Metric Units:
Locate the *M_Window-Casement-Double.rfa* file.

Press **Open**.

16.

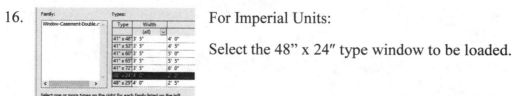

For Imperial Units:

Select the 48" x 24″ type window to be loaded.

For Metric Units:

From the Type Selector drop-down list, select the 1400 x 600mm size for the M_Window-Casement-Double window.

17. The family is already loaded in the project but doesn't have this size. Select the second option to add the additional type.

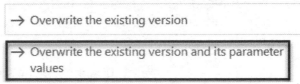

You are trying to load the family Window-Casement-Double, which already exists in this project. What do you want to do?

→ Overwrite the existing version

→ Overwrite the existing version and its parameter values

18. Place the window **6′-6″** [**3000 mm**] from the inner left wall.

Right click and select **Cancel** to exit the command.

19. The arrows appear on the exterior side of the window. If the window is not placed correctly, left click on the arrows to flip the orientation.

20. Select the window so it highlights.

21. Select the **Array** tool under the Modify panel.

22. Select the midpoint of the window as the basepoint for the array.

Endpoint

23.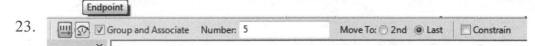

Enable Group and Associate. This assigns the windows placed to a group and allows you to edit the array.

Set the array quantity to **5** on the options bar located on the bottom of the screen. Enable **Last**.

Enabling Constrain ensures that your elements are placed orthogonally.

Array has two options. One option allows you to place elements at a set distance apart. The second option allows you to fill a distance with equally spaced elements.

We will fill a specified distance with five elements equally spaced.

24.

Pick a point **49'-0"** [**14,935.20**] from the first selected point to the right.

25.

You will see a preview of how the windows will fill the space.

Press **ENTER** to accept.

Cancel out of the command.

26. Select the **Measure** tool on the Quick Access toolbar.

27. 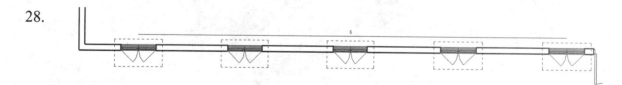 Check the distance between the windows and you will see that they are all spaced equally.

28.

Window around the entire array to select all the windows.

The array count will display.

29. Use the **Mirror→Draw Mirror Axis** tool to mirror the windows to the other side of the wall opposite the door.

30.

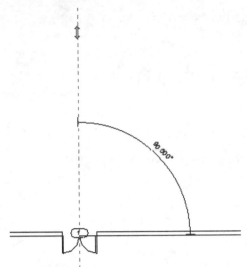

Select the center of the door as the start point of the mirror axis.

Move the cursor upwards at a 90 degree angle and pick a point above the door.

31. Left pick anywhere in the graphics window to complete the command.

You will get an error message, and the windows will not array properly if you do not have the angle set to 90 degrees or your walls are different lengths.

32. Switch to a **3D** View.

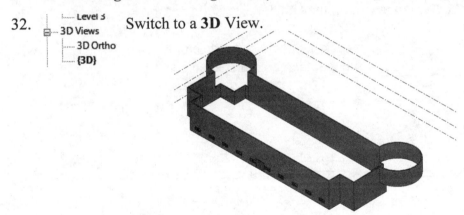

33.

Set the Model Graphics Style to **Consistent Colors**.

We have created a conceptual model to show a client.

34. Save the file as *ex2-6.rvt*.

Exercise 2-7
Creating a Conceptual Mass

Drawing Name: New Conceptual Mass
Estimated Time: 60 minutes

This exercise reinforces the following skills:

- ❑ Masses
- ❑ Levels
- ❑ Load from Library
- ❑ Aligned Dimension
- ❑ Flip Orientation

1. Close any open files.

2. Use the Application Menu and go to **New→Conceptual Mass**.

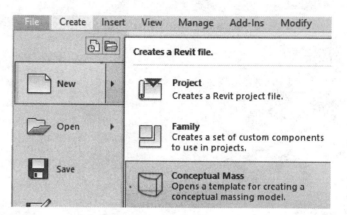

2.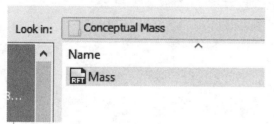

Select the **Mass** template.

Press **Open**.

3. Activate the **South Elevation**.

4.   On the Create ribbon, select the **Level** tool.

5.   Place a Level 2 at **50′ 0″**.

    You can type in 50 as a listening dimension to position the level or place the level, then modify the elevation value by selecting the dimension.

6. Activate the **Level 1** floor plan.

7. Activate the Create ribbon.

    Select the **Plane** tool from the Draw panel.

8. Draw a vertical plane and a horizontal plane to form a box.

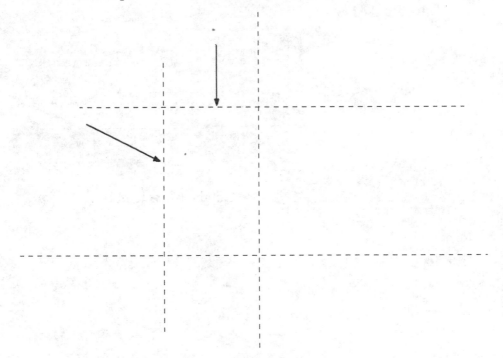

9.

100' - 0"

60' - 0"

Use the ALIGNED DIMENSION tool to add dimensions to the reference planes.

To modify the dimensions, select the plane, then modify the temporary dimension. The permanent dimension will automatically update.

Set the horizontal dimension to 100′ 0″ overall.

Set the vertical dimension to 60′ 0″ overall.

Remember the permanent dimensions are driven by the values of the temporary dimensions. To set the temporary dimension values, select the reference planes.

10.

Load Family

Load from Library

Activate the Insert ribbon.

Select Load Family from **Library→Load Family**.

11.

ProgramData
Autodesk
RVT 2020
Libraries
US Imperial
Mass

Browse to the **Mass** folder.

12.

Mass

Name

RFA Arch
RFA Barrel Vault
RFA Box
RFA Cone
RFA Cylinder
RFA Diagnostic Tripod-1 point
RFA Diagnostic Tripod-2 point
RFA Diagnostic Tripod-3 point
RFA Diagnostic Tripod-4 point
RFA Dome
RFA Gable

Hold down the CTRL Key and select the *Dome*, *Box*, and *Cylinder* files.
Press **Open**.

13.

Component

Model

Activate the Create ribbon.

Select the **Component** tool on the Model panel.

14. 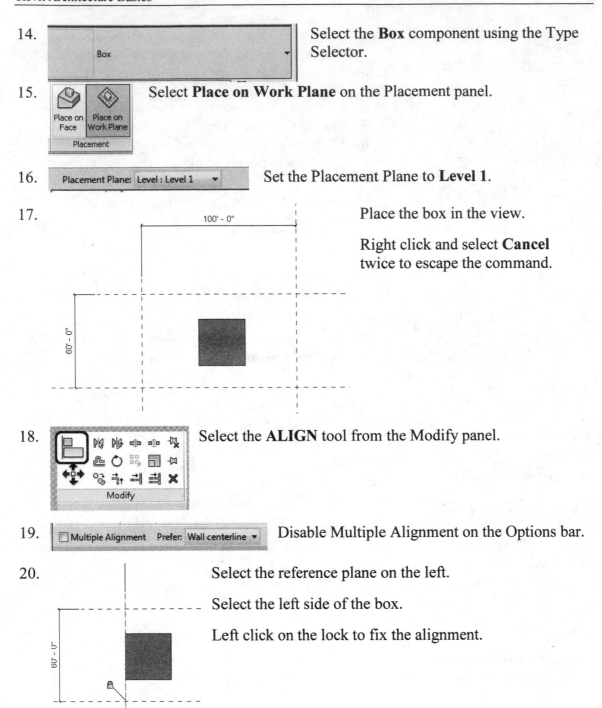 Select the **Box** component using the Type Selector.

15. Select **Place on Work Plane** on the Placement panel.

16. Set the Placement Plane to **Level 1**.

17. Place the box in the view.

Right click and select **Cancel** twice to escape the command.

18. Select the **ALIGN** tool from the Modify panel.

19. Disable Multiple Alignment on the Options bar.

20. Select the reference plane on the left.

Select the left side of the box.

Left click on the lock to fix the alignment.

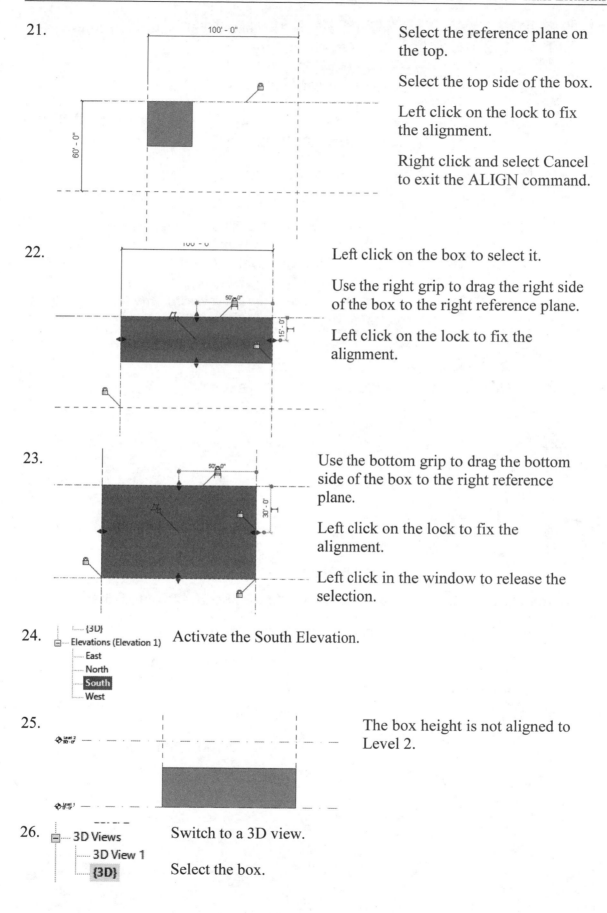

21. Select the reference plane on the top.

Select the top side of the box.

Left click on the lock to fix the alignment.

Right click and select Cancel to exit the ALIGN command.

22. Left click on the box to select it.

Use the right grip to drag the right side of the box to the right reference plane.

Left click on the lock to fix the alignment.

23. Use the bottom grip to drag the bottom side of the box to the right reference plane.

Left click on the lock to fix the alignment.

Left click in the window to release the selection.

24. Activate the South Elevation.

25. The box height is not aligned to Level 2.

26. Switch to a 3D view.

Select the box.

27.

Dimensions	
Width	100' 0"
Height	50' 0"
Depth	60' 0"
Identity Data	
Image	

Change the Height to 50' 0" in the Properties pane.

Press the **Apply** button.

Press ESC to release the box selection or left click anywhere in the window.

28. Floor Plans
— Level 1
— **Level 2**

Activate the **Level 2** floor plan.

29.

Underlay	
Range: Base Level	Level 1
Range: Top Level	Level 2
Underlay Orientation	Look down

With Level 2 highlighted, scroll down the Properties pane to the Underlay category.
Set the Range: Base Level to **Level 1**.

If you don't set the underlay correctly, you won't see the box that was placed on Level 1.

30. 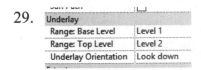 Component

Model

Activate the Create ribbon.

Select the **Component** tool on the Model panel.

31.

Properties	×
Cylinder	▾

Select the **Cylinder** component using the Type Selector.

32.

Dimensions	
Radius	25' 0"
Height	5' 0"

In the Properties Pane:

Set the Radius to **25′ 0″**.
Set the Height to **5′ 0″**.

If you need to adjust the position of the cylinder: Set the Offset to -5′ 0″.

This aligns the top of the cylinder with the top of the box.

33. Place on Face | Place on Work Plane

Placement

Select **Place on Work Plane** on the Placement panel.

34. Placement Plane: Level : Level 2 ▾

Set the Placement Plane to **Level 2**.

35.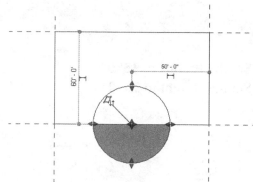

Place a cylinder at the midpoint of the bottom edge of the box.

Right click and select **Cancel** twice to escape the command.

36.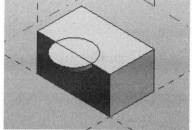

Switch to a 3D View.

37.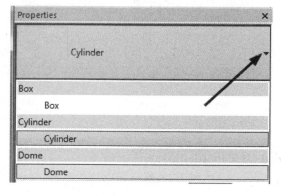

You should see the box as well as the cylinder.

38.

Activate **Level 2**.

39. Activate the Create ribbon.

Select the **Component** tool on the Model panel.

40. Select the **Dome** component using the Type Selector.

41. In the Properties Pane:

Set the Radius to **15′ 0″**.
Set the Height to **20′ 0″**.

42. 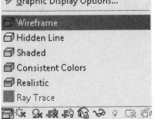 Select **Place on Work Plane** on the Placement panel.

43. Placement Plane: Level : Level 2 ▼   Set the Placement Plane to **Level 2**.

44.

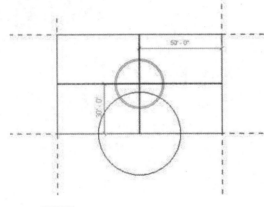

Place the dome so it is centered on the box.

Right click and select Cancel twice to exit the command.

45. 3D Views
      3D View 1
      {3D}

Switch to a 3D View.

46. Graphic Display Options...
    Wireframe
    Hidden Line
    Shaded
    Consistent Colors
    Realistic
    Ray Trace

Switch to a **Wireframe** display.

47.

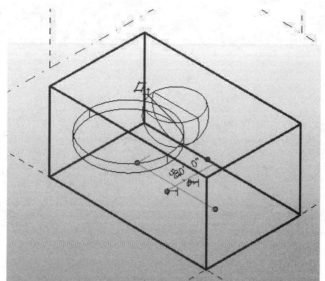

The dome is upside down.

Select the dome.

48.

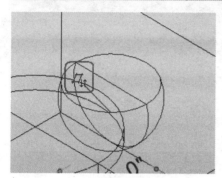

Left click on the orientation arrows to flip the dome.

49. In the Properties pane:

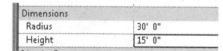

Dimensions	
Radius	30' 0"
Height	15' 0"

Change the Radius to **30′ 0″**.
Change the Height to **15′ 0″**.
Press **Apply**.

50.

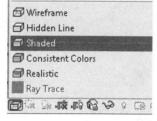

Change the display back to **Shaded**.

51. Save as *ex2-7.rfa*.

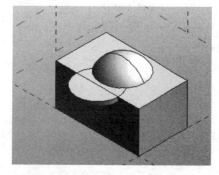

Exercise 2-8
Using a Conceptual Mass in a Project

Drawing Name: New
Estimated Time: 20 minutes

This exercise reinforces the following skills:

- ❑ Masses
- ❑ Load from Library
- ❑ Visibility/Graphics

1. Close any open files.

    Use the Application Menu and go to
    New→Project.

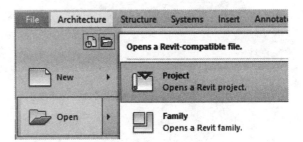

    Press **OK** to accept the Architectural Template.

2. Type **VV** to launch the Visibility/Graphics dialog.

    Enable **Mass** visibility on the Model Categories tab.

    Press **OK**.

3. Switch to a South Elevation.

4. Set Level 2 to **50′ 0″**.

    To change the dimension value, just left click on the dimension and type the new dimension.

5. Activate Level 1.

6. Activate the **Insert** Ribbon.

    Select **Load Family** from the Load from Library panel.

7. 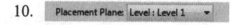 Locate *ex2-7.rfa*.

   Press **Open**.

8. Activate the Architecture ribbon.

   Select **Component→Place a Component** from the Build panel.

9. Select **Place on Work Plane** from the Placement panel.

10. Set the Placement Plane to **Level 1**.

11. Click to place the mass in the view.

    Right click and select Cancel to exit the command.

12. Switch to a 3D view.

13. Type **VV** to launch the Visibility/Graphics dialog.

    Enable **Mass** visibility on the Model Categories tab.

    Press **OK**.

    Right click and select **Zoom to Fit** to see the placed mass family.

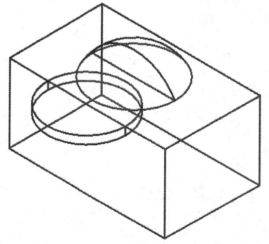

Notice that visibility settings are view-specific. Just because you enabled mass visibility in one view does not mean masses will be visible in all views.

14. 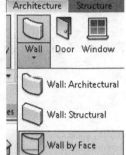 Select the **Wall by Face** tool from the Architecture ribbon.

15.

Select the four sides of the box, the outside arc of the cylinder and the two sides of the dome to place walls.

Right click and select CANCEL.

16. 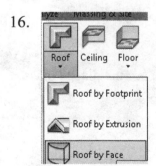 Select the **Roof by Face** tool on the Build panel on the Architecture ribbon.

17.

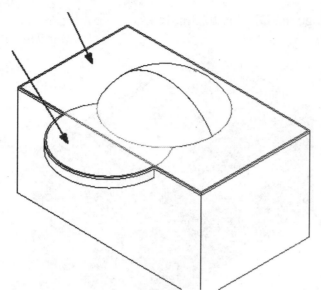

Select the top face of the box and the top half of the cylinder.

18.

Create
ı Roof
tion

Select **Create Roof** on the ribbon.

19. The roof is placed. Right click and select **Cancel**.

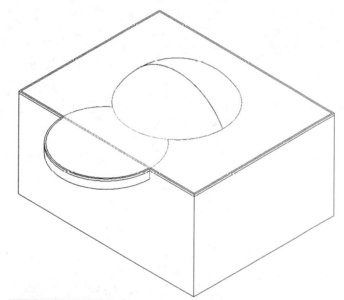

20.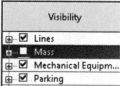

Type **VV** to launch the Visibility/Graphics dialog.

Disable **Mass** visibility on the Model Categories tab.

Press **OK**.

21.

⚘ Graphic Display Options...
🗗 Wireframe
🗗 Hidden Line
🗗 Shaded
🗗 Consistent Colors
🗗 Realistic
⛁ Ray Trace

Change the Display to **Shaded**.

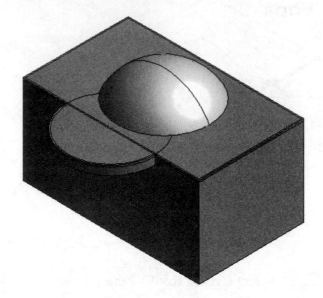

22. Save as *ex2-8.rvt.*

Additional Projects

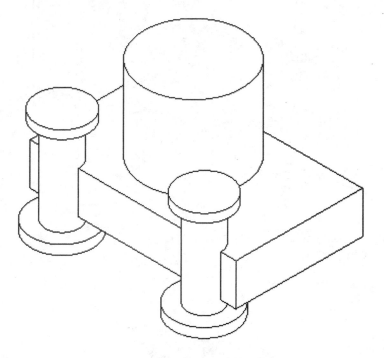

1) Create a conceptual mass family like the one shown.

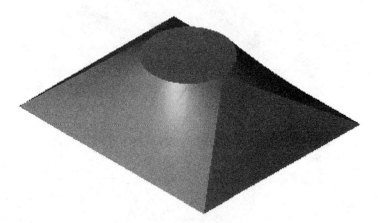

2) Make the shape shown using a Blend. The base is a rectangle and the top is a circle located at the center of the rectangle.

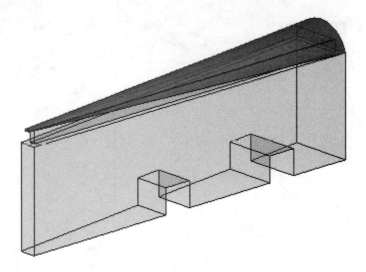

3) Make a conceptual mass family as shown.
   Use Solid Void to create the two openings.

4) Design a pergola using 4″ x 4″ posts and beams.

Lesson 2 Quiz

True or False

1. Masses can be created inside Projects or as a Conceptual Mass file.
2. Forms are always created by drawing a sketch, selecting the sketch, and clicking ⬚ Create Form.
3. In order to see masses, Show Mass must be enabled.
4. Masses are level-based.
5. You can modify the dimensions of conceptual masses.

Multiple Choice [Select the Best Answer]

6. Faces on masses can be converted to the following:

   A. Walls
   B. Roofs
   C. Floors
   D. Doors
   E. A, B, and C, but NOT D

7. You can adjust the distance a mass is extruded by:

   A. Editing the temporary dimension that appears before a solid form is created.
   B. Using the ALIGN tool.
   C. Using the 3D drag tool.
   D. Using the Properties pane located on the left of the screen.
   E. All of the above

8. To insert a conceptual mass family in a project:

   A. Use the INSERT tool.
   B. Use the PLACE COMPONENT tool.
   C. Use the BLOCK tool.
   D. Use the MASS tool.

9. Masses are hosted by:

   A. Projects
   B. Work Planes
   C. Families
   D. Files

10. Mass Visibility can be controlled using:

    A. The SHOW MASS tool
    B. Display Settings
    C. Object Settings
    D. View Properties

11. Each mass is defined by _____ .

    A. A single profile sketch
    B. Multiple profile sketches
    C. Properties
    D. Materials

12. Revit comes with many pre-made mass shapes. Select the mass shape NOT available in the Revit mass library:

    A. BOX
    B. ARC
    C. CONE
    D. TRIANGLE-RIGHT

ANSWERS:

    1) T; 2) T; 3) T; 4) T; 5) T; 6) E; 7) E; 8) B; 9) B; 10) A; 11) A; 12) B

Lesson 3
Floor Plans

> ➤ Put a semi-colon between snap increments, not a comma.
> ➤ If you edit the first grid number to 'A' before you array, Revit will automatically increment them alphabetically for you: A, B, C, etc.
> ➤ You will need to 'Ungroup' an element in an array before you can modify any of the properties.
> ➤ To keep elements from accidentally moving, you can select the element and use **Pin Objects** or highlight the dimension string and lock the dimensions.
> ➤ You can purge the unused families and components in your project file in order to reduce the file space. Go to **File→Purge Unused**.
> ➤ Revit creates stairs from the center of the run, so it may be helpful to place some reference planes or detail lines defining the location of the starting points for the runs of any stairs.
> ➤ Floor plans should be oriented so that North is pointing up or to the right.
> ➤ The direction you draw your walls (clockwise or counterclockwise) controls the initial location of the exterior face of the wall. Drawing a wall from left to right places the exterior on the top. Drawing a wall from right to left places the exterior on the bottom. When you highlight a wall, the blue flip orientation arrows are always adjacent to the exterior side of the wall.

Throughout the rest of the text, we will be creating new types of families. Here are the basic steps to creating a new family.

1. Select the element you want to define (wall, window, floor, stairs, etc.).
2. Select **Edit Type** from the Properties pane.
3. Select **Duplicate**.
4. Rename: Enter a new name for your family type.
5. Redefine: Edit the structure, assign new materials, and change the dimensions.
6. Reload or Reassign: Assign the new type to the element.

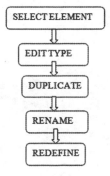

What is a Floor Plan?

A floor plan is a scaled diagram of a room or a building viewed from above. The floor plan may show an entire building, a single building level, or a single room. It may include dimensions, furniture, casework, appliances, electrical outlets, plumbing fixtures or anything else necessary to provide information. A floor plan is not a top view or a bird's eye view. It is drawn to scale. Most plan views represent a cut section of room, 4' above the floor.

Building projects typically start with a site plan. A site plan designates a building site or lot and involves grading the earth and placing the building foundation or pad. Some building projects may assume a pre-existing foundation or pad.

For most building projects, it is normal to start the project by laying out a grid. The grid lines allow references to be made in terms of position/location of various elements in the project. A surveyor will set out the grid lines on the site to help the construction crew determine the placement of structural columns, walls, doors, windows, etc.

Exercise 3-1
Placing a Grid

Drawing Name: default.rte [DefaultMetric.rte]
Estimated Time: 30 minutes

This exercise reinforces the following skills:

- ❑ Units
- ❑ Snap Increments
- ❑ Grid
- ❑ Array
- ❑ Ungroup
- ❑ Dimension
- ❑ Dimension Settings

1.   Start a new project using the default template.

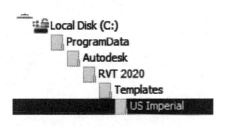

If you did not add the default template to the drop down list, you can start a new project from the Applications Menu and then select default.rte from the Templates folder.

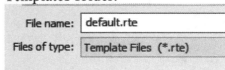

2.   Activate **Level 1**.

3.   Activate the **Manage** ribbon.

      Under **Settings**, select **Snaps**.

4.   Set the snap increments for length to **40′; 25′; 5′ [1500;7620;1220]**.

Press **OK**.   *Use a semi-colon between the distance values.*

6.   Activate the **Architecture** ribbon.

7.   Select **Grid** under the Datum panel.

8.   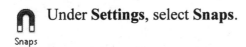 Pick on the left side of the graphics window to start the grid line.

9.   Pick a point above the first point.

Right mouse click and select Cancel twice to exit the Grid mode.

Your first grid line is placed.

10.   Select the Grid text to edit it.

11.   Change the text to **A**.

Left click in the graphics window to exit edit mode.

If you edit the first grid letter before you array, Revit will automatically increment the arrayed grid elements.

12.   Select the grid line so it is highlighted.

Select the **Array** tool under the Modify panel.

Modify

13.

Disable **Group and Associate**. This keeps each grid as an independent element.
Set the Number to **8**.
Enable Move To: **2ⁿᵈ**.
Enable **Constrain**. This keeps the array constrained orthogonally.

14.   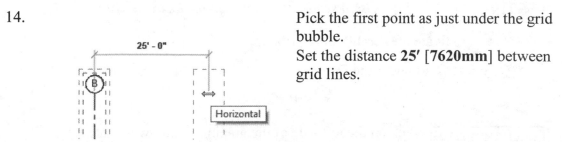 Pick the first point as just under the grid bubble.
Set the distance **25′** [**7620mm**] between grid lines.

15. 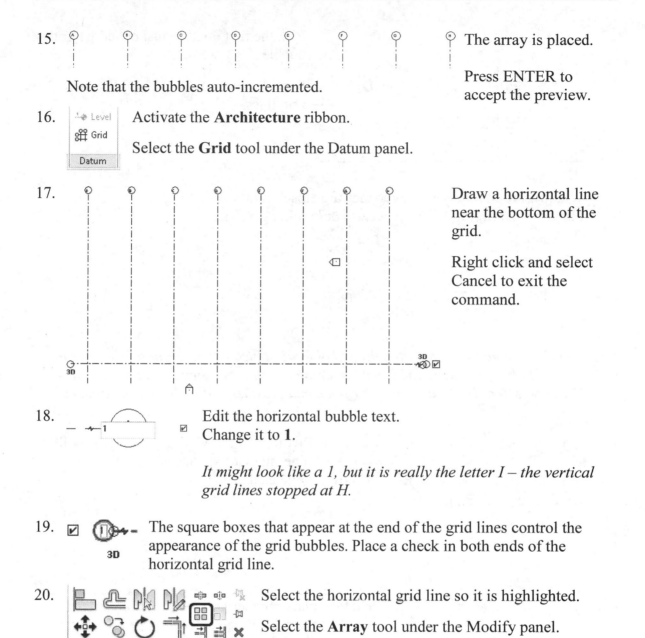 The array is placed.

Press ENTER to accept the preview.

Note that the bubbles auto-incremented.

16. Activate the **Architecture** ribbon.

Select the **Grid** tool under the Datum panel.

17. Draw a horizontal line near the bottom of the grid.

Right click and select Cancel to exit the command.

18. Edit the horizontal bubble text. Change it to **1**.

It might look like a 1, but it is really the letter I – the vertical grid lines stopped at H.

19. The square boxes that appear at the end of the grid lines control the appearance of the grid bubbles. Place a check in both ends of the horizontal grid line.

20. Select the horizontal grid line so it is highlighted.

Select the **Array** tool under the Modify panel.

21. Disable **Group and Associate**.
Set the Number to **3**.
Enable Move To: **2nd**.
Enable **Constrain**.

22.
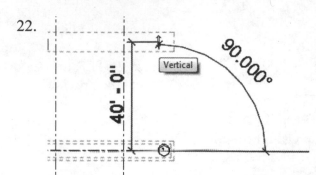

Pick the first point as just under the grid bubble.

Set a distance of **40′ [12200]** between grid lines.

23. Window around the gridlines so they are all selected.
Use the FILTER tool, if necessary, to only select gridlines.

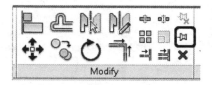

Category:	Count:
☑ Elevations	3
☑ Grids	11
☑ Views	4

If you see Model Groups listed instead of Grids it means you forgot to disable the Group and Associate option when you created the array. Simply select the grid lines and select Ungroup from the ribbon and the lines will no longer be grouped.

Select the PIN tool on the Modify panel. This will fix the gridlines in place so they can't accidentally be moved or deleted.

24. Save the file as *ex3-1.rvt*.

 If you delete a grid line and then place another grid line, the numbering picks up where you left off. For example, if you delete grid line 3 and then add a new grid line, the new grid line will be labeled 4.

Exercise 3-2
Placing Walls

Drawing Name: ex3-1.rvt
Estimated Time: 20 minutes

This exercise reinforces the following skills:

- Walls
- Mirror
- Filter
- Move

1.   Open or continue working in *ex3-1.rvt*.

2.   **Architecture**   Activate the **Architecture** ribbon.

3.   Type **VG** to launch the Visibility/Graphics dialog.
Select the **Annotations Categories** tab.

4.   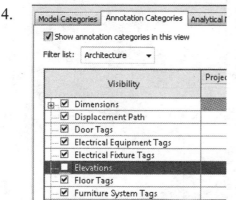   Turn off the Visibility of Elevations in the Visibility/Graphics dialog by unchecking the box.

Press **OK**.

This turns off the visibility of the elevation tags in the display window. This does not DELETE the elevation tags or elevations.

Many users find the tags distracting.

5.   Select the **Wall** tool from the Build panel on the Architecture ribbon.

6.   

Enable **Chain**.

This eliminates the need to pick a start and end point.

7.   Place walls as shown.

   To enable the grid bubbles on the bottom, select the grid and place a check in the small square.

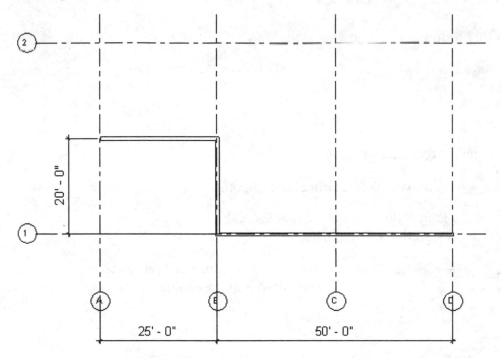

Dimensions are for reference only. Do not place dimensions.

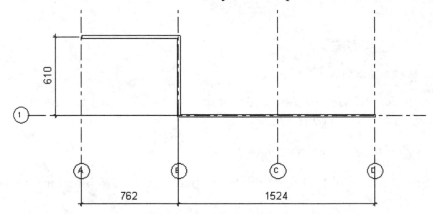

This layout shows units in centimeters.

8.   Window about the walls you just placed.

9.   Select the **Filter** tool.

10. You should just see Walls listed.

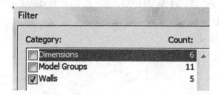

If you see other elements listed, uncheck them and this will select only the walls.

Press **OK**.

Many of my students use a crossing instead of a window or have difficulty selecting just the walls. Use the Filter tool to control your selections and potentially save a lot of time.

11.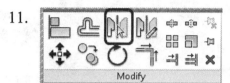

Select the **Mirror→Pick Axis** tool under the Modify Panel.

The Modify Walls ribbon will only be available if walls are selected.

12. | Modify | Walls | ☑ Copy |

Verify that **Copy** is enabled on the Options bar.

13.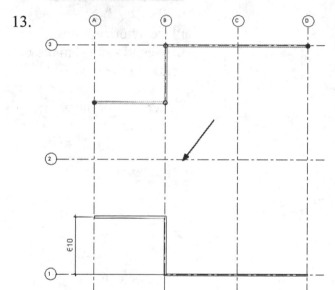

Select the grid line labeled 2 as the mirror axis.

The walls should mirror over.

14. | Architecture |

Select the **Wall** tool from the Build panel on the Architecture ribbon.

15. Draw a vertical wall to close the west side of the building.

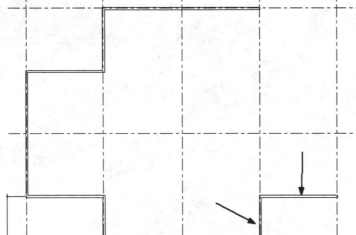

16. Draw two more walls.

Draw a small vertical wall.
Place a horizontal wall between D and E in the center of the bay.

Cancel out of the wall command.

17. Select all the walls except for the horizontal wall located between D and E grids.

You can select by holding down the Ctrl key or use FILTER.

782   1524

610

18. 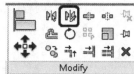 Select **Mirror→Draw Axis** from the Modify panel.

19. 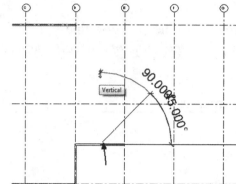 Select the midpoint of the small horizontal wall as the start point for the mirror axis.

Make sure you hold that axis at 90 degrees.

Left click to complete drawing the axis.

Cancel out of the command.

20. 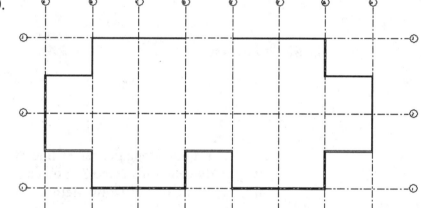 Your building should look like this.

21. Select the **Trim** tool from the Modify Panel on the Modify ribbon.

22. Select the two upper horizontal walls. The trim tool will join them together. Cancel out of the command.

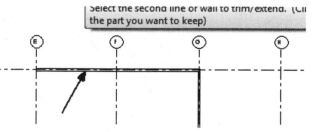

23. Activate the **Architecture** ribbon.

Select the **Grid** tool from the Datum panel on the Architecture ribbon.

24. Select the **Pick** mode.

25. Offset: 20' 0" ☐ Lock   Set the Offset to **20'** on the Options bar.

26. 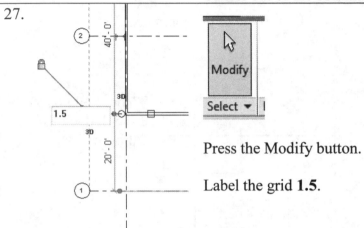   Select Grid 1.

You should see a preview of the new grid line.
Add a grid line along the center line of the walls between Grids 1 and 2.

Cancel out of the command.

27.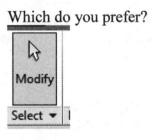

Press the Modify button.

Label the grid **1.5**.

28. 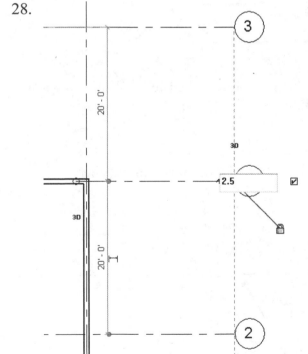   Add a grid line along the center line of the walls between Grids 2 and 3. You can use the Offset method or Draw Method.

Which do you prefer?

Press the Modify button.

Label the grid **2.5**.

29.

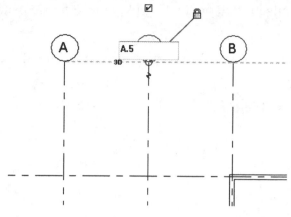

Add a vertical grid line centered between Grids A and B.

You can enter a listening distance of 12' 6" [3810] to place or pick Grid A and offset using a value of 12' 6" on the Option bar.

Press the Modify button.

Label the grid **A.5**.

30.

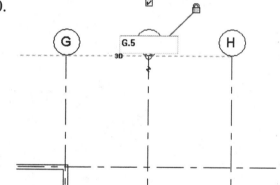

Add a vertical grid line centered between Grids G and H.

You can enter a listening distance of 12' 6" [3810] to place.

Press the Modify button.

Label the grid **G.5**.

31.

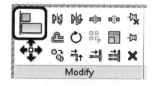

Activate the Modify Ribbon.
Select the **Align** tool on the Modify Panel.

32. ☑ Multiple Alignment   Prefer: Wall centerline ▼   Enable **Multiple Alignment**.

Set **Wall centerline** as the preference in the Options bar.

33.

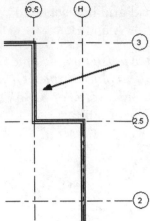

Shift the walls located on the G grid to the G.5 grid using ALIGN.

Select the G.5 grid first, then select the center line of the walls.

Right click and select CANCEL.

34.

Shift the walls located on the B grid to the A.5 grid using ALIGN.

Select the A.5 grid first, then select the center line of the walls.

Right click and select CANCEL twice to exit the command.

35. Save as *ex3-2.rvt*.

When drawing walls, it is important to understand the *location line*. This is the layer from which the wall is drawn and controlled. The location line controls the position of the wall for flipping. It also sets the wall's position when wall thickness is modified. If you change the wall type, say from brick to wood stud, the stud location will be maintained. You can always select the walls and use 'Element Properties' to change the location line. The new user should experiment with this function to fully understand it.

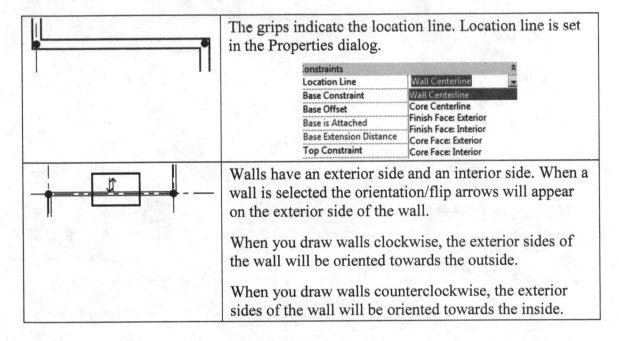

	The grips indicate the location line. Location line is set in the Properties dialog.
	Walls have an exterior side and an interior side. When a wall is selected the orientation/flip arrows will appear on the exterior side of the wall.
	When you draw walls clockwise, the exterior sides of the wall will be oriented towards the outside.
	When you draw walls counterclockwise, the exterior sides of the wall will be oriented towards the inside.

Starting from an AutoCAD file

On some projects, you will receive an AutoCAD file, either from the architect or a subcontractor. In the next exercise, you learn how to bring in an AutoCAD file and use it to create a Revit 3D model.

Exercise 3-3
Converting an AutoCAD Floor plan

Drawing Name: autocad_floorplan.dwg
Estimated Time: 30 minutes

This exercise reinforces the following skills:

- ❑ Import CAD
- ❑ Duplicate Wall Type
- ❑ Wall Properties
- ❑ Trim
- ❑ Orbit

Metric units and designations are indicated in brackets [].
There is a video for this lesson on my Moss Designs YouTube channel:
https://www.youtube.com/user/MossDesigns

1.   Go to **New→Project**.

2.   Press **OK** to use the default template.

3.   Activate the **Insert** ribbon.

4.   Select the **Import CAD** tool from the Import Panel.

5.   Locate the *autocad_floor_plan.dwg*.
   This file is included in the downloaded Class Files.

6.

   Set Colors to **Preserve**.
   Set Layers to **All**.
   Set Import Units to **Auto-Detect**.
   Set Positioning to: **Auto- Center to Center**.
   Press **Open**.

7.

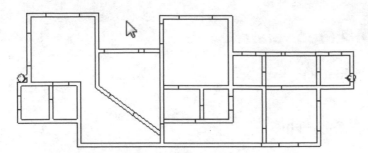

Pick the imported CAD data so it highlights.

8.

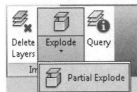

Select the **Partial Explode** tool under the Import Instance panel.

9.

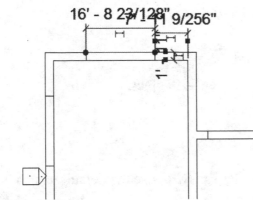

Select a line so it is highlighted.

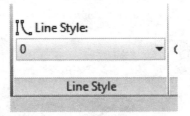

Note the Line Style listed in the Properties pane and on the ribbon.

10.

Activate the **Manage** ribbon.

Under Settings→Additional Settings:

Select **Line Styles**.

11.

Category	Line Weight Projection	Line Color	Line Pat
⊟ Lines	1	RGB 000-166-000	Solid
0	1	Black	Solid
0wall	1	Black	Solid
0wallthick	4	Black	Solid
<Area Boundary>	6	RGB 128-000-255	Solid

Left click on the + symbol next to Lines to expand the list.
Locate the Line Styles named **0**, **0wall**, and **0wallthick**.

12.

Category	Projection	Line Color	Line Pat
⊟ Lines	1	RGB 000-166-000	Solid
0	1	Cyan	Solid
0wall	1	Cyan	Solid
0wallthick	4	Cyan	Solid

Change the color of **0**, **0wall**, and **0wallthick** line styles to **Cyan**.

Click on the color and then select Cyan from the dialog box.
Press **OK**.

13. 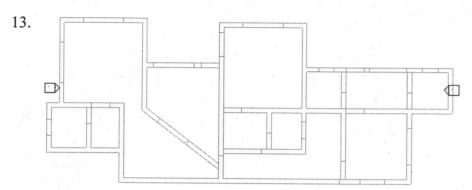 The imported lines will change color to Cyan.

14. Select **Project Units** from the Manage ribbon.

15. Left click on the **Length** button.

16. Set the Rounding **To the nearest ¼"**.

    Press **OK** twice to close the dialogs.

17. Select the **Measure** tool from the Quick Access toolbar.

18. Measure the wall thickness.

    It should read 1'-11".

19. Activate the **Architecture** ribbon.

20. Select the **Wall** tool under the Build panel.

    We need to create a wall style that is 1'-11" thick.

21. Select **Edit Type** from the Properties pane.

22. Select **Duplicate**.

23. Change the Name to **Generic – 1' 11"** [Generic – 584.2mm].

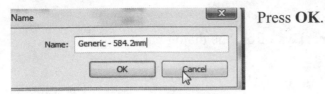

Press **OK**.

24. 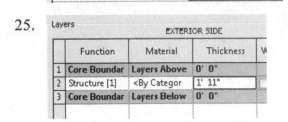 Select **Edit** next to Structure.

25.

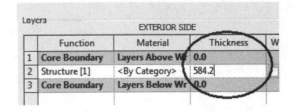

Change the thickness to **1' 11"** [**584.2**].

Press **OK**.

26. Press **OK** to exit the Properties dialog.

27. 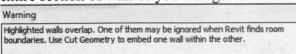 Select the **Pick Lines** tool under the Draw panel.

28. Location Line: Finish Face: Ext ▼   On the Options bar, select **Finish Face: Exterior** for the Location Line.

This means you should select on the exterior side of the wall to place the wall correctly.

29. When you select the line, a preview will appear to indicate the wall orientation. To switch the wall orientation, press the SPACE bar or move the cursor to make a different selection. Select so the dashed line appears inside the wall.

30. Move around the model and select a section of each wall. Do not try to select the entire section of wall or you will get an error message about overlapping walls.

> **Warning**
> Highlighted walls overlap. One of them may be ignored when Revit finds room boundaries. Use Cut Geometry to embed one wall within the other.

Delete any overlapping walls.

31.

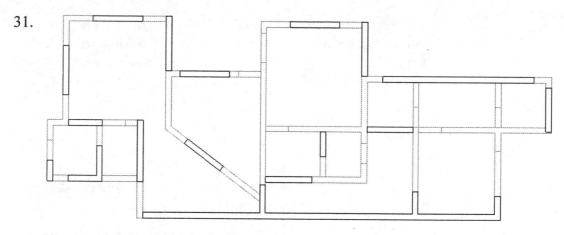

Your model should look similar to this. Note only partial walls have been placed on many of the sections.

32. Activate the View ribbon.

Select **3D View** from the Create panel.

33. The model shows the walls placed.

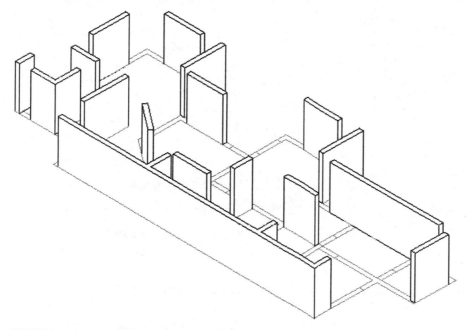

34. Use the Trim tool to extend and trim the walls to create the floor model.

35. When you mouse over the second selection on the TRIM command, you will see a preview of how the walls will connect. If the preview does not look proper, shift your mouse slightly until the preview appears correct.

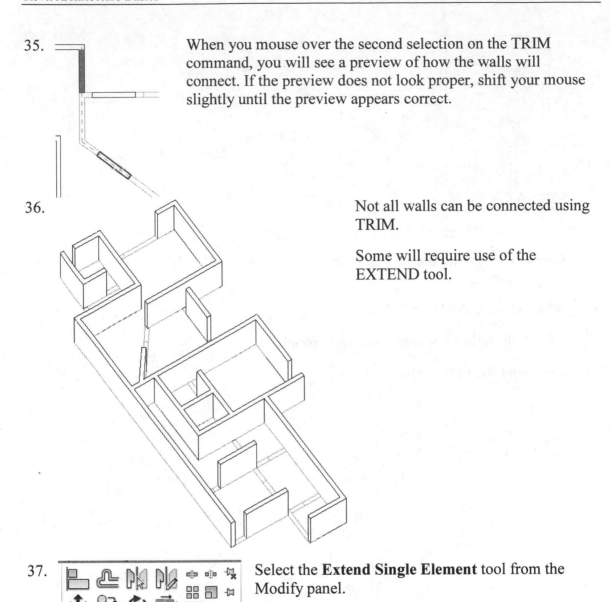

36. Not all walls can be connected using TRIM.

Some will require use of the EXTEND tool.

37. Select the **Extend Single Element** tool from the Modify panel.

To use this tool, select the face of the wall you want to extend TO first, then the wall you want to extend.

38. Orbit the model so you can see where to trim and extend walls.

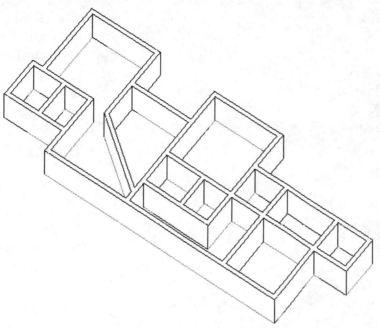

You will still see the colored lines from the original AutoCAD file. They are there to help you check that you placed the walls correctly.

39.   Save as *ex3-3.rvt*.

Wall Properties

Drawing Name: ex3-2.rvt
Estimated Time: 40 minutes

This exercise reinforces the following skills:

- ❑ Walls
- ❑ Filter
- ❑ Wall Properties
- ❑ Flip Orientation
- ❑ Join

> *[ ] brackets are used to indicate metric units or designations.*

1. Open or continue working in *ex3-2.rvt*.

2. Select the exterior walls using **Filter**.
   Use crossing to select the floor plan.

3. ▽   Select the **Filter** tool.
   Filter

4. 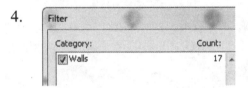   Uncheck all the objects except for Walls.

   Press **OK**.
   This way, only walls are selected.

5. ▾ ⊞ Edit Type   Select **Edit Type** in the Properties panel.

6. [Duplicate...]   Select **Generic – 8″** [**Generic – 200 mm**] under Type.

   Select **Duplicate**.

7. 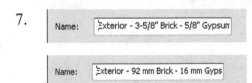   For the name, enter:

   Exterior - 3-5/8″ Brick - 5/8″ Gypsum
   [**Exterior- 92mm Brick -16mm Gypsum**].

   Press **OK**.

8. | Parameter | Value |
   |-----------|-------|
   | **Construction** | |
   | Structure | Edit... |

   Select **Edit** for Structure.

9. [<< Preview]   Enable the Preview button to expand the dialog so you can see the preview of the wall structure.

10. [Insert]   Select the **Insert** Button.

In the dialog, notice that the bottom of the list indicates toward the interior side and the top of the list indicates toward the exterior side of the wall.

11. Select the second line in the Function column.

Select **Finish 1 [4]** from the drop-down list.

12. Press the **Insert** button until you have seven layers total.

Arrange the layers as shown.

Assign the Functions as shown:

Layer 1: Finish 1 [4]   Layer 5: Structure [1]
Layer 2: Thermal/Air Layer   Layer 6: Core Boundary
Layer 3: Substrate [2]   Layer 7: Finish 2 [5]
Layer 4: Core Boundary

13. Select the **Material** column for Layer 1.

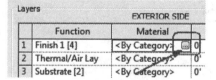

14. In the Materials search box, type **bri**.

15. Any materials with 'bri' as part of the name or description will be listed.

Highlight **Brick, Common** from the Name list.

16. Note that we can set how we want this material to appear when we render and shade our model.

Enable **Use Render Appearance**.
We only need to enable this for the outside materials. The inner materials are not visible during rendering.

17.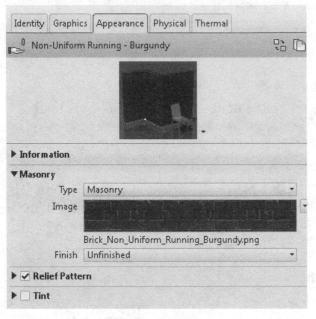

Left mouse click on the fill pattern in the Surface Pattern field.

18. Select the **Appearance** tab.

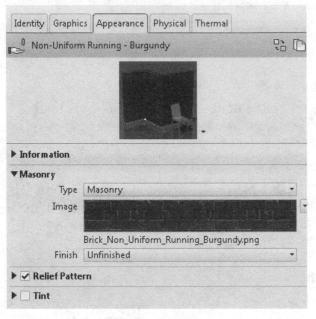

Note the image of how the brick will appear when the wall is rendered.

Press **OK**.

19.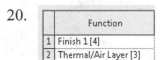

Change the Thickness to **3 5/8″ [92mm]**.

Function	Material	Thickness
1 Finish 1 [4]	Brick, Common	92.0

20.

Locate Layer 2 and verify that it is set to Thermal/Air Layer. Select the **Material** column.

Function
1 Finish 1 [4]
2 Thermal/Air Layer [3]

21.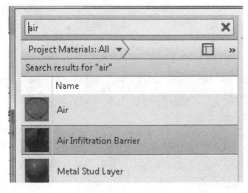

On the Materials tab:

Type **air** in the search field.

Select **Air Infiltration Barrier**.

Press **OK**.

22.

	Function	Material	Thickness
1	Finish 1 [4]	Brick, Common	0' 3 5/8"
2	Thermal/Air Layer [3]	Air Infiltration Barrier	0' 1"

EXTERIOR SIDE

	Function	Material	Thickness
1	Finish 1 [4]	Brick, Common	92.0
2	Thermal/Air Layer [3]	Air Infiltration Barrier	25.4

Set the Thickness of **Layer 2** to **1″** [**25.4**].

23.

	Function	Material	Thickness
1	Finish 1 [4]	Brick, Common	0' 3 5/8"
2	Thermal/Air Layer [3]	Air Infiltration Barrier	0' 1"
3	Substrate [2]	Plywood, Sheathing	0' 0 1/2"

	Function	Material	Thickness
1	Finish 1 [4]	Brick, Common	92.0
2	Thermal/Air Layer [3]	Air Infiltration Barrier	25.4
3	Substrate [2]	Plywood, Sheathing	17.2

Set Layer 3 to Substrate [2], **Plywood, Sheathing**; Thickness to ½″ [**17.2**].

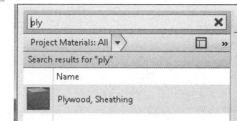

24.

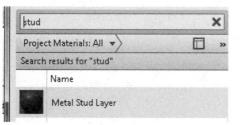

Set the Material for Structure [1] to **Metal Stud Layer**.

Press **OK**.

25.

	Function	Material	Thickness
1	Finish 1 [4]	Brick, Common	0' 3 5/8"
2	Thermal/Air Layer [3]	Air Infiltration Barrier	0' 1"
3	Substrate [2]	Plywood, Sheathing	0' 0 1/2"
4	**Core Boundary**	**Layers Above Wrap**	**0' 0"**
5	Structure [1]	Metal Stud Layer	0' 6"

	Function	Material	Thickness
1	Finish 1 [4]	Brick, Common	92.0
2	Thermal/Air Layer [3]	Air Infiltration Barrier	25.4
3	Substrate [2]	Plywood, Sheathing	17.2
4	**Core Boundary**	**Layers Above Wrap**	**0.0**
5	Structure [1]	Metal Stud Layer	152.4

Set the Thickness for Layer 4 Structure [1] to **6″** [**152.4**].

The Core Boundary Layers may not be modified or deleted.

These layers control the location of the wrap when walls intersect.

26.

	Function
1	Finish 1 [4]
2	Thermal/Air Layer [3]
3	Substrate [2]
4	**Core Boundary**
5	Structure [1]
6	**Core Boundary**
7	Finish 2 [5]

Select the Material column for Layer 7: Finish 2 [5].

27. 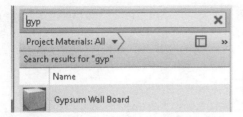 Select **Gypsum Wall Board**.

28. 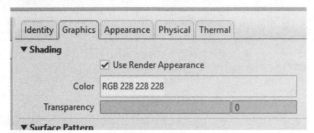 Select the Graphics tab.

Enable **Use Render Appearance**.

Press **OK**.

29. 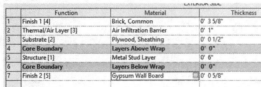 Set Layer 7 to **Finish 2 [5]**, **Gypsum Wall Board**; Thickness to **5/8″ [16mm]**.

Press **OK**.

	Function	Material	Thickness
1	Finish 1 [4]	Brick, Common	92.0
2	Thermal/Air Layer [3]	Air Infiltration Barrier	25.4
3	Substrate [2]	Plywood, Sheathing	17.2
4	**Core Boundary**	**Layers Above Wrap**	**0.0**
5	Structure [1]	Metal Stud Layer	152.4
6	**Core Boundary**	**Layers Below Wrap**	**0.0**
7	Finish 2 [5]	Gypsum Wall Board	16

30.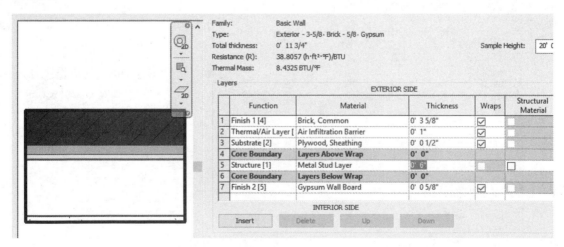

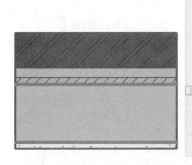
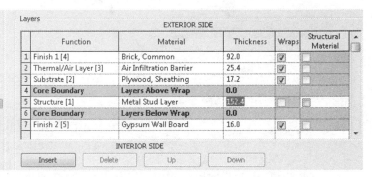

Layers

	Function	Material	Thickness	Wraps	Structural Material
			EXTERIOR SIDE		
1	Finish 1 [4]	Brick, Common	92.0	☑	☐
2	Thermal/Air Layer [3]	Air Infiltration Barrier	25.4	☑	☐
3	Substrate [2]	Plywood, Sheathing	17.2	☑	☐
4	**Core Boundary**	**Layers Above Wrap**	**0.0**		
5	Structure [1]	Metal Stud Layer	152.4	☐	☐
6	**Core Boundary**	**Layers Below Wrap**	**0.0**		
7	Finish 2 [5]	Gypsum Wall Board	16.0	☑	☐

INTERIOR SIDE

Insert | Delete | Up | Down

Set Layer 1 to Finish 1 [4], Brick, Common, Thickness: **3-5/8″** [**92**].
Set Layer 2 to Thermal/Air Layer [3], Air Infiltration Barrier, Thickness: **1″** [**25.4**].
Set Layer 3 to Substrate [2], Plywood, Sheathing, Thickness to ½″ [**17.2**].
Set Layer 4 to Core Boundary, Layers Above Wrap, Thickness to **0″**.
Set Layer 5 to Structure [1], Metal Stud Layer, Thickness to **6″** [**152.4**].
Set Layer 6 to Core Boundary, Layers Below Wrap, Thickness to **0″**.
Set Layer 7 to Finish 2 [5], Gypsum Wall Board, Thickness to **5/8″** [**16**].
Press **OK** to exit the Edit Assembly dialog.

31.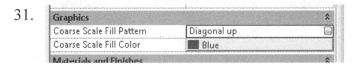

Set the Coarse Scale Fill Pattern to **Diagonal up**.

Set the Coarse Fill Color to **Blue**.

32. Press **OK** to exit the Properties dialog.
Because your walls were selected when you started defining the new wall style, your walls now appear with the new wall style.

33. If you zoom in, you will see that the coarse fill pattern is Diagonal Up and is Color Blue.

This only appears if the Detail Level is set to Coarse.

34.  Set the Detail Level to **Medium**.

You will not see the hatching or any change in line width unless the Detail Level is set to Medium or Fine.

35. We now see the wall details, but the brick side is toward the interior and the gypsum board is towards the exterior for some walls. In other words, the walls need to be reversed or flipped.

How do we know the walls need to be flipped?

36. When we pick the walls, the orientation arrows are located adjacent to the exterior side of the wall.

The orientation arrows are the blue arrows that are activated when an object is selected.

37. 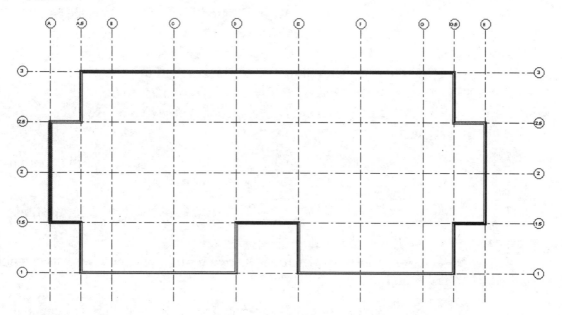 Select the wall, right click and select **Change wall's orientation** or just click the blue arrows.

38. Go around the building and flip the orientation of the walls so that the brick is on the outside.

If you have problems selecting a wall instead of a grid line, use the TAB key to cycle through the selection.

39. Save the file as *ex3-4.rvt*.

Exercise 3-5
Add Level 1 Interior Walls

Drawing Name: ex3-4.rvt
Estimated Time: 10 minutes

This exercise reinforces the following skills:
A video is available for this exercise on my Moss Designs YouTube channel:
https://www.youtube.com/user/MossDesigns

- ❑ Wall
- ❑ 3D View
- ❑ Visibility
- ❑ Wall Properties

1.   Open or continue working in *ex3-4.rvt*.

2.   ⊟ Floor Plans   Activate **Level 1**.
        Level 1
        Level 2
        Site

3.   Select the **Wall** tool on the Build panel under the Architecture ribbon.
     Wall

4.   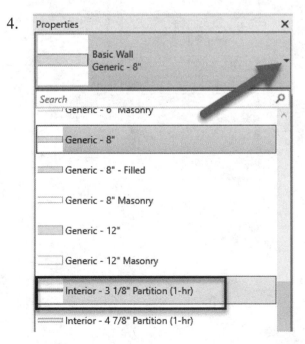   Scroll down the list of wall styles available under the Type Selector.

     Locate the **Interior- 3 1/8" Partition (1-hr)** style.

For metric, look for the **Interior-138mm Partition (1-hr)** style.

5.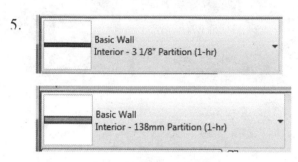

In the Properties panel, select **Interior: 3 1/8″ Partition (1-hr)** [**Basic Wall: Interior - 138mm Partition (1-hr)**].

6. Place interior walls as shown.

Imperial Units

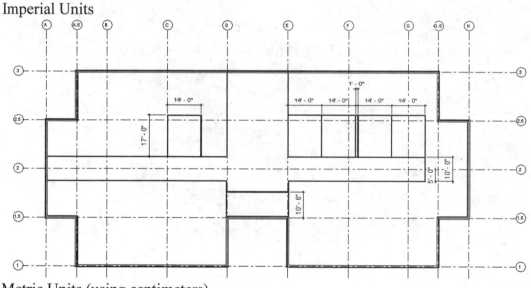

Metric Units (using centimeters)

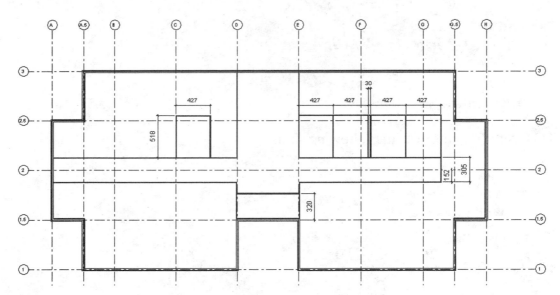

7. Activate the View ribbon.
   Select the **3D View** tool.

   We see our first floor interior walls.

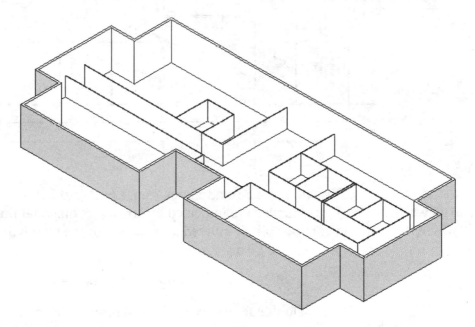

8. Save the file as *ex3-5.rvt*.

Exercise 3-6
Add Level 2 Interior Walls

Drawing Name: ex3-5.rvt
Estimated Time: 20 minutes

This exercise reinforces the following skills:

- ❑ View Properties
- ❑ Wall
- ❑ Wall Properties
- ❑ 3D View

1.   Open *ex3-5.rvt.*

2.   Activate **3D View**.

Some of our interior walls will be continuous from Level 1 to Level 2.
The walls that will be extended are indicated by arrows. A diagonal line indicates
where the walls need to be split or divided as a portion of the wall will be
continuous and a portion will be a single story.

3.   Switch view to a **South Elevation** view.

Double left click in the browser window on the South
Elevation view.

Elevations (Building Elevation)
    East
    North
    South
    West

4.    Select the level line.

Right click and select **Maximize 3D Extents**.

This will stretch the level to cover the entire building.

Repeat for the other level.

5. 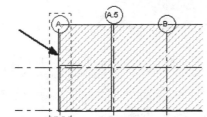   Select the first grid line labeled **A**.

6. 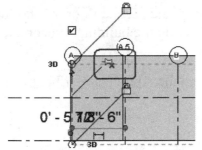   Lift click on the pin to unpin the gridline.

You have to unpin the element before it can be modified.

7. 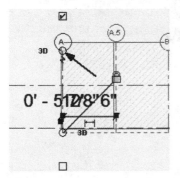   Use the grip indicated to drag the grid line above the building.

8. All the grid lines which were created as part of the array adjust. If you forgot to uncheck Group and Associate when you created the array, you will not be able to adjust the grids without either UNGROUPING or editing the group.

To ungroup, simply select one of the gridlines and select Ungroup from the ribbon.

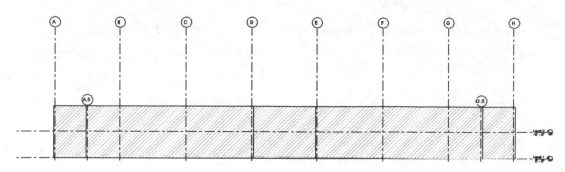

9. 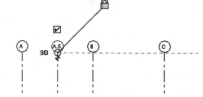 Use the grip to adjust the location of the remaining grid lines.

10.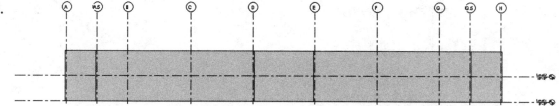

The current model has level lines for the ground plane and second floor, but not for the roof. We will add a level line for the roof to constrain the building's second level and establish a roof line.

11. Select the **Level** tool on the Datum panel on the Architecture ribbon.

12. Draw a level line **10′ [4000** mm] above Level 2.

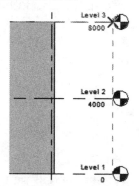

13. Rename Level 3 by double left clicking on the level name. Rename to **Roof Line**.

14. Would you like to rename corresponding views?

☐ Do not show me this message again [Yes] [No]

Press **Yes**.

This renames the views in the Project Browser to Roof Line.

15. Switch back to the Level 1 Floor plan view.

16. Activate the **Modify** ribbon.

Select the **Split** tool.

17. ☐ Delete Inner Segment   Uncheck **Delete Inner Segment** on the Options bar.

18. Split the walls where portions will remain only on level 1 and portions will be continuous up to the roof line.

19. Split the walls at the intersections indicated.

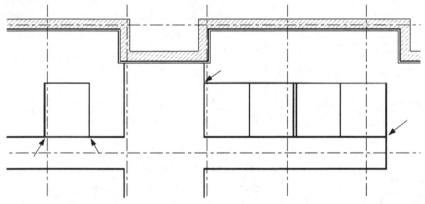

20. Holding down the Control key, pick the walls indicated.

21.

Walls (12)	▾	Edit Type
Constraints		⌃ ▲
Location Line	Wall Centerline	
Base Constraint	Level 1	
Base Offset	0' 0"	≡
Base is Attached	☐	
Base Extension Distance	0' 0"	
Top Constraint	Up to level: Roof Line	
Unconnected Height	20' 0"	

On the Properties pane:

Set the Top Constraint to **Up to Level: Roof line**.

Press **OK**.

By constraining walls to levels, it is easier to control their heights. Simply change the level dimension and all the walls constrained to that level will automatically update.

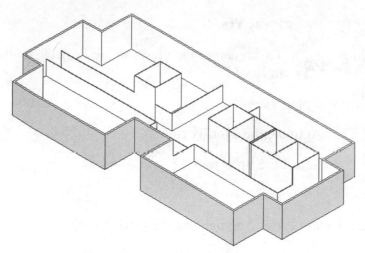

Walls that extend through multiple levels should be created out of just one wall, not walls stacked on top of each other. This improves model performance and minimizes mistakes from floor to floor. For a stair tower, create the wall on the first floor and set its top constraint to the highest level.

22.

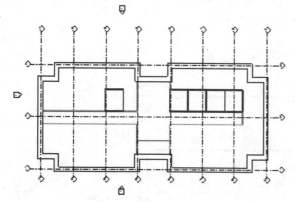

Activate the **Level 2** view.

It can be disconcerting to try to create elements on the second level (Level 2) when you can see the interior walls in the first level (Level 1).

23.

On the Properties Pane:
Locate the Underlay category.
Set the Range: Base Level to Level 2.

This means that the user will only see what is visible on Level 2.

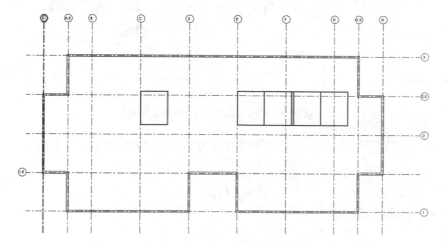

24. Save the file as *ex3-6.rvt*.

Global Parameters

Global Parameters are project-based. They can be used to control material settings or used to define dimensions. For example, if you want to ensure that your project complies with building codes so that all hallways are a minimum of 8' wide, you can define a global parameter and apply it to any hallway dimensions that need to meet that code.

What makes this especially powerful is that if the building code changes or the building specification changes, the dimensions can update simply by changing the single global parameter.

The Reveal Constraints tool will show dimensions that have had global parameters assigned to them. It will display an additional red dimension with a label below the original dimension.

I collaborate with a large group using BIM 360 on building layouts. The CAD Manager will create a sheet in the project which lists all the "rules" or global parameters in use in the project. He then sets the project up so that the sheet is the first view displayed for anybody opening the project.

You can specify the view that Revit displays by default when a project is opened.

The default setting for a starting view is <Last Viewed>, that is, whichever view was active the last time the model was closed.

When a project is shared across a collaboration team, the specified starting view is applied to all local models. This view is opened when the central or any local model is opened, and when any team member uses the Open dialog to detach from the central or create a local model.

Best practices for starting views:

- You can reduce the amount of time required to open a model by choosing a simple view as the starting view.

- You might create a special view that contains important project information and notices that you want to share with team members. Use this view as the starting view so that team members always see this information when they open the project.

To specify a starting view:

1.   Open the model.

2.   Click Manage tab ➤ Manage Project panel ➤ 🖼 Starting View.

3.   In the Starting View dialog, specify the starting view, and click OK.

        When Revit loads this model, the specified view is opened.

Adding Doors

Doors are loadable families. This means doors are external files which are inserted into a building project. Revit comes with a library of door families which you can load and insert into a building project. Each door family may have several different size combinations (Height and Width). You may also download door families from various websites, such as bimobject.com or revitcity.com. If you do a search for Revit families or Revit library, a long list of possible sources for content will be displayed.

Be sure to check out http://revit.autodesk.com/library/Archive2009/html/. *These are older Revit families, but there is a lot of excellent and useful content here and it is all free. Hint: If you select the file cabinet at the top of the screen, you can download the entire directory at once!*

Exercise 3-7
Add Doors

Drawing Name: ex3-6.rvt
Estimated Time: 10 minutes

This exercise reinforces the following skills:

- ❑   Door
- ❑   Load From Library
- ❑   Global Parameters

1.   Open or continue working in *ex3-6.rvt*.

2.   Activate the **Level 1** view.

3.   Home   Activate the **Architecture** ribbon.

4.   Select the **Door** tool from the Build panel.

     Door

5.   Select **Single-Flush 36″ x 84″ [M_Single-Flush: 0915 × 2134mm]** from the drop-down.

6.   Enable **Tag on Placement** on the ribbon.

     This will automatically add door tags to the doors as they are placed.

7.   Place the doors as shown. **Verify that you are on Level 1.**
     Remember if you press the space bar that will flip the door orientation before you click to place.

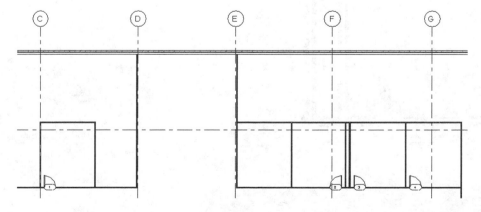

8.   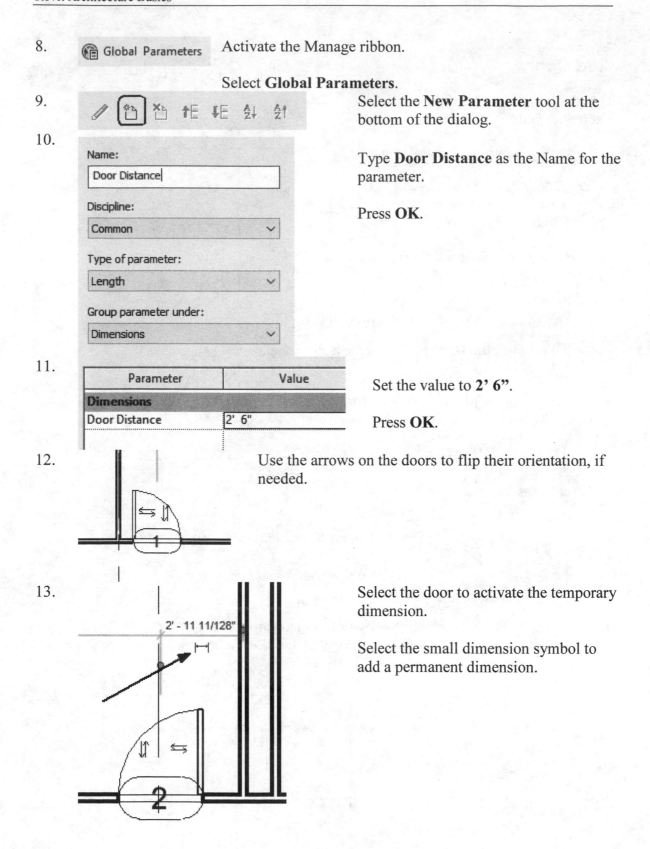 Global Parameters   Activate the Manage ribbon.

                     Select **Global Parameters**.

9.   Select the **New Parameter** tool at the bottom of the dialog.

10.

Name:

Door Distance

Discipline:

Common

Type of parameter:

Length

Group parameter under:

Dimensions

Type **Door Distance** as the Name for the parameter.

Press **OK**.

11.

Parameter	Value
Dimensions	
Door Distance	2' 6"

Set the value to **2' 6"**.

Press **OK**.

12. Use the arrows on the doors to flip their orientation, if needed.

13. Select the door to activate the temporary dimension.

2' - 11 11/128"

Select the small dimension symbol to add a permanent dimension.

14. 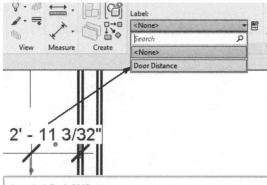 Select the permanent dimension.

Notice on the ribbon there is a label dimension panel.

Select the **Door Distance** from the drop-down.

15. 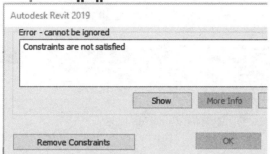 Select **Remove Constraints** if this error dialog appears.

16. Repeat for all the doors.

 Door numbers are often keyed to room numbers – 101A, etc. To edit the door tag, just click on the number and it will change to blue and you can renumber it. You can also change the door number from the Properties dialog box of the door or from the door schedule. A change in any of those locations will update automatically on all other views.

---

17. Select the **Door** tool on the Architecture ribbon.

18. 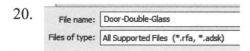 Select **Load Family** from the Mode panel.

19. Browse to the *Doors* folder.

20. Locate the *Door - Double-Glass. Rfa [M_Double-Glass .rfa]* file under the *Doors* folder and **Open**.

21.

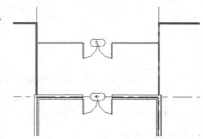

Select size **72" x 78" [1800 x 2000mm]** type to load.

Press **OK**.

Types:

Type	Width	H
	(all) ▾	(
1700 x 2000m	5' 6 119/128"	6' 6 189/256"
1700 x 2050m	5' 6 119/128"	6' 8 181/256"
1700 x 2100m	5' 6 119/128"	6' 10 173/256"
1800 x 1950m	5' 10 111/128"	6' 4 99/128"
1800 x 2000m	5' 10 111/128"	6' 6 189/256"
1800 x 2050m	5' 10 111/128"	6' 8 181/256"
1800 x 2100m	5' 10 111/128"	6' 10 173/256"

right for each family listed on the left   OK

22.

Place two double glass entry doors at the middle position of the entry walls.

The double doors should swing out in the path of egress for accuracy. The interior doors should typically swing into the room space not the hall.

23. Activate **Level 2**.

24. Home   Activate the **Architecture** ribbon.

25. 🚪 Select the **Door** tool from the Build panel.
Door

26. Select **Single-Flush 36″ x 84″ [M_Single-Flush: 0915 × 2134mm]** from the drop-down.

27. Place the doors as shown.

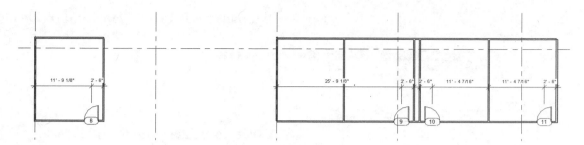

28. Use the Door Distance global parameter to position the doors in each room.

    We will be adjusting the room sizes in a later exercise.

29. Save the file as *ex3-7.rvt*.

Exercise 3-8
Define a Starting View

Drawing Name: ex3-7.rvt
Estimated Time: 20 minutes

This exercise reinforces the following skills:

- ❏ Add a Sheet
- ❏ Load a Sheet template
- ❏ Add Notes
- ❏ Global Parameters
- ❏ Define a Starting View

1. Open or continue working in *ex3-7.rvt*.

2. In the Project browser: Highlight the Sheets category. Right click and select **New Sheet.**

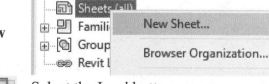

3. Select the Load button.

4. Browse to the *Titleblocks* folder and select the **B 11 x 17 Horizontal** title block.

    Press **Open**.

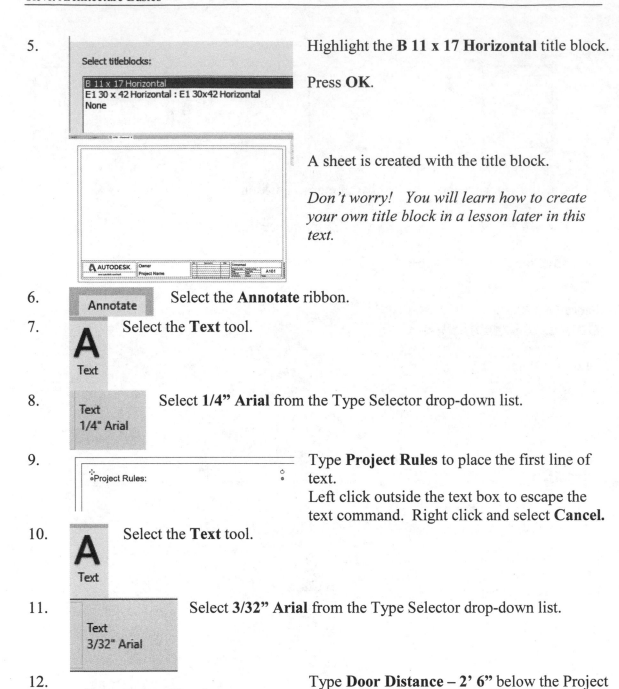

5.   Highlight the **B 11 x 17 Horizontal** title block.

Press **OK**.

A sheet is created with the title block.

Don't worry!   You will learn how to create your own title block in a lesson later in this text.

6.   Select the **Annotate** ribbon.

7.   Select the **Text** tool.

8.   Select **1/4" Arial** from the Type Selector drop-down list.

9.   Type **Project Rules** to place the first line of text.
Left click outside the text box to escape the text command. Right click and select **Cancel.**

10.   Select the **Text** tool.

11.   Select **3/32" Arial** from the Type Selector drop-down list.

12.   Type **Door Distance – 2' 6"** below the Project Rules: header.

13. 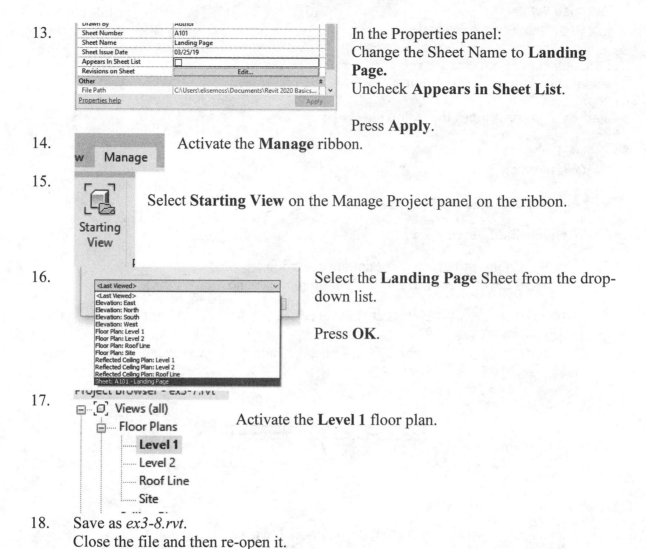 In the Properties panel:
Change the Sheet Name to **Landing Page.**
Uncheck **Appears in Sheet List**.

Press **Apply**.

14. Activate the **Manage** ribbon.

15. Select **Starting View** on the Manage Project panel on the ribbon.

16. Select the **Landing Page** Sheet from the drop-down list.

Press **OK**.

17. Activate the **Level 1** floor plan.

18. Save as *ex3-8.rvt*.
Close the file and then re-open it.
19. Notice it re-opens to the Landing Page.

If you want the project to open to the last active view, change the Starting View to <Last Viewed>, which is the default.

Exercise 3-9
Exploring Door Families

Drawing Name: ex3-8.rvt
Estimated Time: 60 minutes

This exercise reinforces the following skills:

- ❑ Door
- ❑ Hardware

1. Open or continue working in *ex3-8.rvt*.

2. Activate the **Level 1** view.

3. Zoom into the exterior double glass door.
   Select the door to activate the Properties pane.

4. 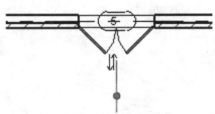 Note that you can adjust the Swing Angle for the door.

   Change it to 45 degrees and see how it updates in the display window.

   Note that the other double glass door doesn't change. The swing angle is controlled by each instance, not by type.

5. Select the **Door** tool on the Architecture ribbon.

   Door

6. Select **Load Family** from the Mode panel.

   Load Family

7. Browse to the *Doors/Commercial* folder.

8. 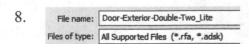 Locate the *Door-Exterior-Double-Two_Lite. Rfa [M_ the Door-Exterior-Double-Two_Lite.rfa]* file under the *Doors* folder and **Open**.

   File name: M_Double-Glass 2.rfa

9.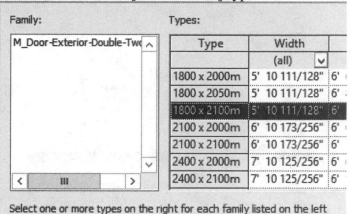

Select the 72" x 82" [1800 x 2100] type.

Press **OK**.
Press ESC to exit the command.
Replace the outside exterior door with the new door type.

Select the door and then select the new door type using the Type Selector drop-down list.

10.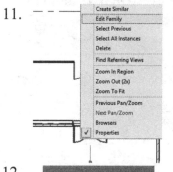

11. Highlight the door.

Right click and select **Edit Family**.

12.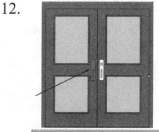

Select the left door hardware.

| Exterior Pull |
| Without Lock |

Note that the Properties panel designates it as an Exterior Pull Without Lock.

13. Expand the Doors folder under Families in the Browser.

You see the different families that are used in the door including the hardware.

14. 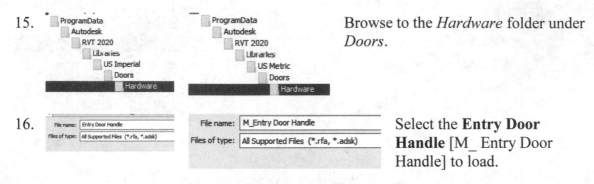 Activate the Insert ribbon and select **Load Family**.

15. Browse to the *Hardware* folder under *Doors*.

16. 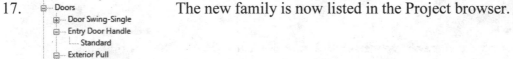 Select the **Entry Door Handle** [M_ Entry Door Handle] to load.

Press **Open**.

17. The new family is now listed in the Project browser.

18. 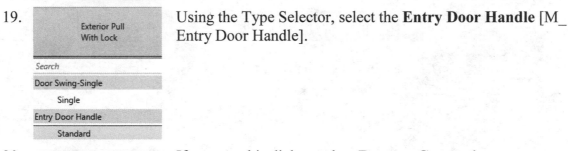 Select the door hardware on the right.

19. Using the Type Selector, select the **Entry Door Handle** [M_ Entry Door Handle].

20. If you see this dialog, select **Remove Constraints**.

21.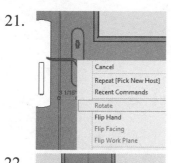

Select the hardware.
Right click and select **Flip Hand**.

22.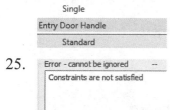

The door handle will flip to the correct orientation.

23.

Select the hardware on the left.

24.

Using the Type Selector, select the **Entry Door Handle** [M_ Entry Door Handle].

25.

If you see this dialog, select **Remove Constraints**.

26.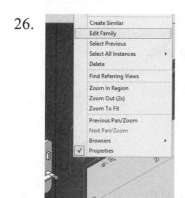

Select the handle on the right.

Right click and select **Edit Family**.

27. Select **Family Types** on the ribbon.

28. Change the Function to **Exterior**.

29. Notice that there is only one family type defined.

30. Select **New** to create a new family type.

31. Type **Standard No Lock** in the Name field.

    Press **OK**.

32. The new type is listed in the Type name list.

33. Under Parameters:
    Select **New Parameter.**

34. Enable **Family Parameter**.
    Type **Lock** in the Name field.
    Enable **Type**.
    Set Discipline to **Common**.
    Set Type of Parameter to **Yes/No**.
    Set Group Parameter under to **Construction**.
    Press **OK**.

    Press **OK** to close the dialog box.

35. Select the cylinder representing the lock.

36. Select the small gray box to the right of the Visible field.

37. Highlight **Lock** and press **OK**.

38. 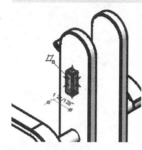 Orbit the model around.

    Select the rectangle representing the lock.

39. Select the small gray box to the right of the Visible field.

40. Highlight **Lock** and press **OK**.

41. Launch the Family Types dialog again.

42. Select the **Standard** type selected under the Type Name:

43. Enable **Lock**.

44. 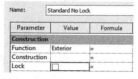 With Standard No Lock type selected:
    Uncheck **Lock**.

    Press **OK**.

45.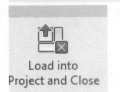
   Save As *Entry Door Handle-2.rfa* in your exercise folder.

46. Select **Load into Project and Close** from the ribbon.

47. Load into the door family.

48. Press ESC to exit the command which is active.

   Select the left side door handle.

49. Use the Type Selector to set the Type to Standard No Lock.

50. Use the Type Selector to set the hardware on the right to *Entry Door Handle-2-Standard*.
   Use the Type Selector to set the hardware on the left to *Entry Door Handle-2-Standard No Lock*.

51. Save as *Door-Exterior-Double-Two_Lite 2.rfa*

52. Load into the ex3-8.rvt project and close.

53. Escape out of the active command.

   Replace the exterior door with the new version using the type selector.

54. Activate the South elevation.

55. 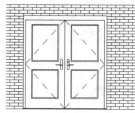 The door is displayed with the new hardware.

Hardware is only displayed when the detail level is set to Fine.

56.  Change the Display detail to see how the door display changes.

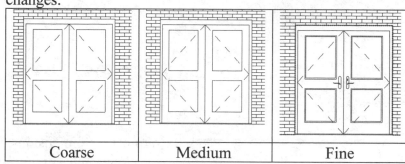

Coarse	Medium	Fine

57. Save as *ex3-9.rvt*.

Stairs

There are two basic methods to create a set of stairs in Revit. You can use a "run" which involves defining the stair type and then drawing a line by selecting two points. Revit will automatically generate all the stair geometry based on the location and distance between the two points. In the second method, you use a sketch. In the sketch, you must define the boundaries of the stairs – that is the inside and outside edges – and the riser lines. The distance between the risers determines the width of the tread.

The most common mistakes I see users make when sketching stairs are:

- Creating a closed boundary – the boundary lines should be open. The boundaries define where the stringers are for the stairs.

- Placing risers on top of each other. If you place a riser on top of a riser, you will get an error.

- Making the riser lines too short or too long. The riser line endpoints should be coincident to the inside and outside boundary lines.

You can assign materials to the stairs based on the stair type. Stairs are system families, which means they are unique to each project.

Exercise 3-10
Adding Stairs

Drawing Name: ex3-9.rvt
Estimated Time: 40 minutes

This exercise reinforces the following skills:

   ❑ Stairs

1. Open *ex3-9.rvt.*

2. 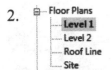   Activate **Level 1** Floor Plan.

3. 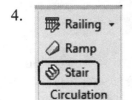   To turn off the dimensions, type **VV** to launch the Visibility/Graphics dialog.
Select the **Annotation Categories** tab.
Disable **Dimensions** to turn off the visibility of dimensions.
Close the dialog.

4. Select the **Stair** tool under the Circulation panel on the Architecture ribbon.

5. Highlight **Run**.

Select the **Straight Run** option.

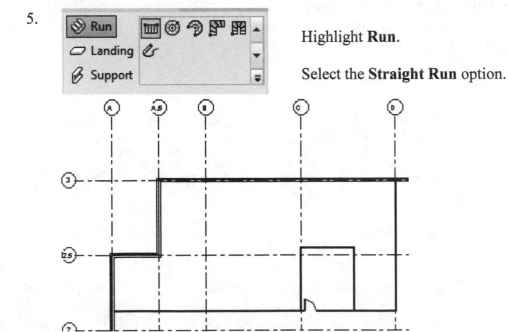

Our first set of stairs will be located in the room at Grids C-2.5.

6.
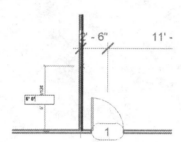

On the Options bar:

Set the Location line to Exterior Support: Left.
Set the Offset to 0' 0".
Set the Actual Run Width to 3' 0".
Enable Automatic Landing.

7.

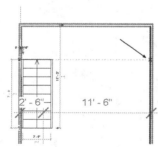

Start the stairs 6' 0" above the door by selecting the inside wall face and typing in the dimension.

8.

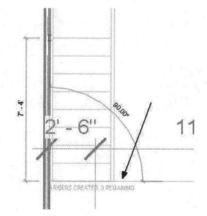

Drag the mouse straight up.

Notice at the start point where you selected Revit generates a riser count.

When you see a riser count of 9, left click to complete the first run of stairs.

9.

Select a start point for the second run of stairs.

This point should be located directly across from the end point of the first run and on the inside face of the wall.

10.

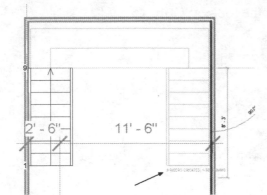

Drag your mouse straight down until you see a Revit prompt indicating that all the risers have been defined.

Left click to complete the stairs.

11.   Select the **Green Check** on the ribbon to finish the stairs.

12. The stairs are completed.

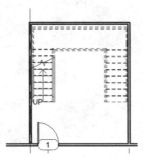

13. 3D Views Activate the **3D View**.
    {3D}

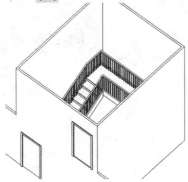

Railing was added on both sides of the stairs.

We will correct this in the next exercise.

14. Press F8 to access the orbit tool.

15. Save as *ex3-10.rvt*.

Exercise 3-11
Creating a Handrail on a Wall

Drawing Name: ex3-10.rvt
Estimated Time: 20 minutes

This exercise reinforces the following skills:

- Railings
- Railing Types

1. Open *ex3-10.rvt.*

2.   Floor Plans   Activate **Level 1**.
       Level 1
       Level 2
       Roof Line
       Site

3. Hover your mouse over the Railing on the left side of the stairs.

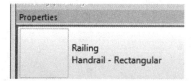

Left click to select the railing.

If you have difficulty selecting the railing, use the TAB key to cycle the selection or use the FILTER tool. You can check that you have selected the railing because it will be listed in the Properties pane.

4. Select **Edit Path** under the Mode panel.

5. Delete the lines indicated.

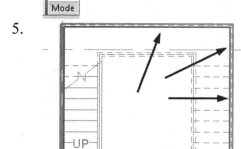

6. 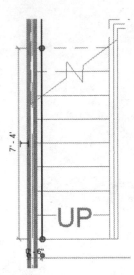 Move the two remaining lines six inches [152.4 mm] away from the wall.

7. 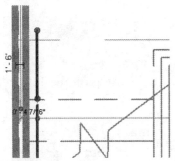 Adjust the top vertical line so it is 1′ 6″ [457.2 mm] above the top riser.

8. 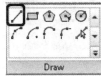 Select the **LINE** tool from the Draw panel.

9. 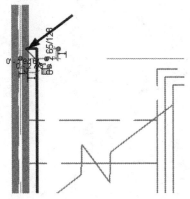 Add a short horizontal line at the top of the railing. This is the bar on the handrail that attaches to the wall.

10. 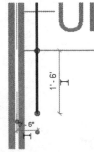 Add a vertical line extending below the bottom riser a distance of 1′ 6″ [457.2 mm].

11. 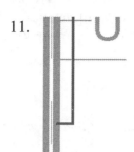 Add another horizontal line that attaches to the wall at the bottom of the railing.

12. 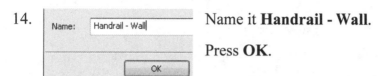 Select **Edit Type** from the Properties pane.

13. Duplicate... Select **Duplicate**.

14. Name: Handrail - Wall

Name it **Handrail - Wall**.

Press **OK**.

OK

15. 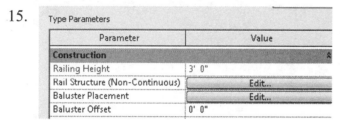 Press **Edit** next to Baluster Placement.

16. 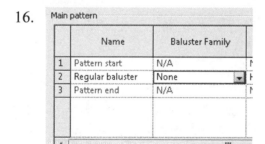 On Row 2 of the **Main Pattern** table, set the Regular baluster to **None** in the Baluster Family column.

17. 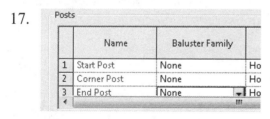 In the **Posts** table, set all the posts to **None** in the Baluster Family column.

18.

	Name	Baluster Family
1	Pattern start	N/A
2	Regular baluster	None
3	Pattern end	N/A

Main pattern

Break Pattern at: Each Segment End ⌄

Justify: Beginning ⌄ Excess Length Fill : No

☐ Use Baluster Per Tread On Stairs   Balusters Per Tread: 2

Posts

	Name	Baluster Family	Base	Base offset	
1	Start Post	None	Host	0' 0"	To
2	Corner Pos	None	Host	0' 0"	To
3	End Post	None	Host	0' 0"	To

Corner Posts At: Each Segment End ⌄

Verify that you have set the balusters to **None** in both areas.

Many of my students will set it to None in one area and not the other.

Press **OK** twice to exit all dialogs.

19. Select the **Green Check** under Mode to exit Edit Mode.

If you get an error message, check the sketch. It must be a single open path. All line segments must be connected with no gaps and no intersecting lines or overlying lines.

20. Switch to a 3D view.

21. Inspect the railing.

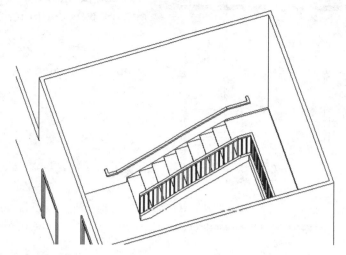

22. 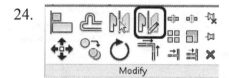 Activate **Level 1**.

23. 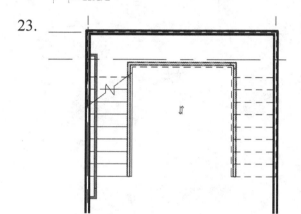 Select the handrail you just created.

24. 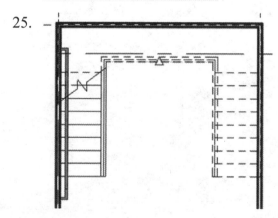 Select **Mirror→Draw Axis** from the Modify panel.

25. Locate the midpoint on the landing and left click at the start of the axis.

26. Select a point directly below the midpoint to end the axis definition.

You don't see anything – that's because the handrail was placed on the second level.

27. 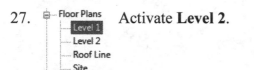 Activate **Level 2**.

28. You see the handrail.

29. Switch to a 3D view.

30. Inspect the railing.

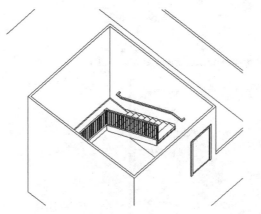

31. Save as *ex3-11.rvt*.

Exercise 3-12
Modifying the Floor Plan – Skills Review

Drawing Name: ex3-11.rvt
Estimated Time: 30 minutes

This exercise reinforces the following skills:

- ❑ 3D View
- ❑ Copy
- ❑ Align
- ❑ Floor
- ❑ Railing

1.   Open *ex3-11.rvt.*

2.   Activate the **North** Elevation.

3.   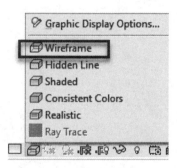 Set View Properties to **Wireframe**.

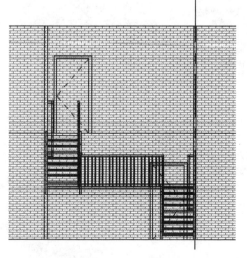

We see that our stairs in elevation look pretty good.

4.   Activate **Level 2**.

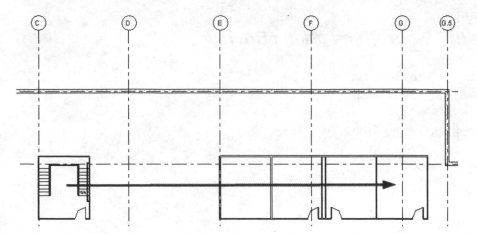

We want to copy the stairs, railings, and landing to the stairwell indicated.

5. Window around the walls, stair and railings.

6. Select the **Filter** tool.

Filter

7. | Use **Filter** to ensure that only the Stairs and Railings are selected.

Press **OK**.

Category:	Count:
☐ Dimensions	1
☑ Railings	3
☑ Railings: Top Rails	2
☑ Stair Paths	1
☑ Stairs	1
☑ Stairs: Landings	1
☑ Stairs: Runs	2
☑ Stairs: Supports	8
☐ Walls	1

Check All

Check None

8. 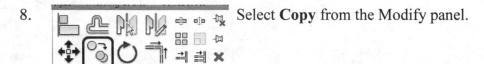 Select **Copy** from the Modify panel.

Modify

The Modify→Copy can only be used to copy on the same work plane/level.

9.

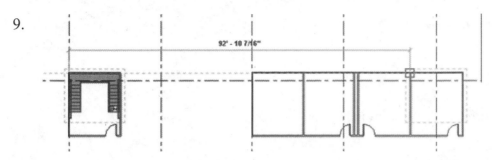

Select the top left corner of the existing stairwell as the basepoint.

Select the top left corner of the target stairwell as the target location.
Left click to release.

(Instead of copying the stair, you can also repeat the process we used before to create a stair from scratch.)

10.  Activate **3D View**.

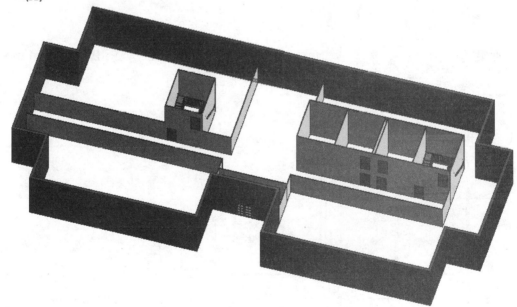

You should see a complete stair with landing and railing in both stairwells.

11. Save as *ex3-12.rvt*.

Exercise 3-13
Defining a 2-hr Wall

Drawing Name: ex3-12.rvt
Estimated Time: 10 minutes

This exercise reinforces the following skills:

- Split
- Wall Properties

Stairwell walls are usually 2-hour walls. The walls, as currently defined, are 1-hour walls. We need to load 2-hr walls into the project from the Project Browser.

1. Open *ex3-12.rvt.*

2.   Activate the **Level 2 Floor Plan**.

3.   Select the **Split** tool from the Modify panel on the Modify ribbon.

4. Split the walls at the intersections indicated.

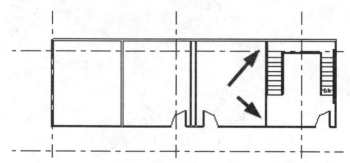

5. Select the stairwell walls indicated.

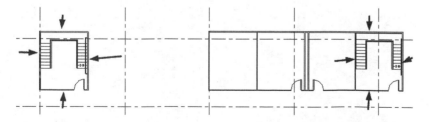

These walls will be defined as 2-hr walls.

6.   You can window around both areas while holding down the CTRL key.
Filter

Then select the Filter tool from the ribbon.

7.
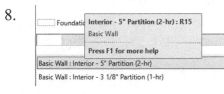
Uncheck all the elements EXCEPT walls.

Press **OK**.

8.

On the Properties pane, select the **Interior - 5″ Partition (2-hr) [Basic Wall: Interior - 135mm Partition (2-hr)]** from the Type Selector drop down list.

Left click in the window to clear the selection.

9.   Save the file as *ex3-13.rvt*.

Exercise 3-14
Adding an Elevator

Drawing Name: ex3-13.rvt
Estimated Time: 40 minutes

This exercise reinforces the following skills:

- ❑   3D View
- ❑   Delete
- ❑   Wall Properties
- ❑   Families

1.   Open *ex3-13.rvt*.

2.   Floor Plans
     — **Level 1**
     — Level 2
     — Roof Line
     — Site

Activate the **Level 1 Floor Plan**.

3.   Load Family

Activate the **Insert** ribbon.

Select **Load Family** from the Load from Library panel.

4. 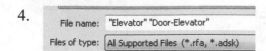 Locate the *Elevator* and the *Door-Elevator* in the downloaded Class Files.

Press **Open**.

5. The elevator is now available in the Project browser for your current project.

 To locate it, look for the folder called Specialty Equipment and expand that folder.

6. Select the *6' 10" x 5' 2" family*.

Hold down the left mouse button and drag into the display window.

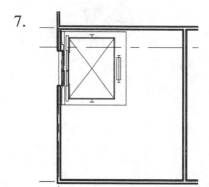

7. 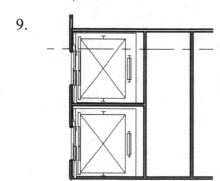 The elevator will appear on the cursor when you hover over a wall.

The elevator is wall-hosted, so needs a wall to be placed.

Press the SPACE bar to rotate the elevator to orient it properly.

Left pick to place.

8. 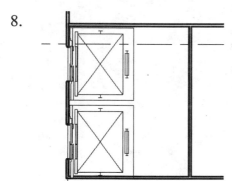 Place two elevators.

Right click and select CANCEL to exit the command.

9. Add an **Interior – 3 1/8" Partition (1-hr) [Interior-138mm Partition (1-hr)]** wall behind and to the right of the two elevators as shown.

Add a wall between the two elevator shafts. **Use the midpoint of the wall behind the elevators to locate the horizontal wall.**

Constrain both walls to the Roof Line level.

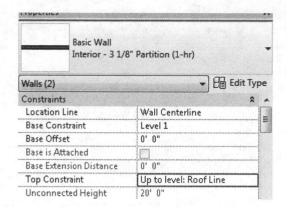

Basic Wall		
Interior - 3 1/8" Partition (1-hr)		

Walls (2)   Edit Type

Constraints		
Location Line	Wall Centerline
Base Constraint	Level 1
Base Offset	0' 0"
Base is Attached	☐
Base Extension Distance	0' 0"
Top Constraint	Up to level: Roof Line
Unconnected Height	20' 0"

10. Activate the **Manage** ribbon.

11. 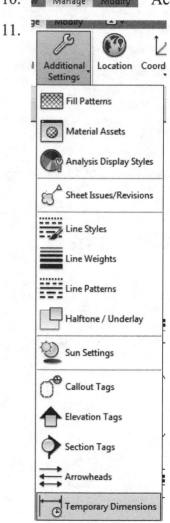 Go to **Settings→Additional Settings→Temporary Dimensions**.

- Fill Patterns
- Material Assets
- Analysis Display Styles
- Sheet Issues/Revisions
- Line Styles
- Line Weights
- Line Patterns
- Halftone / Underlay
- Sun Settings
- Callout Tags
- Elevation Tags
- Section Tags
- Arrowheads
- Temporary Dimensions

12. 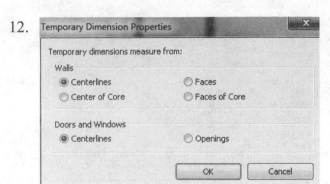 Enable **Centerlines** for Walls.

Enable **Centerlines** for Doors and Windows.

Press **OK**.

We turned off the visibility of dimensions earlier. Let's create a Level 1 floor plan which displays the dimensions.

13. 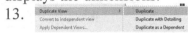 Highlight the Level 1 view.
Right click and select **Duplicate View→Duplicate with Detailing.**
This copies the view along with any annotations, such as dimensions.

14.  Right click on the copied view and select **Rename**.

15.  Change the name to **Level 1 – Annotations Visible**.

16. 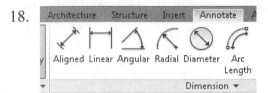 Type **VV** to launch the Visibility/Graphics dialog.
Activate the Annotation Categories tab.

Enable **Dimensions**.
Press **OK** to close the dialog.

17. Annotate Activate the **Annotate** ribbon.

18. Select the **Aligned** tool under Dimension.

Make sure visibility of dimensions is turned on in the Visibility/Graphics dialog.

19. 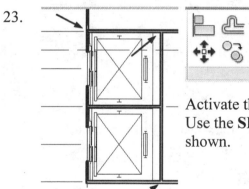   Select the center of the wall.
Select the center line of the elevator.
Select the center line of the wall.
Left pick to place the continuous dimension.

Repeat for the other elevator.

20. 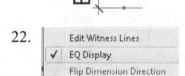   Select the dimension.
Left click on the EQ symbol to set the dimensions equal and position the elevator centered in the space.

21.   The dimension display changes to show EQ symbols.

22.      To display dimensions:
Select the extension line of the dimension.

Uncheck the EQ Display to show dimensions.

23.   Activate the **Modify** ribbon.
Use the **SPLIT** tool and split the walls at the intersections shown.

24.   Elevator shaft walls are 2-hour, usually CMU.

Select the elevator shaft walls.

Filter   Use the **Filter** tool to ensure only the walls are selected.

25. 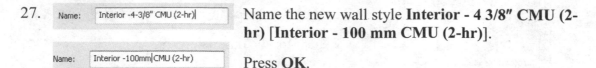 Select **Edit Type** from the Properties pane.

26. [Duplicate...] Select the **Duplicate** button.

27. Name: [Interior -4-3/8" CMU (2-hr)|]   Name the new wall style **Interior - 4 3/8″ CMU (2-hr) [Interior - 100 mm CMU (2-hr)]**.

    Name: [Interior -100mm|CMU (2-hr)]   Press **OK**.

28.

Type Parameters	
Parameter	Value
Construction	
Structure	[Edit...]

Select **Edit** next to Structure.

29.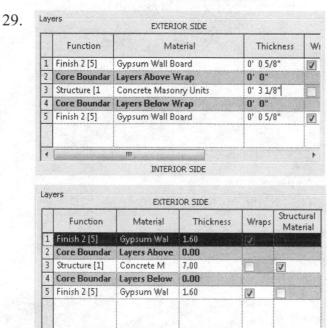

Set Layer 1 to Finish 2 [5], **Gypsum Wall Board, 5/8″ [1.6 mm]**.
Set Layer 2 to Core Boundary.
Set Layer 3 to Structure [1], **Concrete Masonry Units, 3 1/8″ [7 mm]**.
Set Layer 4 to Core Boundary.
Set Layer 5 to Finish 2 [5], **Gypsum Wall Board, 5/8″ [1.6 mm]**.

Press **OK**.

30.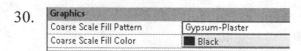

Set the Coarse Scale Fill Pattern to **Gypsum-Plaster**.

Press **OK**.

31. Set the properties of the elevator shaft walls to the new style: **Interior - 4 3/8″ CMU [Interior-100 mm CMU]**.

Use **Modify→ALIGN** to adjust the walls.

32. ☐ Multiple Alignment   Prefer: Wall faces ▼   Set the Option bar to **Prefer Wall faces**.

33. Use the ALIGN tool to adjust the walls so the faces are flush. *You can use the TAB key to tab through the selections.*

Re-adjust the position of the elevators to ensure they remain centered in the shaft.

You can click on the EQ toggle on the aligned dimension to update.

34. ⊟ 3D Views
    └ {3D}   Switch to a 3D view.

Verify that all the walls appear properly, with walls for the second floor constrained as needed.

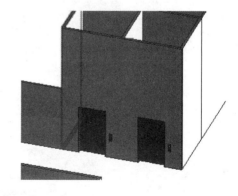

Next, we add the elevator doors for the second floor.

35. ⊟ Floor Plans
    ├ Level 1
    ├ **Level 2**
    ├ Roof Line
    └ Site   Activate **Level 2**.

36. Activate the **Architecture** ribbon.

Door   Select the **Door** tool.

37. 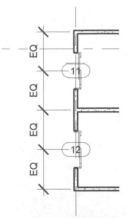 Verify that the Door-Elevator is selected on the Properties palette.

Door-Elevator
12000 x 2290 Opening

38. Place the doors so they are centered in each shaft.

39. 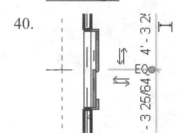 Switch to a 3D view to verify the placement of the elevator doors.

If the doors don't look proper, return to the Level 2 view.

40. You can use the orientation/flip arrows to re-orient the door or flip the side location for the elevator buttons on the Properties panel.

Graphics	
Elevator Call Button - Right Side	☑
Elevator Call Button - Left Side	☐
Materials and Finishes	

41. Save the file as *ex3-14.rvt*.

➢ The Snap Settings you use are based on your Zoom factor. In order to use the smaller increment snap distances, zoom in. To use the larger increment snap distances, zoom out.
➢ When placing a component, use the spacebar to rotate the component before placing.
➢ You can purge any unused families from your project by using Manage→ Settings→Purge Unused.

Exercise 3-15
Load Family

Drawing Name: ex3-14.rvt
Estimated Time: 15 minutes

This exercise reinforces the following skills:

- ❑ Load Family
- ❑ Space Planning

All of the content required in this exercise is included in the downloaded Class Files.

1. Open or continue working in *ex3-14.rvt*.

2. Activate **Level 1**.

3. Select the **Component→Place a Component** tool from the Build panel on the Architecture ribbon.

4. Select **Load Family** from the Mode panel.

5. Locate the following files in the downloaded exercises for this text.

   - Urinal w Screen
   - Counter Top w Sinks
   - Dispenser-Towel
     Toilet Stall-3D w Toilet
   - Toilet Stall-Accessible-Front-3D w Toilet
   - Mirror
   - Trash Receptacle

   You can select more than one file by holding down the CTRL key.
   Press **Open**.

6. 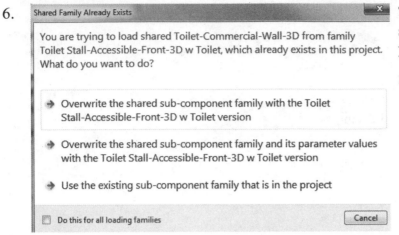 The toilet stall families share sub-components, so you will see this message.

Select the top option.

7. Under the Properties Pane:

Change Element Type:

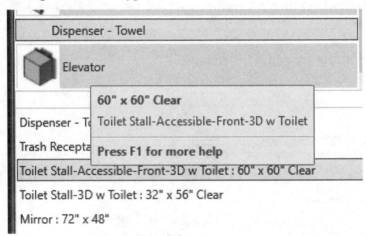

Select the **Toilet Stall-Accessible-Front-3D w Toilet**.

You need to select the 60" x 60" Clear family.

8. 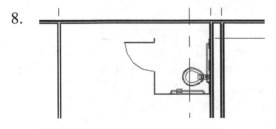 The toilet will be visible when the cursor is over a wall.
Click to place. Do not try to place it exactly. Simply place it on the wall.

Use the SPACE bar to orient the family.

9. 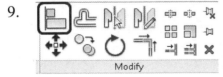 Use the ALIGN tool from the Modify **ribbon** to align the cubicle with the wall.

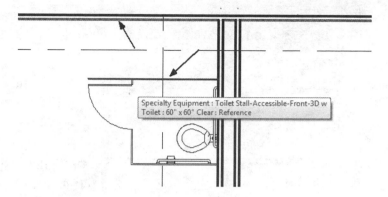

10. 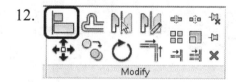 Scroll down the Project Browser.

Locate the **Toilet Stall-3D w Toilet 36″ x 60″ Clear** under Specialty Equipment.

11. Place two toilet stalls by dragging into the window. Use the SPACE bar to orient the family.

You won't see the toilets unless you hover over the wall.

12. Use the ALIGN tool from the Modify **ribbon** to align the cubicles.

Make sure multi-alignment is disabled in the Options bar.

13. 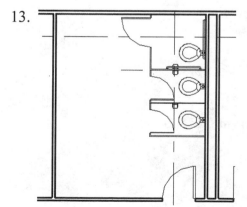 The toilet stalls should all be nestled against each other.

14. 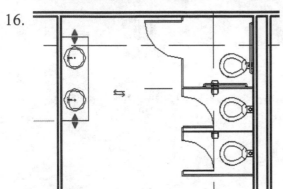 Select the **Component→Place a Component** tool from the Build panel on the Architecture ribbon.

15. Select the **Counter Top w Sinks 600 mm x 1800 mm Long** from the Type drop down list on the Properties pane.

16. Place the **Counter Top w Sinks 600 mm x 1800 mm Long** in the lavatory.

Notice if you pull the grips on this family you can add additional sinks.

17. 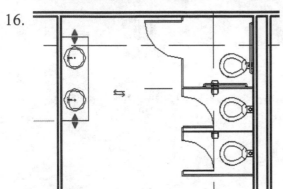 Select the **Component→Place a Component** tool from the Build panel on the Architecture ribbon.

18. Select the **Dispenser - Towel** from the Type drop down list on the Properties pane.

19. Place the **Dispenser – Towel.**

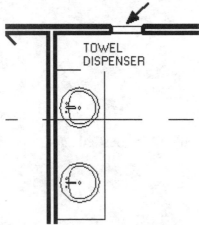

20. 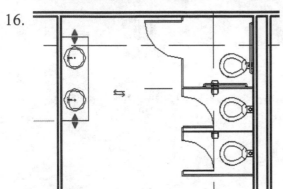 Select the **Component→Place a Component** tool from the Build panel on the Architecture ribbon.

21. 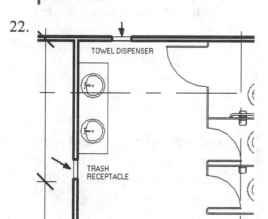 Select the **Trash Receptacle** from the Type drop down list on the Properties pane.

22. Place the **Trash Receptacle next to the sink.**

23. Select the **Component→Place a Component** tool from the Build panel on the Architecture ribbon.

24. 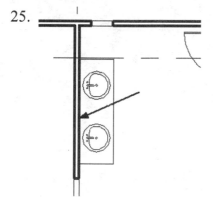 Select the **Mirror: 72″ x 48″** from the Type drop down list on the Properties pane.

25. Place the mirror above the sink.

26.

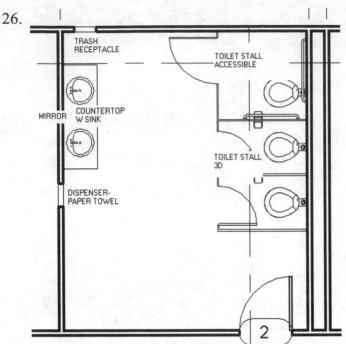

Place the components as shown.

You can place components by selecting them from the drop down list on the Option bar or by dragging and dropping from the browser.

You can flip the orientation of the component by pressing the SPACE bar before you pick to place it.

27. Save the file as *ex3-15.rvt*.

When you duplicate a view it uses the copied view name as the prefix for the new view name and appends Copy #; i.e. '[View Name] Copy 1' instead of naming the Duplicate View 'Copy of [View Name]'. This makes it easier to sort and locate duplicated views.

Exercise 3-16
Adjust Room Sizes

Drawing Name: ex3-15.rvt
Estimated Time: 15 minutes

This exercise reinforces the following skills:

- ❑ Pin
- ❑ Global Parameters
- ❑ Floor Plan
- ❑ Dimensions

Verify that the rooms are the same size and that the plumbing space is 1'-0" wide. Pin the walls into position to ensure they don't move around on you.

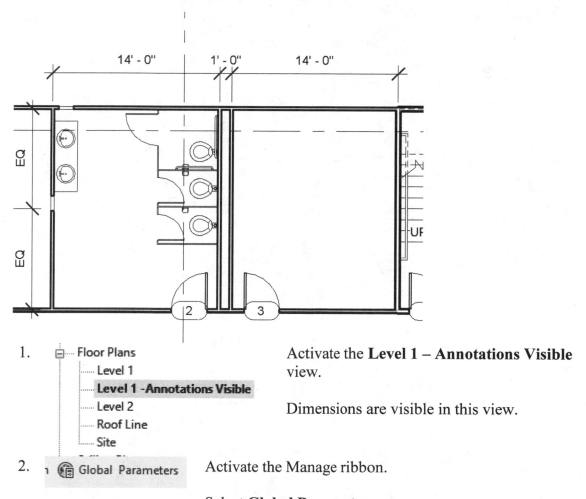

1.   Floor Plans
        Level 1
        Level 1 -Annotations Visible
        Level 2
        Roof Line
        Site

Activate the **Level 1 – Annotations Visible** view.

Dimensions are visible in this view.

2.   Global Parameters

Activate the Manage ribbon.

Select **Global Parameters**.

3.   Select the **New Parameter** tool.

4.

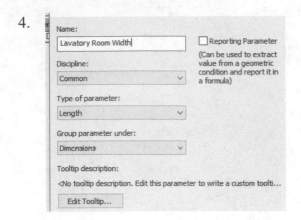

Type **Lavatory Room Width** for the Name.

Press **OK**.

5.

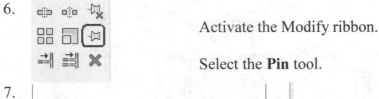

Parameter	Value
Dimensions	
Door Distance	2' 6"
Lavatory Room Width	14' 0"

Set the Width to **14' 0"**.

Press **OK**.

6.

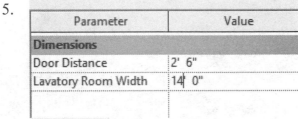

Activate the Modify ribbon.

Select the **Pin** tool.

7.

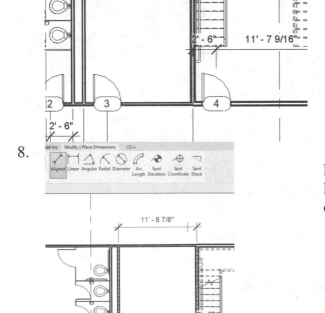

Select the left side wall of the stairwell to pin or fix the wall into position.

By placing a pin, you ensure that the wall will not move around when you change any dimensions.

Cancel out of the PIN tool.

8.

Place a horizontal dimension on the right lavatory walls using the ALIGNED dimension tool.

9.

Select the dimension.

Select the **Lavatory Room Width** label from the ribbon.

The walls will adjust.

10.

Move the wall for the women's lavatory to the left.

11.

Select the ALIGNED dimension tool.

12.

Place an aligned horizontal dimension on the two lavatory walls.

13.

Select the left wall to activate the temporary dimension.

Change the temporary dimension to **1' 0".**

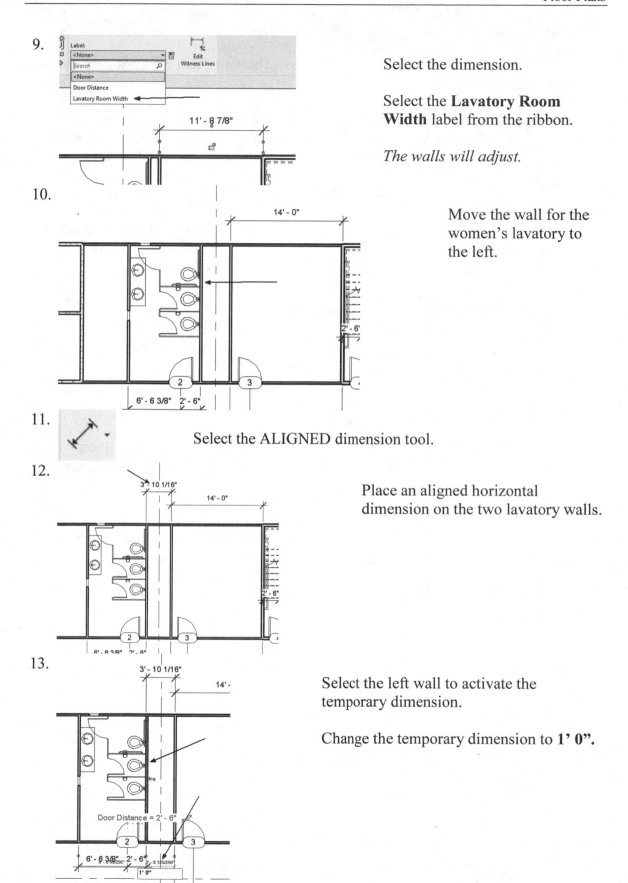

14.

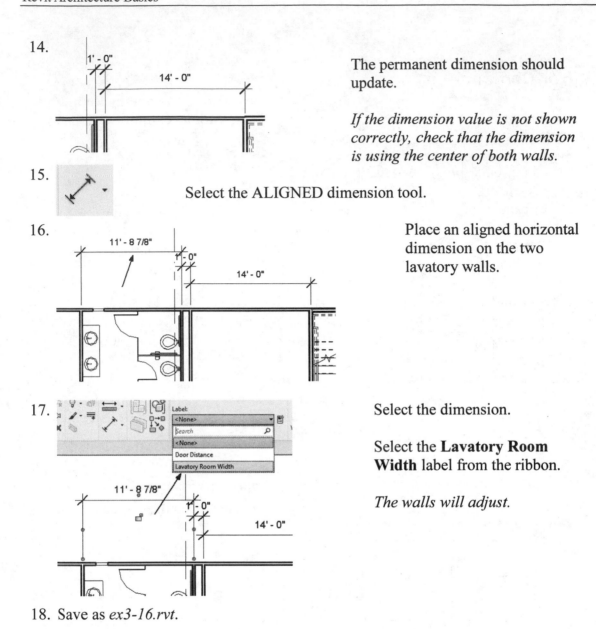

The permanent dimension should update.

If the dimension value is not shown correctly, check that the dimension is using the center of both walls.

15. Select the ALIGNED dimension tool.

16. Place an aligned horizontal dimension on the two lavatory walls.

17. Select the dimension.

Select the **Lavatory Room Width** label from the ribbon.

The walls will adjust.

18. Save as *ex3-16.rvt*.

EXTRA: *Add the Lavatory Room Width dimension to the Landing Page sheet.*

Exercise 3-17
Mirror Components

Drawing Name: ex3-16.rvt
Estimated Time: 15 minutes

This exercise reinforces the following skills:

- ❏ Mirror
- ❏ Align
- ❏ Filter
- ❏ Wall Properties

In order for the Mirror to work properly, the rooms need to be identical sizes.
This is a challenging exercise for many students. It takes practice to draw a mirror axis
six inches from the end point of the wall. If you have difficulty, draw a detail line six
inches between the 1' gap used for plumbing, and use the Pick Axis method for Mirror.

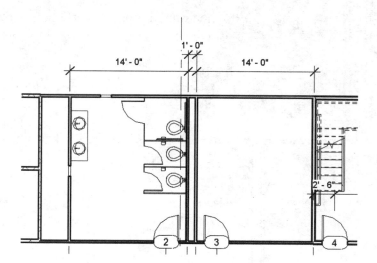

1. Open or continue working in *ex3-16.rvt*.

2. Activate **Level 1 – Annotations Visible**.

3.

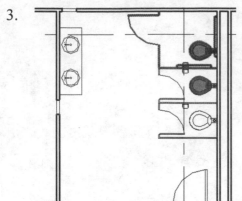

Hold down the Control key and select the top two stalls or just window around them.

They should highlight in red.
Select the countertop with sinks, the trash receptacle, the mirror and the towel dispenser.

Everything is selected EXCEPT one toilet stall.

4.

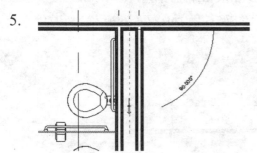

Select the **Mirror→Draw Mirror Axis** tool under the Modify panel.

5.

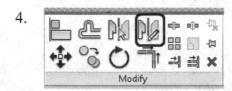

Start a line at the midpoint into the wall space.

Then pick a point directly below the first point selected.

6.

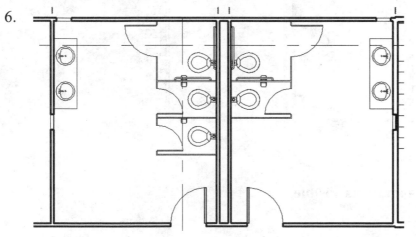

The stalls are mirrored.

7.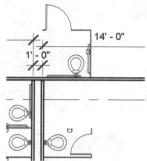

If the ADA stall lands outside the lavatory, select the ADA toilet family.

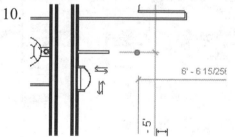

Select **Pick New Host** from the ribbon.
Then, select the wall where you want the ADA toilet stall to be placed.
Use the **ALIGN** tool to position the stall properly.

8.

```
Plumbing Fixtures
  Sink - Vanity - Round
  Toilet-Commercial-Wall-3D
    15" Seat Height
    19" Seat Height
  Urinal (1)
  Urinal w Screen
    Urinal w Screen
```

You should have loaded the *Urinal w Screen* into your project browser from the *plumbing fixtures* folder.

Drag and drop into the men's lavatory or use the Place Component tool.

9.

```
No Tag Loaded                                    x

There is no tag loaded for Plumbing Fixtures. Do
you want to load one now?


                              [ Yes ]   [ No ]
```

If you see this dialog, press **No**.

10. *You can use the orientation flip arrows to flip the orientation of the privacy screen.*

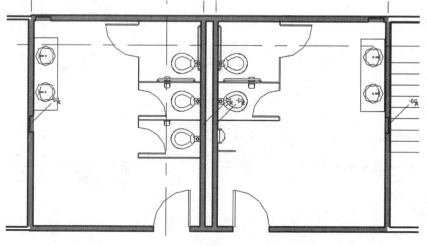

The walls surrounding both lavatories should be set to the correct wall type - **Interior - 5″ Partition (2-hr)** [**Interior - 135mm Partition (2-hr)**].

```
Basic Wall
Interior - 5" Partition (2-hr)
```

11. Use **FILTER** to only select walls.

Filter

Category:	Count:
☐ Casework	2
☐ Door Tags	2
☐ Doors	2
☐ Plumbing Fixtures	1
☐ Railings: Top Rails	1
☐ Specialty Equipment	11
☐ Stairs: Supports	2
☑ Walls	6

12. Set the walls to the correct type:

Basic Wall
Interior - 5" Partition (2-hr)

Interior - 5″ Partition (2-hr) [Interior - 135mm Partition (2-hr)] using the Properties pane type selector.

13.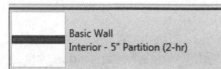
Modify

Use ALIGN from the Modify panel on the Modify ribbon to align the lavatory walls.

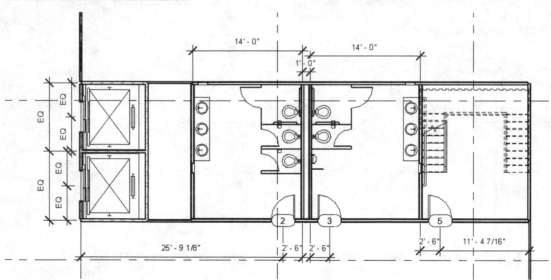

14. Save the file as *ex3-17.rvt*.

To re-select the previous selection set, hit the left arrow key.

Exercise 3-18
Create a 3D View

Drawing Name: ex3-17.rvt
Estimated Time: 10 minutes

This exercise reinforces the following skills:

- 3D view
- Section Box
- Properties

1. Open *ex3-17.rvt*.

2. 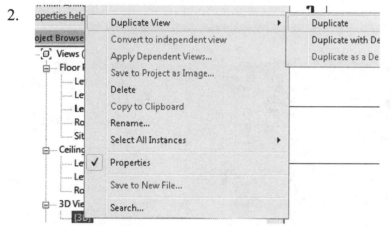 Highlight the 3D view in the Project Browser.

    Right click and select
    Duplicate View→ Duplicate.

3.  Right click on the new 3D view and select **Rename**.

4. Type **3D - Lavatory**.

    Press **OK**.

5. Scroll down the Properties pane to the Extents section.

    Enable **Section Box**.

    Press **OK**.

6.

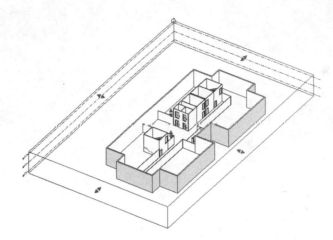

A clear box appears around the model. Left click to select the box.

There are grips indicated by arrows on the box.

7.

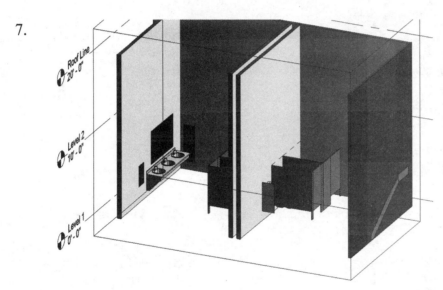

Select the grips and reduce the size of the box, so only the lavatories are shown.

Use the **SPIN** tool to rotate your view so you can inspect the lavatories.

If you hold down the SHIFT key and middle mouse button at the same time, you can orbit the model.

8.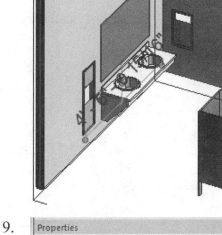

If you need to adjust the elevation/height of the mirror:

Select the mirror.

9.

In the Properties pane:

Set the elevation to **4′ 0″**.

Press **Apply**.

Repeat for the other mirror.

10. Save as *ex3-18.rvt.*

Copying Lavatory Layouts

Drawing Name: ex3-18.rvt
Estimated Time: 10 minutes

This exercise reinforces the following skills:

- ❏ Filter
- ❏ Group
- ❏ Copy Aligned

1. Open *ex3-18.rvt.*

2. Activate **Level 1**.

3. Select the components used in the lavatory layout.

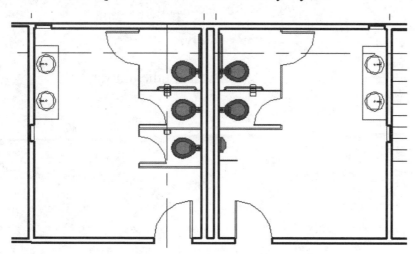

4.   ▽ **Hint: Use the Filter tool to select only the**
    Filter **Casework, Plumbing Fixtures and Specialty**
    Equipment.

Category:	Count:
☑ Casework	2
☐ Dimensions	3
☐ Door Tags	2
☐ Doors	2
☑ Plumbing Fixtures	1
☐ Railings: Top Rails	1
☑ Specialty Equipment	9
☐ Stairs: Supports	2
☐ Walls	9

5. Select the **Create Group** tool from the Create panel.

6. 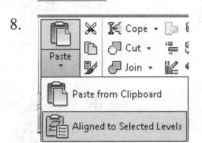 In the Name field, type **Lavatory**.

Press **OK**.

7. 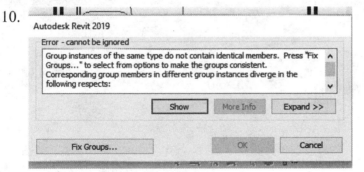 Select **Copy to Clipboard** from the Clipboard panel.

8. Select **Paste→Aligned to Selected Levels** from the Clipboard panel.

9. Select **Level 2**.

Press **OK**.

10. Select **Fix Groups**, if you see this dialog.

11. 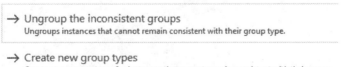 Select **Ungroup the inconsistent groups**.

12. Switch to **Level 2**.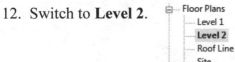

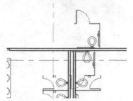

The lavatory
should be placed.

Once again, the ADA toilet stalls are acting up. Select each
stall and use the Pick New Host option to locate the stalls
correctly.

Then use the ALIGN tool to position them properly in each
lavatory.

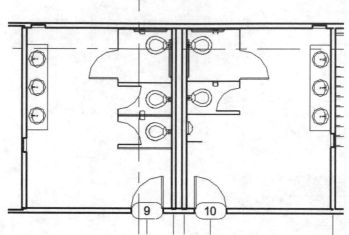

13. Save the file as *ex3-19.rvt*.

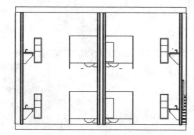

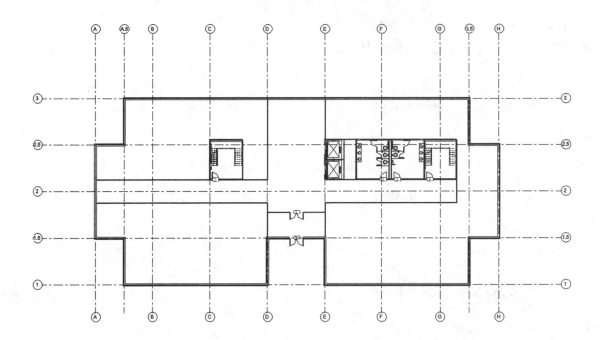

Level 1

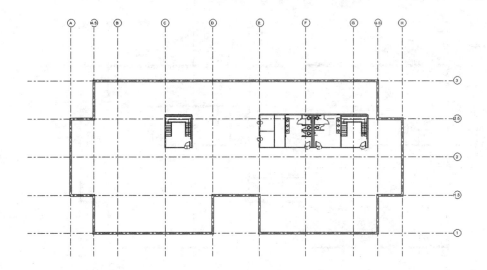

Level 2
Create a duplicate view of Level 2 using the Duplicate with Detailing option.
Rename the view Level 2 – Annotations Visible.
Turn off the visibility of elevations and dimensions in the Level 2 view.
Turn on the visibility of elevations and dimensions on the Level 2 – Annotations Visible
view.

Exercise 3-20
Add a Door to a Curtain Wall

Drawing Name: ex3-19.rvt
Estimated Time: 30 minutes

This exercise reinforces the following skills:

- ❑ Curtain Wall
- ❑ Modify Wall Curtain Wall
- ❑ Elevations
- ❑ Grid Lines
- ❑ Load Family
- ❑ Properties
- ❑ Type Selector

1.   Open *ex3-19.rvt*.

2.
Floor Plans
— Level 1
— Level 1 -Annotations Visible
— Level 2
— Level 2 - Annotations Visible
— Roof Line
— Site
    Activate **Level 1** Floor Plan.

3.   Select the East Wall.

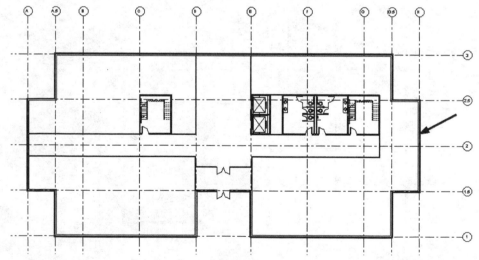

4.
Ext| **Storefront**
| Curtain Wall
| Press F1 for more help
Curtain Wall : Storefront
    Select the **Storefront** Curtain Wall using the Type Selector on the Properties pane.

5.

Visibility
☑ Elevations
☑ Floor Tags
☑ Furniture System Tags

Type **VV** to launch the Visibility/Graphics dialog.
Enable **Elevations**.
Press **OK**.

6.

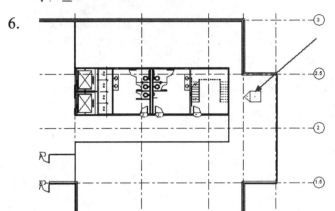

Locate your East Elevation marker. In this case the marker is inside the building.

7.

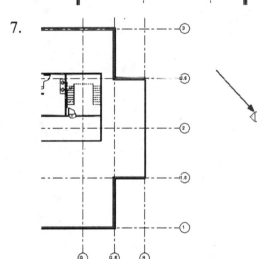

Place your cursor over the square part of the elevation marker. Hold down your left mouse button. Drag the marker so it is outside the building.

8.

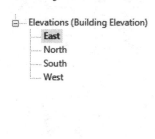

Left click on the triangle part of the elevation. A blue line will appear to indicate the depth of the elevation view.

Move the blue line so it is outside the building.

Double left click on the triangle to open the East Elevation view or select East elevation in the Project Browser.

Elevations (Building Elevation)
    East
    North
    South
    West

9.

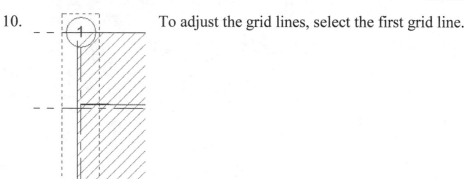

This is the east elevation view when the marker is outside the building and the depth is outside the building.

10. To adjust the grid lines, select the first grid line.

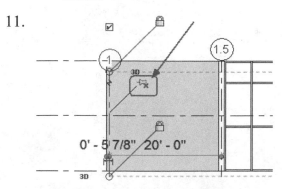

11. Unpin the grid line if it has a pin on it.

0' - 5 7/8" 20' - 0"

12. 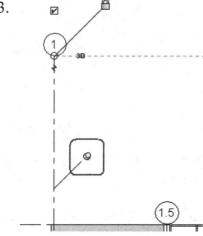 Select the grid line again.

Drag the bubble using the small circle above the building model.

13. 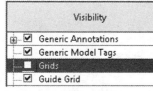 Left click to re-enable the pin.

14. 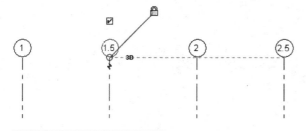 The group will adjust. Drag the remaining gridlines so the bubbles are aligned with the grid line group.

15.

Visibility
⊞ ☑ Generic Annotations
☑ Generic Model Tags
■ Grids
☑ Guide Grid

Type VV to launch the Visibility/Graphics dialog. Select the Annotations tab. Turn off visibility of grids.

Press **OK**.

16. 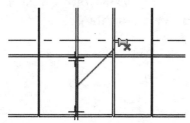 Select the center mullion.
Use the TAB key to cycle select.

Unpin the selection.

Right click and select **Delete** or press the **Delete** key on your keyboard to delete.

17. 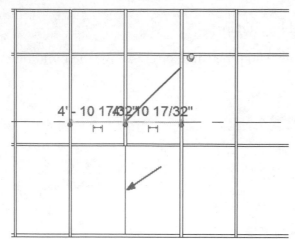   Select the grid line.

18.   Select the **Add/Remove Segments** tool from the ribbon.

Add/Remove
Segments
Curtain Grid

19. 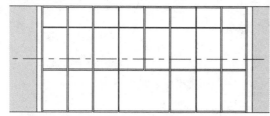   Select the grid line to be removed.

Left click below the building to accept.

There will now be a wide pane to be used to create a door.

20. 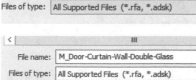   Select the **Load Family** tool from the Load from Library panel on the Insert ribbon.

Load   Load as
Family Group
Load from Library

21.    Select the Imperial Library folder on the left pane of the Load Family dialog.

Imperial Lib...

Imperial Det...

22.

| File name: | Door-Curtain-Wall-Double-Glass |
| Files of type: | All Supported Files (*.rfa, *.adsk) |

| File name: | M_Door-Curtain-Wall-Double-Glass |
| Files of type: | All Supported Files (*.rfa, *.adsk) |

Locate the **Door-Curtain-Wall-Double-Glass [M_ Door-Curtain-Wall-Double-Glass]** family from the *Doors* folder.
Press **Open**.

23.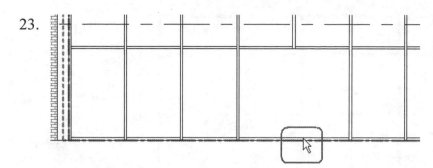

Hover your cursor over the edge of the panel.

DO NOT CLICK.

24.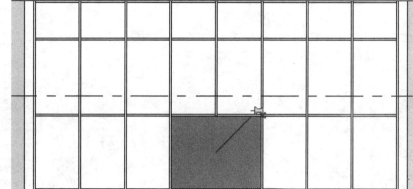

Press the tab.

Repeat until only the glass pane is highlighted.

25.

You should see **System Panel Glazed** in the Properties pane.

It is grayed out because it is pinned which prevents editing.

26. 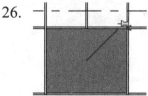 Unpin the glazing.

27. From the Properties pane:
Assign the **Door-Curtain-Wall-Double-Glass [M_ Door-Curtain-Wall-Double-Glass]** door to the selected element.

Left click to complete the command.

28. You will now see a door in your curtain wall.
Remember you won't see the hardware unless the Detail Level is set to Fine.

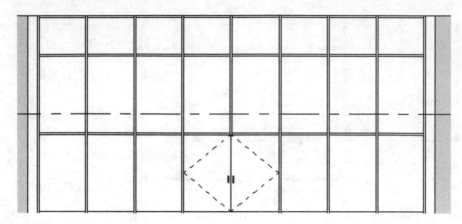

29. Save as *ex3-20.rvt*.

Exercise 3-21
Modifying a Curtain Wall

Drawing Name: ex3-20.rvt
Estimated Time: 30 minutes

This exercise reinforces the following skills:

- ❑ Use Match to change an Element Type
- ❑ Curtain Wall
- ❑ Curtain Wall Grids
- ❑ Curtain Wall Doors

Several students complained that the door was too big and asked how to make it a smaller size.

1.   Open *ex3-20.rvt.*

2.   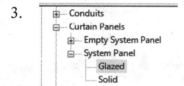   Switch to the **East** elevation view.

3.      Scroll down the **Project Browser**.

    Expand the **Curtain Panel** category.
    Expand the **System Panel** category.

    Locate the **Glazed** style.

4.   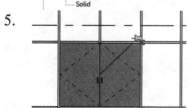   Right click on the **Glazed** System Panel.
    Select **Match**.

5.   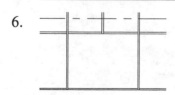   Select the Curtain Wall door.

6.   *The door is replaced with glazing.*

    Right click and select **Cancel**.

7. Activate the Architecture ribbon.

    Select the **Curtain Grid** tool.

8. 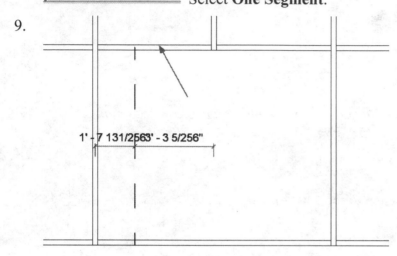 On the ribbon:

    Look for the Placement panel.

    Select **One Segment**.

9. Hover your mouse over the horizontal mullion on the left side of the pane to preview the placement of the curtain grid.

1' - 7 131/256" 3' - 3 5/256"

Note that you use the horizontal mullion to place a vertical grid and a vertical mullion to place a horizontal grid.

10. 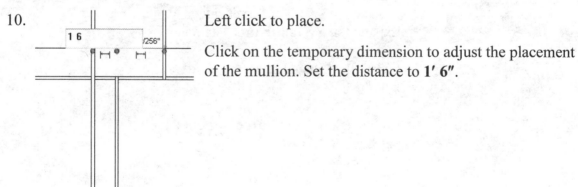 Left click to place.

    Click on the temporary dimension to adjust the placement of the mullion. Set the distance to **1' 6"**.

11.

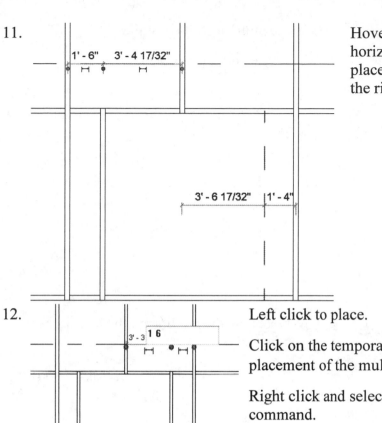

Hover your mouse over the horizontal mullion to preview the placement of the curtain grid on the right side.

12.

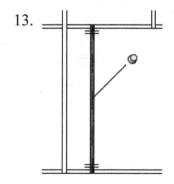

Left click to place.

Click on the temporary dimension to adjust the placement of the mullion. Set the distance to **1′ 6″**.

Right click and select **Cancel** to exit the command.

13.

Left click on the left vertical mullion.

Note that on the top and bottom there is a cross-hatch icon.

This is used to edit how the mullion joins/intersects the horizontal mullions.

Left click on the top cross-hatch.

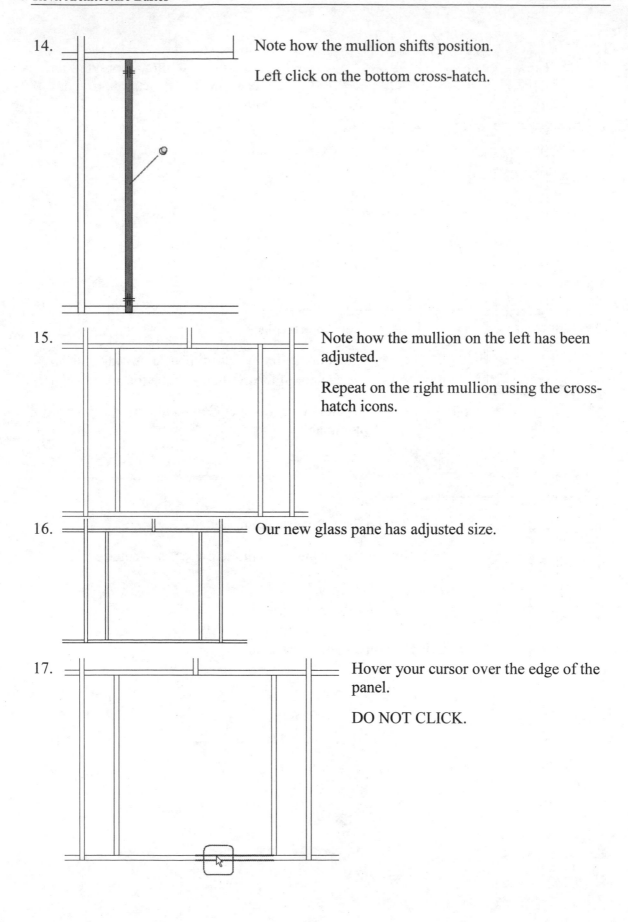

14.   Note how the mullion shifts position.

Left click on the bottom cross-hatch.

15.   Note how the mullion on the left has been adjusted.

Repeat on the right mullion using the cross-hatch icons.

16.   Our new glass pane has adjusted size.

17.   Hover your cursor over the edge of the panel.

DO NOT CLICK.

18.  Press the tab.

Repeat until only the glass pane is highlighted.

Left click to select the glass pane.

19. You should see **System Panel Glazed** in the Properties pane.

20. 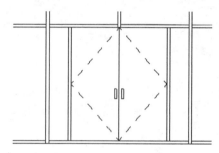 From the Properties pane:
Assign the Curtain Wall Dbl Glass door to the selected element using the Type Selector.

Left click to complete the command.

21. You will now see a door in your curtain wall.
The hardware is only visible if the Detail Level is set to **Fine**.

22. Save as *ex3-21.rvt*.

Exercise 3-22
Curtain Wall with Spiders

Drawing Name: ex3-21.rvt
Estimated Time: 15 minutes

This exercise reinforces the following skills:

- ❑ Curtain Walls
- ❑ System Families
- ❑ Properties

1. Open *ex3-21.rvt*.

2.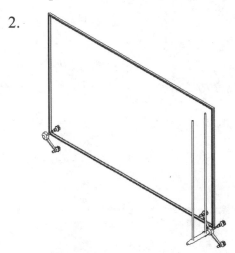
We are going to replace the existing panels in the curtain wall with a panel using spiders.

In order to use the panel with spiders, we need to load the family into the project.

3.
Go to the **Insert** ribbon. Select **Load Family**.

4.
File name:	spider clips
Files of type:	All Supported Files (*.rfa, *.adsk)

Browse to where you are storing the exercise files. Select the *spider clips.rfa*.
Press **Open**.

5.
Select the curtain wall so that it is highlighted.

6.   [⊞ Edit Type]   Select **Edit Type** from the Properties panel.

7.   [Duplicate...]   Select **Duplicate**.

8.   Name: [Curtain Wall with Spiders|]   Rename *Curtain Wall with Spiders.*
                                                 Press **OK**.

9.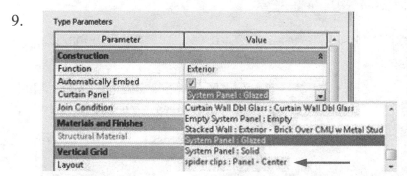

Locate the Curtain Panel parameter.

Select the drop-down arrow.

Select the spider clips: Panel - Center family.

10.

Parameter	Value
Construction	⁀
Function	Exterior
Automatically Embed	☑
Curtain Panel	spider clips : Panel - Center ▾
Join Condition	Vertical Grid Continuous
Materials and Finishes	⁀

You should see the spider clips panel family listed as assigned to the Curtain Panel.

11.

Vertical Mullions	⁀
Interior Type	Rectangular Mullion : 2.5" x 5" rec
Border 1 Type	Rectangular Mullion : 2.5" x 5" rec
Border 2 Type	Rectangular Mullion : 2.5" x 5" rec
Horizontal Mullions	⁀
Interior Type	Rectangular Mullion : 2.5" x 5" rec
Border 1 Type	Rectangular Mullion : 2.5" x 5" rec
Border 2 Type	Rectangular Mullion : 2.5" x 5" rec

Scroll down the dialog box and locate the Vertical and Horizontal Mullions.

12.

Adjust for Mullion Size	☐
Vertical Mullions	⁀
Interior Type	Rectangular Mullion : 2.5" x 5" rectangular ▾
Border 1 Type	None
Border 2 Type	Circular Mullion : 2.5" Circular
Horizontal Mullions	Rectangular Mullion : 1" Square
	Rectangular Mullion : 1.5" x 2.5" rectangular
Interior Type	Rectangular Mullion : 2.5" x 5" rectangular
Border 1 Type	Rectangular Mullion : 2.5" x 5" rectangular
Border 2 Type	Rectangular Mullion : 2.5" x 5" rectangular
Identity Data	⁀
Keynote	

Use the drop-down selector to set all the Mullions to **None**.

13.

Vertical Mullions	
Interior Type	None
Border 1 Type	None
Border 2 Type	None
Horizontal Mullions	
Interior Type	None
Border 1 Type	None
Border 2 Type	None

Verify that all the mullions are set to **None**.

Press **OK**.

14.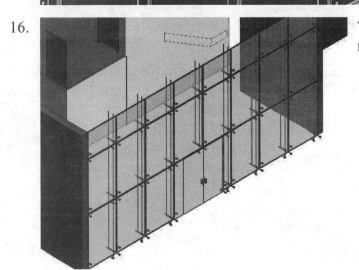

If you see this dialog, press **Delete Mullions**.

15.

If you still see mullions...

Select one of the mullions. Right click and use **Select All Instances→Visible in View**.

Then press the Delete key on your keyboard.

16.

The curtain wall with spiders and no mullions is displayed.

17. Save as *ex3-22.rvt*.

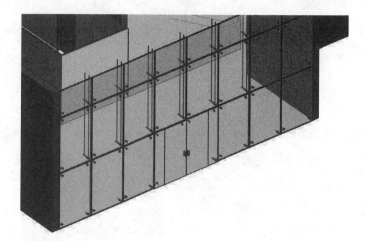

Extra: Replace the bottom panels with the Spider Panel – End Panel family to create a cleaner looking curtain wall.

Hint: Load the Spider Panel – End Panel family and replace each bottom panel with the new family using the Type Selector (similar to placing the curtain wall door). You will need to unpin before you can replace the panel.

Exercise 3-23
Adding Windows

Drawing Name: ex3-22.rvt
Estimated Time: 60 minutes

This exercise reinforces the following skills:

- ❑ Window
- ❑ Window Properties
- ❑ Array
- ❑ Mirror
- ❑ Copy-Move

1.   Open *ex3-22.rvt*.

2.   Switch to a **South** elevation view.

Elevations (Building Elevation)
    East
    North
    South
    West

3.   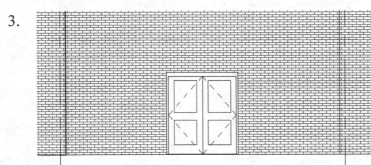   Zoom into the entrance area of the building.

Display is set to Coarse, Hidden Line.

4.   Select the **Architecture** ribbon.

Window   Select the **Window** tool from the Build panel.

5.      Set the Window Type to **Fixed: 36″ x 48″ [M_Fixed: 0915 x 1220mm]**.

6. 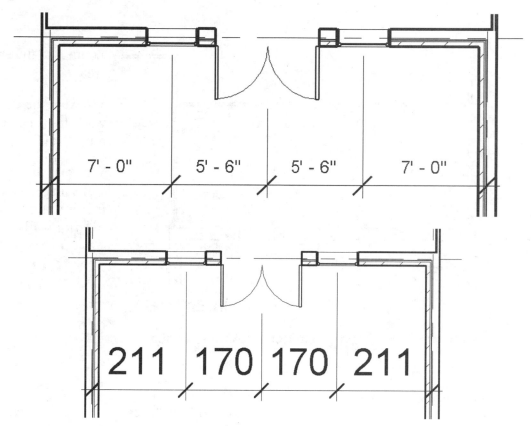 Place two windows **5′ 6″ [170 cm]** from the door's midpoint.

7. In plan view, the windows will be located as shown.

7' - 0"   5' - 6"   5' - 6"   7' - 0"

211 | 170 | 170 | 211

8. 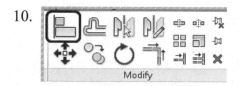 Activate the **South** Elevation.

   North
   South
   West

9. Zoom into the entrance area.

10. Use the **ALIGN** tool to align the top of the windows with the top of the door.
    Select the **Align** tool on the Modify panel on the Modify ribbon.

    Modify

11. ☑ Multiple Alignment   Prefer: Wall faces   Enable **Multiple Alignment** on the Options bar.

12. Select the top of the door, then the top of the left window. Then, select the top of the second window.

    Right click and select **Cancel** twice to exit the command.

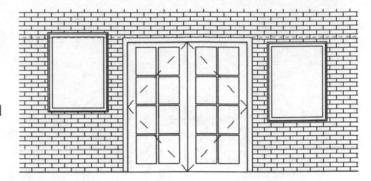

13. 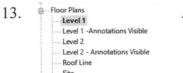 Activate **Level 1**.

14. 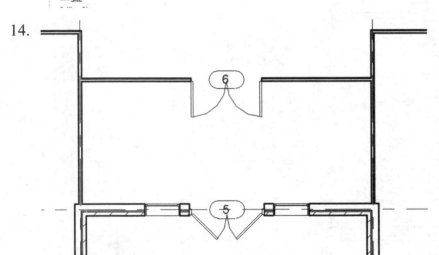   Zoom into the entry area.

    We want the exterior walls to be centered on the grid lines.

    We want the interior walls aligned to the interior face of the exterior walls.

15. 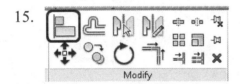   Activate the **Modify** ribbon.

    Select the **ALIGN** tool.

16.    On the Options bar:
    Set Prefer to Wall faces.

17. 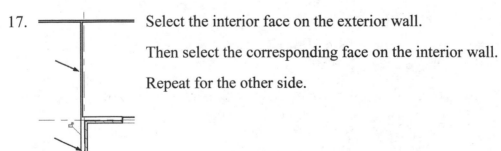   Select the interior face on the exterior wall.

    Then select the corresponding face on the interior wall.

    Repeat for the other side.

18. 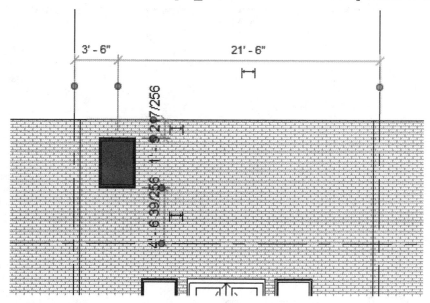 Activate the South elevation.

19. Place a **Fixed: 36″ x 48″** [**M_Fixed: 0915 x 1220mm**] window on the Level 2 wall.

Locate the window 3′ 6″ [110 cm] to the right of grid D.

20. Select the window.
    Select the **Array** tool from the Modify panel.

21. Enable **Group and Associate**.

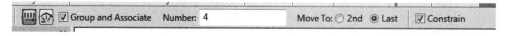

Set the Number to **4**.
Enable Move to: **Last**.
Enable **Constrain**. *This forces your cursor to move orthogonally.*

22. Pick the midpoint of the top of the window as the base points.

Move the cursor 18′ [542 cm] towards the right.
Pick to place.

23. Press ENTER to accept the preview.

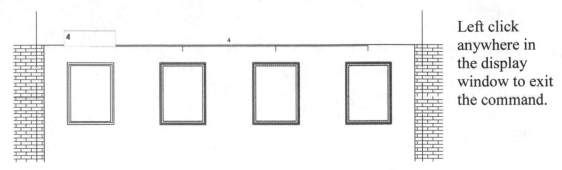

Left click anywhere in the display window to exit the command.

24.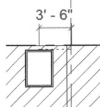

Use the **Measure** tool from the Quick Access tool bar to verify that the last window is 3′ 6″ [1110 cm] from grid E.

Verify that the four windows are equally spaced using the Measure tool.

25.

Select the first window.
Select the window group.

Select **Edit Group** from the Group panel.

26.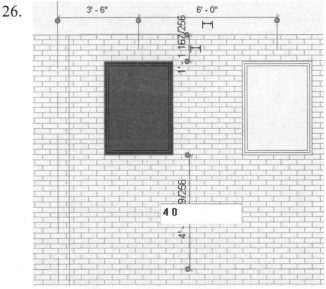

Select the first window.

27.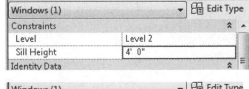

In the Properties pane:
Set the Sill Height to **4′ 0″** [**120 cm**].

28.  Select **Finish** from the Edit group toolbar.

29. The entire group of windows adjusts.

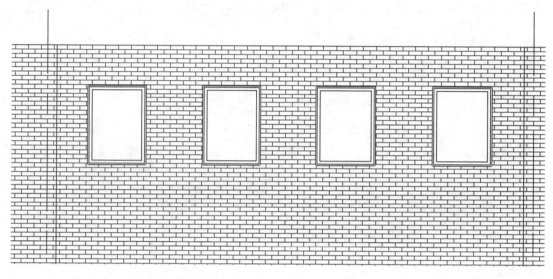

30. Save the file as *ex3-23.rvt*.

Challenge Exercise:

Add additional windows to complete the project.
Use the MIRROR and ARRAY tools to place windows.

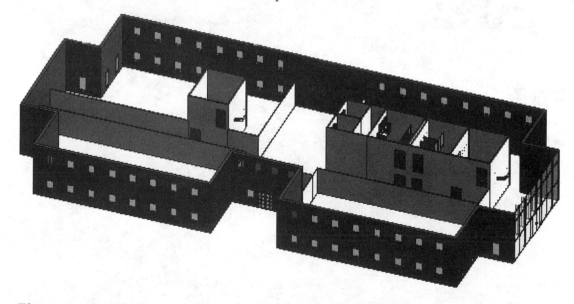

Elements can only be arrayed on the same work face.

Exercise 3-24:
Floor Plan Layout

Drawing Name: ex3-23.rvt
Estimated Time: 20 minutes

This exercise reinforces the following skills:

- ❑ Loading a Title block
- ❑ Add a New Sheet
- ❑ Adding a view to a sheet
- ❑ Adjusting Gridlines and Labels
- ❑ Hiding the Viewport

1.   Open or continue working in *ex3-23.rvt*.

2.   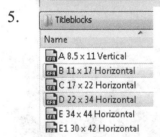   In the browser, highlight **Sheets**.
Right click and select **New Sheet**.

3.   [Load...]   Select **Load**.

4.   Select the *Imperial Library* folder located on the left pane.

5.   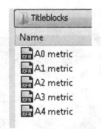

Browse to the *Titleblocks* folder.

Select the **D 22 x 34 Horizontal [A3 metric]** Titleblock.

Press **Open**.

6.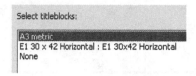

Highlight the **D 22 x 34 Horizontal** [**A3 metric**] Titleblock.

Press **OK**.

7.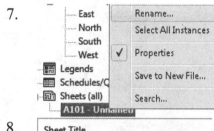

Highlight **A102-Unnamed** in the browser.
Right click and select **Rename**.

8.

Sheet Title	
A102	Number
First Level Floor Plan	Name

Change the Name of the Sheet to **First Level Floor Plan** and then press **OK**.

9. Drag and drop the Level 1 Floor Plan from the browser into the sheet.

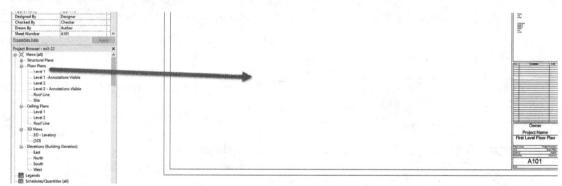

10.
① Level 1
  1/8" = 1'-0"

If you zoom into the View label, you see the scale is set to *1/8" = 1'-0"*.

You can adjust the position of the label by picking on the label to activate the blue grip. Then hold the grip down with the left mouse button and drag to the desired location.

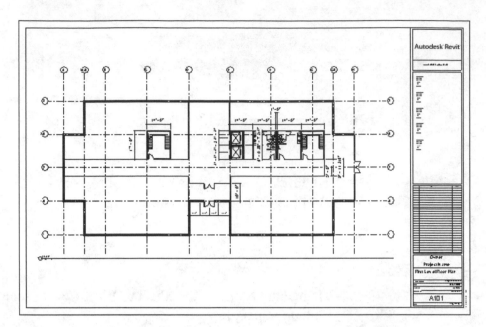

11. *Elevation markers will be printed.*

    To turn off the elevation markers:

    Hover the cursor over the view.
    Right click and select **Activate View**.

This is similar to activating model space in AutoCAD.

12. Type **VV** to launch to the Visibility/Graphics dialog.

    Select the **Annotation Categories** tab.

    Uncheck **Elevations**.

    Press **OK**.

13. 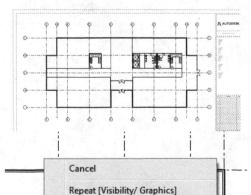 Adjust the position of the grids as needed,

    Unpin one grid and then use the grip next to
    the bubble head to re-position.

14. Hover the cursor over the view.
    Right click and select **Deactivate View**.

    This is similar to activating paper space in AutoCAD.

15.

Other	
File Path	
Drawn By	J. Student
Guide Grid	<None>

In the Properties Pane:
Change **Author** next to Drawn By to your name.

16. The title block updates with your name.

17. Save as *ex3-24.rvt*.

Notes:

Additional Projects

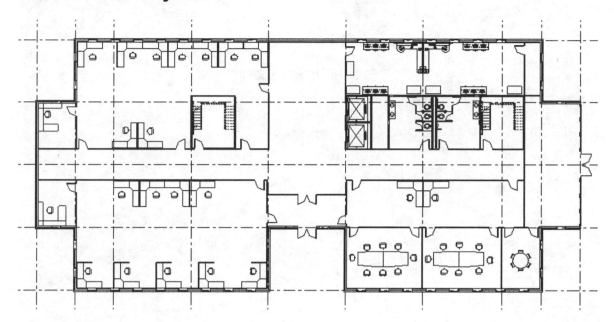

1) Duplicate the Level 1 Floor plan and create an office layout similar to the one shown.

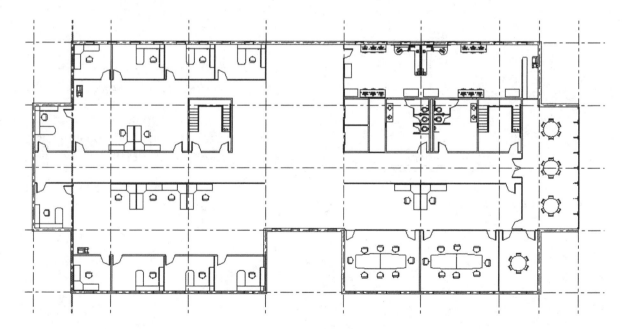

2) Duplicate the Level 2 Floor plan and create an office layout similar to the one shown.

3) On Level 2, in the cafeteria area, add furniture groups using a table and chairs.
   Add vending machines. (Families can be located from the publisher's website.)

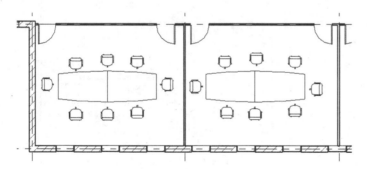

4) Furnish the two conference rooms on Level 2.

Add tables and chairs to the two conference room areas.

Use Group to create furniture combinations that you can easily copy and move.

There are conference table group families available on the publisher's website.

5) Furnish the office cubicles. Families can be found on the publisher's website.

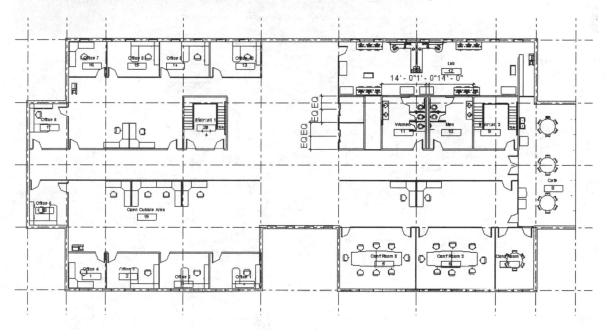

6) Add room tags to the rooms on Level 2.
   Create a room schedule.

<Room Schedule>		
A	**B**	**C**
Number	Name	Area
1	Office 4	136 SF
2	Office 3	187 SF
3	Office 2	181 SF
4	Office 1	183 SF
5	Conf Room 3	470 SF
6	Conf Room 2	479 SF
7	Conf Room 1	233 SF
8	Cafe	708 SF
9	Stairwell 2	229 SF
10	Men	228 SF
11	Women	228 SF
12	Lab	1058 SF
13	Office 10	183 SF
14	Office 9	181 SF
15	Office 8	187 SF
16	Office 7	136 SF
17	Office 6	176 SF
18	Office 5	176 SF
19	Open Cubicle A	6168 SF
20	Stairwell 1	225 SF

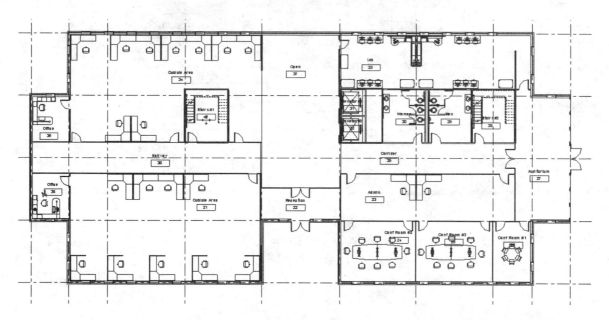

7) Use the Room Separator tool and the Room Tag tool to add rooms and room tags to Level 1.

Lesson 3 Quiz

True or False

1. Changes in the settings for one view do not affect other views.

Multiple Choice

2. When placing doors, which key is pressed to flip orientation?
   Choose one answer.

   A. Ctrl+S
   B. H
   C. Spacebar
   D. F
   E. L or R

3. Which of the following is NOT an example of bidirectional associativity?
   Choose one answer.

   A. Draw a wall in plan and it appears in all other views
   B. Add an annotation in a view
   C. Change a door type in a schedule and all the views update
   D. Flip a section line and all views update

4. Curtain Grids can be defined using all of the following except:
   Choose one answer.

   A. Vertical line
   B. Angled line
   C. Horizontal line
   D. Partial line

5. A stair can consist of all the following EXCEPT:

   A. Runs

   B. Landings

   C. Railings

   D. Treads

6. Which command is used to place a free-standing element, such as furniture?
   Choose one answer.

   A. Detail Component
   B. Load Family
   C. Repeating Detail
   D. Model In Place
   E. Place a Component

7. Select the TWO that are type properties of a wall:
   Choose at least two answers.

   A. FUNCTION
   B. COARSE FILL PATTERN
   C. BASE CONSTRAINT
   D. TOP CONSTRAINT
   E. LOCATION LINE

8. Which is NOT true about placing windows?
   Choose one answer.

   A. Windows require a wall as a host
   B. Windows cut an opening in the wall when placed
   C. The location of the exterior side of the window can be selected
   D. Sill height is adjustable in plan view

9. If you highlight Level 1, right click and select Duplicate View→Duplicate, what
   will the new view name be by default?

   A. Copy of Level 1
   B. Level 1 Copy 1
   C. Level 1 (2)
   D. None of the above

ANSWERS:
   1) T; 2) C; 3) B; 4) B; 5) D; 6) E; 7) A & B; 8) D; 9) B

Lesson 4
Materials

Materials are used:

- For Rendering
- For Scheduling
- To control hatch patterns
- To control appearance of objects

Revit divides materials into TWO locations:

- Materials in the Project
- Autodesk Asset Library

You can also create one or more custom libraries for materials you want to use in more than one project. You can only edit materials in the project or in the custom library – Autodesk Materials are read-only.

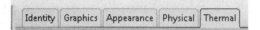

Materials have FIVE asset definitions – Identity, Graphics, Appearance, Physical and Thermal. Materials do not have to be fully defined to be used – YOU DON'T HAVE TO DEFINE ALL FIVE ASSETS. Each asset contributes to the definition of a single material.

You can define Appearance, Physical, and Thermal Assets independently without assigning/associating it to a material. You can assign the same assets to different materials. An asset can exist without being assigned to a material, BUT a material must have at least THREE assets (Identity, Graphics, and Appearance). Assets can be deleted from a material in a project or in a user library, but you cannot remove or delete assets from materials in locked material libraries, such as the Autodesk Materials library, the AEC Materials library, or any locked user library. You can delete assets from a user library, BUT be careful because that asset might be used somewhere!

You can only duplicate, replace or delete an asset if it is in the project or in a user library. The Autodesk Asset Library is "read-only" and can only be used to check-out or create materials.

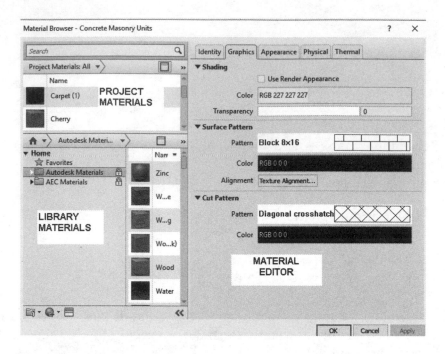

The Material Browser is divided into three distinct panels.
The Project Materials panel lists all the materials available in the project.
The Material Libraries list any libraries available.
The Material Editor allows you to modify the definition of materials in the project or in custom libraries.

The search field allows you to search materials using keywords, description or comments.

The button indicated toggles the display of the Material Libraries panel.

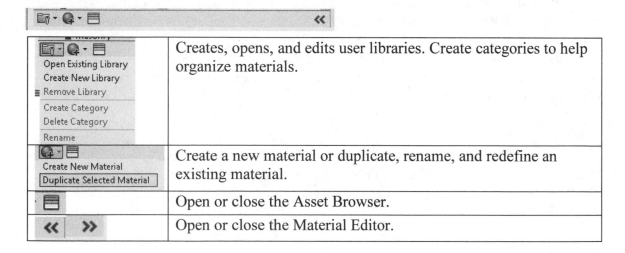

Open Existing Library / Create New Library / Remove Library / Create Category / Delete Category / Rename	Creates, opens, and edits user libraries. Create categories to help organize materials.
Create New Material / Duplicate Selected Material	Create a new material or duplicate, rename, and redefine an existing material.
	Open or close the Asset Browser.
« »	Open or close the Material Editor.

Exercise 4-1

Exercise 4-1
Modifying the Material Browser Interface

Drawing Name: ex3-24.rvt
Estimated Time: 10 minutes

This exercise reinforces the following skills:

- ❑ Navigating the Material Browser and Asset Browser

1.  Activate the **Manage** ribbon.

   Materials   Select the **Materials** tool.

2.   Make the Material Browser appear as shown.

   Toggle the Materials Library pane OFF by selecting

   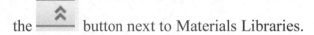 the button next to Materials Libraries.

   Collapse the Material Editor by selecting the >> Arrow on the bottom of the dialog.

3. 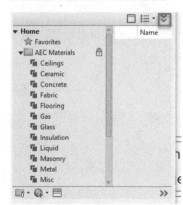 Toggle the Materials Library ON by selecting the double arrows button next to Material Libraries.

4.

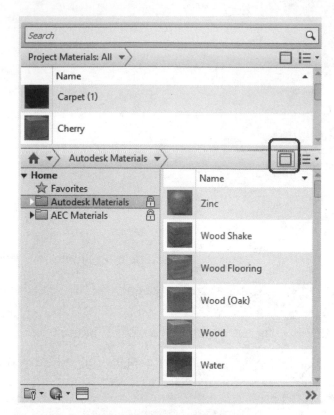

Select the toggle button indicated to hide the library names.

5.

In the Project Materials panel, set the view to **Show Unused**.

Scroll through the list to see what materials are not being used in your project.

6. 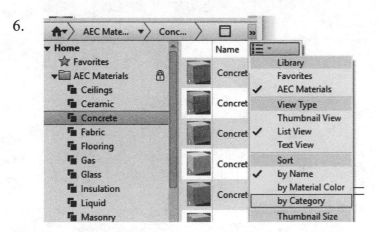   In the Material Library pane, change the view to **sort by Category**.

7.   Close the Material dialog.

The Material Browser has up to five tabs.

The Identity Tab

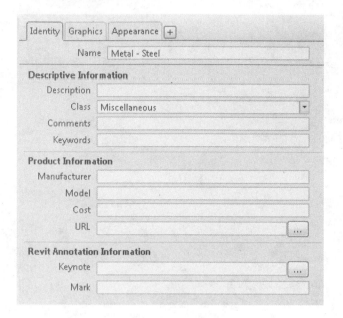

The value in the Description field can be used in Material tags.

Using comments and keywords can be helpful for searches.

Product Information is useful for material take-off scheduling

Assigning the keynote and a mark for the legend makes keynote creation easier.

The Graphics Tab

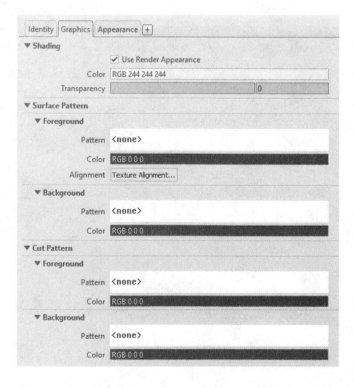

Shading –
Determines the color and appearance of a material when view mode is SHADED.

If you have Use Render Appearance unchecked, then the material will show a color of the material when the display is set to Shaded.

If Use Render Appearance is checked, then the material uses the Render Appearance Color when the display is set to Shaded.

Comparing the appearance of the model when Render Appearance is disabled or enabled:

☑ Use Render Appearance

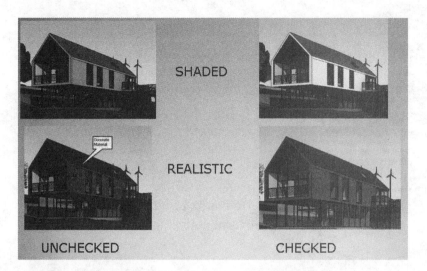

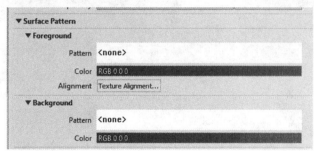

Surface Pattern will be displayed in all the following Display settings: Wireframe/Hidden/Shaded/Realistic/ Consistent Colors.

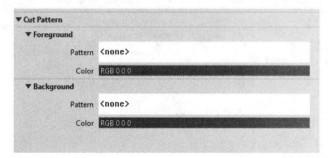

Cut Patterns appear in section views.

The Appearance Tab

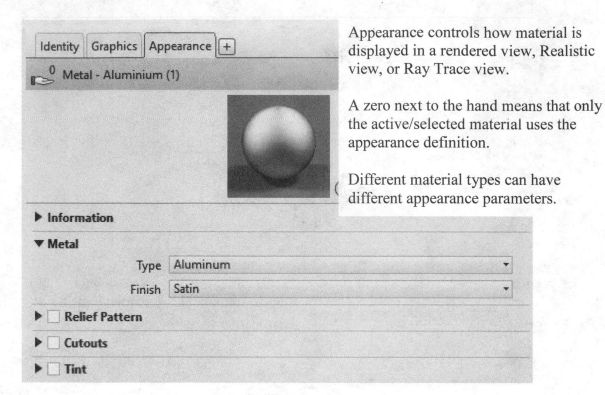

Appearance controls how material is displayed in a rendered view, Realistic view, or Ray Trace view.

A zero next to the hand means that only the active/selected material uses the appearance definition.

Different material types can have different appearance parameters.

The assigned class controls what parameters are assigned to the material. If you create a custom class, it will use generic parameters. The images below illustrate how different classes use different appearance parameters.

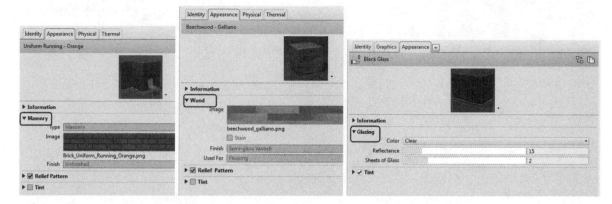

The Physical Tab

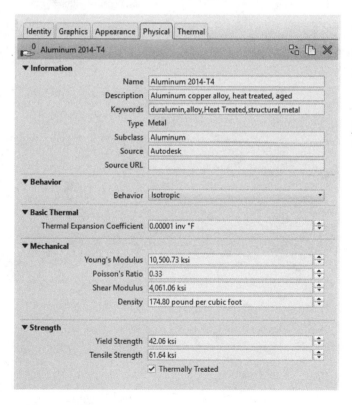

This information may be used for structural and thermal analysis of the model.

You can only modify physical properties of materials in the project or in a user library.

The Thermal Tab

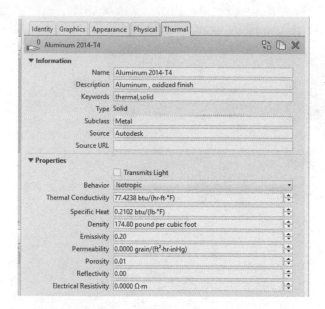

The Thermal tab is only available in Revit. (The Autodesk Material Library is shared by any installed Autodesk software, such as AutoCAD.)

This information is used for thermal analysis and energy consumption.

Material libraries organize materials by category or class.

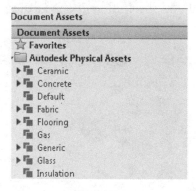

Material structural classes are organized into the following types:
- Basic
- Generic
- Metal
- Concrete
- Wood
- Liquid
- Gas
- Plastic

This family parameter controls the hidden view display of structural elements. If the Structural Material Class of an element is set to Concrete or Precast, then it will display as hidden. If it is set to Steel or Wood, it will be visible when another element is in front of it. If it is set to Unassigned, the element will not display if hidden by another element.

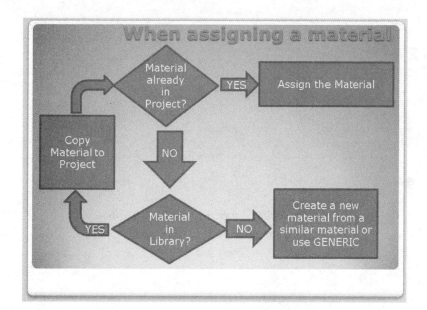

Use this flow chart as a guide to help you decide whether or not you need to create a new material in your project.

Exercise 4-2
Copy a Material from a Library to a Project

Drawing Name: ex3-24.rvt
Estimated Time: 10 minutes

In order to use or assign a material, it must exist in the project. You can search for a material and, if one exists in one of the libraries, copy it to your active project.

This exercise reinforces the following skills:

- ❑ Copying materials from the Materials Library
- ❑ Modifying materials in the project

1. Go to the Manage ribbon.

2. Select the **Materials** tool.

3. Type **oak** in the search field.

4. 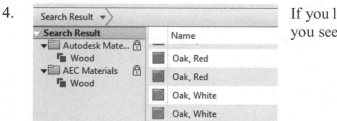 If you look in the Materials Library pane, you see that more oak materials are listed.

5. 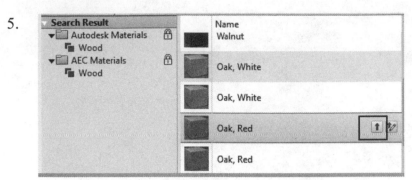 Select the Up arrow to copy the **Oak, Red** material from the library to the active project.

6. The material is now available for use in the project.

7. Type **brick** in the search field.

8. There are several more brick materials in the library.

9. Locate Brick, Adobe and then select the Copy and Edit tool.

This allows you to copy the material to the project and edit it.

10. Activate the Identity tab.

Select the button next to **Keynote**.

11.

Division 04	Masonry
04 05 00	Common Work Results for Masonry
04 20 00	Unit Masonry
04 21 00	Clay Unit Masonry

Highlight **Clay Unit Masonry**. Press **OK**.

12.

Revit Annotation Information

Keynote 04 21 00

Mark

The keynote number auto-fills the text box.

13.

Identity | Graphics | Appearance | Physical | Therma

Shading

☑ Use Render Appearance

Color RGB 191 184 146

Transparency

Activate the Graphics tab.
Enable **Use Render Appearance**.

14.

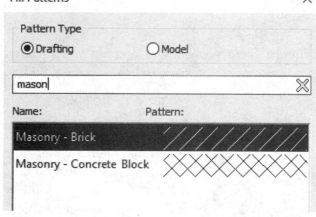

Identity | Graphics | Appearance | Physical | Thermal

▼ Shading

☑ Use Render Appearance

Color RGB 191 184 146

Transparency 0

▼ Surface Pattern

▼ Foreground

Pattern Masonry - Brick

Color RGB 120 120 120

Alignment Texture Alignment...

Left click on **<none>** in the Pattern box under Surface Pattern Foreground and select **Masonry – Brick**.

Press **OK**.

15.

Fill Patterns ✕

Pattern Type

◉ Drafting   ○ Model

mason ✕

Name: Pattern:

Masonry - Brick

Masonry - Concrete Block

You can use the search field to help you locate the desired fill pattern.

16.

▼ Surface Pattern

Pattern Masonry - Brick

Color RGB 120 120 120

Alignment Texture Alignment...

The fill pattern now appears in the Pattern box.

17.

Activate the **Appearance** tab.

Left click on the image name.

18.

Locate the *adobe brick* image file included in the exercise files.
Press **Open**.

19.

Change the Finish to **Unfinished**.

Press **OK**.

20. Close the Materials dialog.

21. Save as *ex4-2.rvt*.

You have modified the material definition in the project ONLY and not in the Materials library.

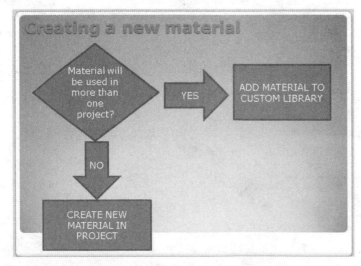

If you need to create a new material, you should first consider whether the material might be used in more than one project.

If the material may be used again, add the material to your custom library.

Steps for creating a new material:

1. Locate a material similar to the one you want, if you can (otherwise, use Generic)
2. Duplicate the material
3. Rename
4. Modify the Identity, Graphics, and Appearance tabs
5. Apply to the object
6. Preview using Realistic or Raytrace
7. Adjust the definition as needed

Exercise 4-3
Create a Custom Material Library

Drawing Name: ex4-2.rvt
Estimated Time: 5 minutes

This exercise reinforces the following skills:

□ Materials Library

1. Go to the Manage ribbon.

2. Select the **Materials** tool.
   Materials

3. At the bottom of the dialog:
   Open Existing Library
   Create New Library
   Remove Library
   Select **Create New Library**.

4. 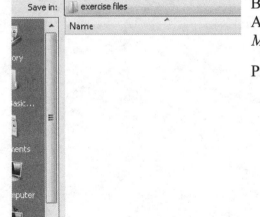 Browse to where you are storing your work. Assign your materials file a name, like *Custom Materials*.
   Press **Save**.

   Save in: exercise files
   Name
   File name: Custom Materials
   Files of type: Library Files (*.adsklib)

5. ▼ Home
   ☆ Favorites
   ▶ AEC Materials
   Custom Materials

   The custom library is now listed in the Libraries pane.

   Note that there is no lock next to the Custom Materials library because this library can be edited.

   Press **OK** to close the Materials dialog.

6. Save the project as *ex4-3.rvt*.

Exercise 4-4
Create Paint Materials

Drawing Name: ex4-3.rvt
Estimated Time: 30 minutes

This exercise reinforces the following skills:

❑   Materials

1.   Go to the Manage ribbon.

2.      Select the **Materials** tool.

3.   Type **Paint** in the search text box.

Select the Create **New Material** tool at the bottom of the dialog.

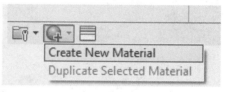

4.   Activate the **Identity** tab.
In the Name field, type **Paint - SW6126NavajoWhite**.
In the Description field, type **Sherwin Williams Paint – Interior**.
Select **Paint** for the Class under the drop-down list.
In the Manufacturer field, type **Sherwin Williams**.
In the Model field, type **SW6126NavajoWhite**.

5.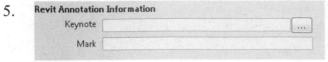

Select the **... browse** button next to Keynote.

6.

Division 09	Finishes
09 22 00	Supports for Plaster and Gypsum Board
09 23 00	Gypsum Plastering
09 24 00	Portland Cement Plastering
09 29 00	Gypsum Board
09 30 00	Tiling
09 51 00	Acoustical Ceilings
09 64 00	Wood Flooring
09 65 00	Resilient Flooring
09 68 00	Carpeting
09 72 00	Wall Coverings
09 73 00	Wall Carpeting
09 81 00	Acoustic Insulation
09 84 00	Acoustic Room Components
09 91 00	Painting
09 91 00.A1	Paint Finish
09 91 00.A2	Semi-Gloss Paint Finish

Select the **09 91 00.A2 Semi-Gloss Paint Finish** in the list.

Press **OK**.

7.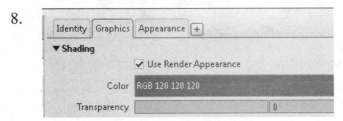

Note the keynote information auto-fills based on the selection.

8.

Activate the Graphics tab.

Place a check on **Use Render Appearance**.

9.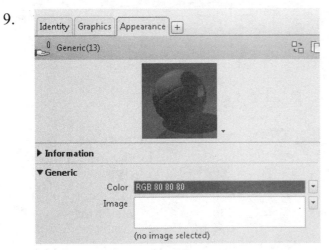

Select the Appearance tab.

Left click on the Color button.

10.

Select an empty color swatch under Custom colors.

11. 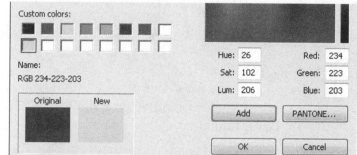 In the Red field, enter **234**.
In the Green field, enter **223**.
In the Blue field, enter **203**.

Press **Add**.

This matches Sherwin William's Navajo White, SW6126.

12. Select the color swatch.

Press **Add**.

The color swatch will be added to the pallet.

Press **OK**.

13.  The preview will update to display the new color.

Press **OK**.

14. Click on the bottom of the Material Browser to add a material.

15. On the Identity tab:

On the Identity tab:

Type **Paint - Interior-SW0068 Heron Blue** for the name of the material.

16.

Enter the Description in the field.

You can use Copy and Paste to copy the Name to the Description field.

Set the Class to **Paint**.
Enter the Manufacturer and Model information.

17.

Select the **...** browse button next to Keynote.

18.

09 91 00	Painting
09 91 00.A1	Paint Finish
09 91 00.A2	Semi-Gloss Paint Finish

Locate the **Semi-Gloss Paint Finish**.
Press **OK**.

19.

The Identity tab should appear as shown.

20.

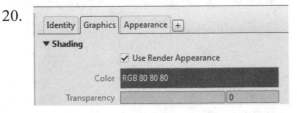

Activate the **Graphics** tab.

Enable **Use Render Appearance**.

21. 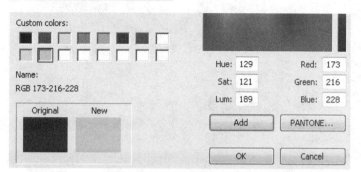 Activate the Appearance tab.

Select the down arrow next to Color to edit the color.

22. Select an empty slot under Custom Colors.

Set Red to 173.
Set Green to 216.
Set Blue to 228.

Press **Add**.

23. Select the custom color.

Press **OK**.

24. The color is previewed.

25. If you return to the Graphics tab, you see that the Color has been updated.

26. Press **OK** to close the Materials dialog.

27. Save as *ex4-4.rvt*.

Exercise 4-5
Add Categories and Materials to a Custom Library

Drawing Name: ex4-4.rvt
Estimated Time: 10 minutes

This exercise reinforces the following skills:

- ❏ Material Libraries
- ❏ Create Library Categories
- ❏ Add Materials to Custom Libraries

1. Go to the Manage ribbon.

2. Select the **Materials** tool.

3. In the Library Pane:

   Highlight the Custom Materials library.
   Right click and select **Create Category**.

4. Type over the category name and rename it to **Paint**.

5. Locate the Navajo White paint in the Project Materials list.

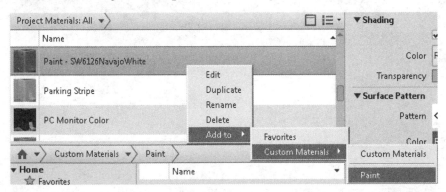

   Right click and select **Add to → Custom Materials → Paint**.

6. 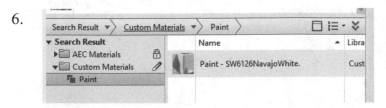 The Paint material is now listed in your Custom Materials library.

The material will now be available for use in other projects.

7. Locate Heron Blue paint in the Project Materials list.

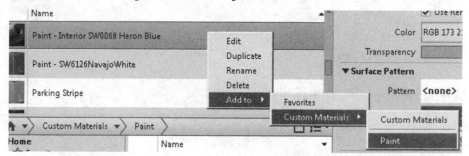

Right click and select **Add to → Custom Materials → Paint**.

8. 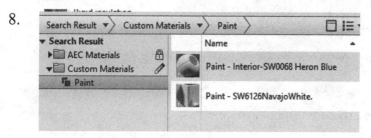 There are now two paint materials available in the Custom Materials library.

9. Close the Materials Manager.

10. Save as *ex4-5.rvt*.

Exercise 4-6
Defining Wallpaper Materials

Drawing Name: ex4-5.rvt
Estimated Time: 15 minutes

This exercise reinforces the following skills:

- ❑ Materials
- ❑ Defining a new class
- ❑ Using the Asset Manager

1.  Go to **Manage→Materials**.

    Materials

2. | wallpaper | ✕ |
    |---|---|

    Type **Wallpaper** for the name of the material.

3. The search term was not found in the document.

    There are no wallpaper materials in the project or the library.

4. Default

    Default Floor

    Locate the **Default** material in the project.

5.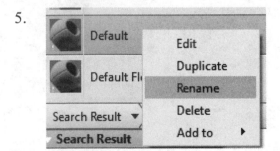

    Default
    Default Fl

    Edit
    Duplicate
    Rename
    Delete
    Add to ▶

    Search Result ▼
    Search Result

    Highlight the **Generic** material.
    Right click and select **Rename**.

6. Wallpaper - Striped

    Rename to **Wallpaper - Striped**.

7.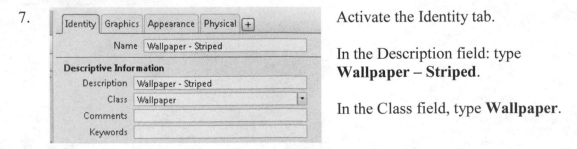

    Activate the Identity tab.

    In the Description field: type **Wallpaper – Striped**.

    In the Class field, type **Wallpaper**.

This adds Wallpaper to the Class drop-down list automatically.

8.  Browse to assign a keynote.

9. 09 72 00   Wall Coverings
        09 72 00.A1   Vinyl Wallcovering

     Locate the **Vinyl Wallcovering** keynote. Press **OK**.

10. The Identity tab should appear as shown.

11. 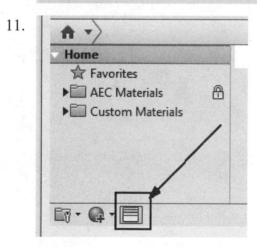 Launch the **Asset Browser** by pressing the button on the dialog.

12. stripes   Type **stripes** in the search field.

13. Highlight Wall Covering.
There are several materials available.

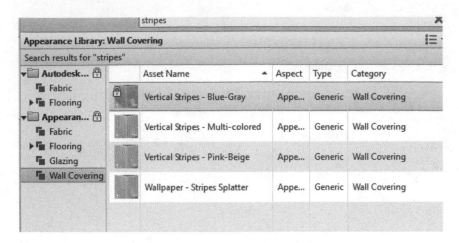

Highlight the **Vertical Stripes – Blue-Gray**.

14. Select the double arrow on the far right to copy the material properties to the Wallpaper-Striped definition.

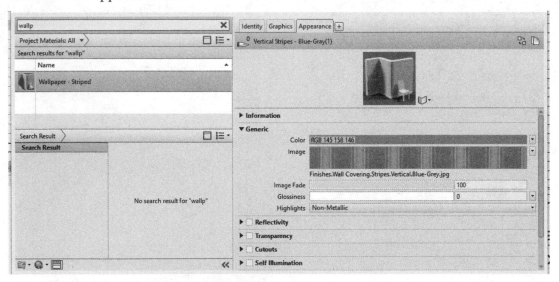

15. Activate the Appearance tab.

Note how the appearance has updated.

16.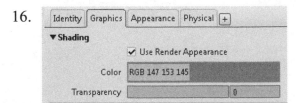

Activate the Graphics tab.

Place a check on Use Render Appearance.

Note that the color has updated for the Render Appearance.

17. Close the Material Browser.

18. Save as *ex4-6.rvt*.

Extra: *Create a Wallpaper – Floral material using the Asset Browser.*

Note: The materials you defined are local only to this project. To reuse these materials in a different project, they should be added to a custom library or use Transfer Project Standards on the Manage ribbon.

Exercise 4-7
Defining Vinyl Composition Tile (VCT)

Drawing Name: ex4-6.rvt
Estimated Time: 30 minutes

This exercise reinforces the following skills:

- ❑ Materials
- ❑ Copy a material
- ❑ Rename a material
- ❑ Use the Asset Browser

1.   ⬤   Go to **Manage→Materials**.

   Materials

2. | laminate | Type **laminate** in the search field.

3.   Highlight the **Laminate - Ivory, Matte** material.

   On the Identity tab:

   Type in the Description field:
   Laminate - Ivory, Matte

   Type in the Class field:
   Laminate.

   Browse for the keynote.

   (Identity / Graphics / Appearance tab dialog)
   Name: Laminate - Ivory, Matte
   Descriptive Information
   Description: Laminate- Ivory, Matte
   Class: Laminate
   Comments:
   Keywords:
   Product Information
   Manufacturer:
   Model:
   Cost:
   URL:
   Revit Annotation Information
   Keynote:
   Mark:

4.   Select the keynote for **Vinyl Composition Tile**.

   Press **OK**.

09 64 00	Wood Flooring
09 65 00	Resilient Flooring
09 65 00.A1	Resilient Flooring
09 65 00.A2	Vinyl Composition Tile
09 65 00.A3	Rubber Flooring

5.

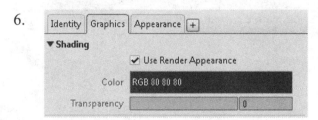

The Identity tab should appear as shown.

6.

On the Graphics tab:
Enable **Use Render Appearance**.

7.

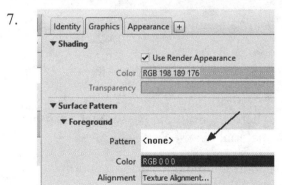

Select the Pattern button under Surface Pattern Foreground to assign a hatch pattern.

8.

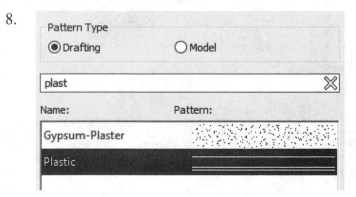

Scroll down and select the **Plastic** fill pattern.

You can also use the Search field to help you locate the fill pattern.

Press **OK**.

9.

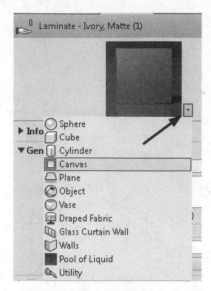

Select the Appearance tab.

In the Preview window:

Select the down arrow.
Highlight Canvas to switch the preview to display
a canvas.

10.

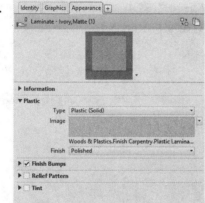

The Appearance tab should look as shown.

11.

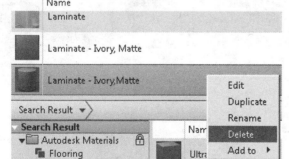

There are multiple Ivory, Matte laminate
materials in the Project.

Highlight the ones that were not
modified.

Right click and press **Delete**.

12.

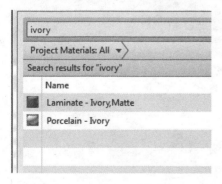

Verify that there is only one Ivory, Matte laminate
material in the project by doing a search on ivory.

13. [tile] Type **tile** in the search box.

14.

Tile, Mosaic, Gray

Tile, Porcelain, 4in

Vinyl Composition Tile

Search Result ▼〉

Edit
Duplicate
Rename
Delete
Add to ▶

Highlight the **Tile, Mosaic, Gray**.

Right click and select **Edit**.

15.

Identity | Graphics | Appearance | Physical | Thermal

Name Tile, Mosaic, Gray

Descriptive Information

Description Mosiac tile, gray

Class Generic

Comments

Keywords

Carpet
Ceramic
Concrete

Product Information

Manufacturer

Model

Cost

URL

Earth
Fabric
Gas
Generic
Glass
Laminate
Liquid

On the Identity tab:

Add a Description.

Change the Class to **Laminate**.

16.

Descriptive Information

Description Mosiac tile, gray

Class Laminate

Comments

Keywords VCT, gray

Add **VCT, gray** to the Keywords list.

17.

09 65 00   Resilient Flooring
  09 65 00.A1   Resilient Flooring
  09 65 00.A2   Vinyl Composition Tile
  09 65 00.A3   Rubber Flooring

Assign **Vinyl Composition Tile** to the keynote.

18.

Identity | Graphics | Appearance | Physical | Thermal

Name Tile, Mosaic, Gray

Descriptive Information

Description Mosiac tile, gray

Class Laminate

Comments

Keywords VCT, gray

Product Information

Manufacturer

Model

Cost

URL [...]

Revit Annotation Information

Keynote 09 65 00.A2   [...]

Mark

The Identity tab should appear as shown.

19.

On the Graphics tab:

Enable **Use Render Appearance**.

20.

Select the Pattern button under Surface Pattern Foreground to assign a hatch pattern.

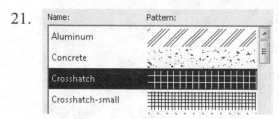

21.

Scroll down and select the **Crosshatch** fill pattern.

Press **OK**.

22.

The Graphics tab should appear as shown.

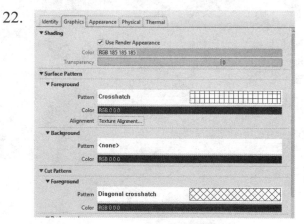

23. [TILE]   Type **TILE** in the search field.

24.  Scroll down the Document materials and locate the Vinyl Composition Tile.

Right click and select **Rename**.

25. 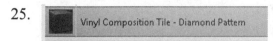 Name the new material **Vinyl Composition Tile - Diamond Pattern**.

26. Activate the Identity tab.

In the Description field: Type **VCT-Diamond Pattern**.
Set the Class to **Laminate**.
Enter **VCT, Diamond** for Keywords.

Assign **09 65 00.A2** as the Keynote.

27. Activate the Graphics tab.

Enable **Use Render Appearance** under Shading.

28. Left click in the Surface Pattern Foreground area.

29.

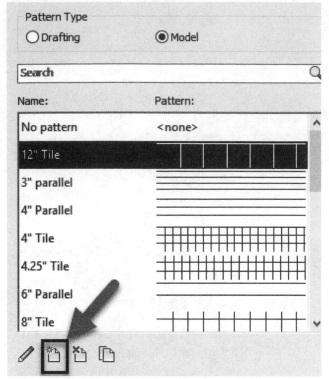

Select the **New** button.

30.

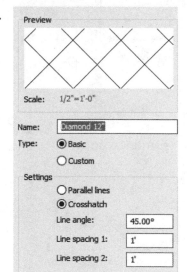

Enable **Basic**.

In the Name field, enter **Diamond 12″** [**Diamond 300**].

Set the Line Space 1: to **1'** [**300**].
Set the Line Space 2: to **1'** [**300**].
Enable **Crosshatch**.

Press **OK**.

31.

Ceiling 24x48

Diamond 12″

HerringBone 2x6

Pattern Type
○ Drafting   ◉ Model

| No Pattern | OK | Cancel |

The new hatch pattern is listed.

Highlight to select and press **OK**.

32.

The new hatch pattern is displayed.

33. Launch the Asset Browser.

34. Type diamond in the search box at the top of the Asset Browser dialog.

35. Locate the **Diamonds1** property set.

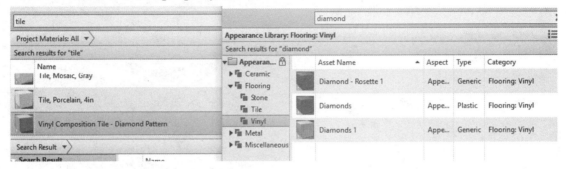

At the far left, you should see a button that says replace the current asset in the editor with this asset.

36.

The Appearance tab will update with the new asset information.

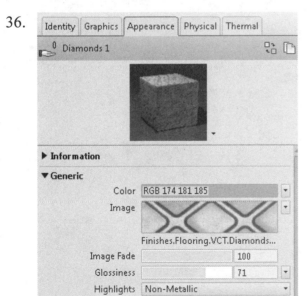

37. 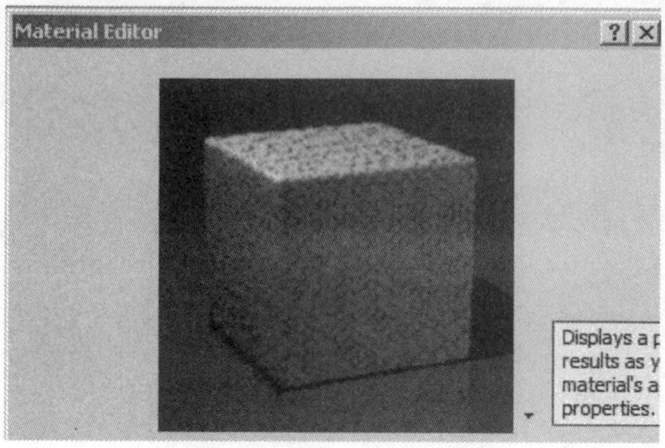
Material Editor
Displays a p
results as y
material's a
properties.

The Material Browser will update.

Close the Asset Browser.

Press **OK.**

38. Save as *ex4-7.rvt*.

Exercise 4-8
Define a Glass Material

Drawing Name: ex4-7.rvt
Estimated Time: 20 minutes

This exercise reinforces the following skills:

- ❑ Materials
- ❑ Use the Asset Browser
- ❑ Duplicate a Material

1.   ⬤ Go to **Manage→Materials**.
    Materials

2.   `glass`   Type glass in the search field.

3.   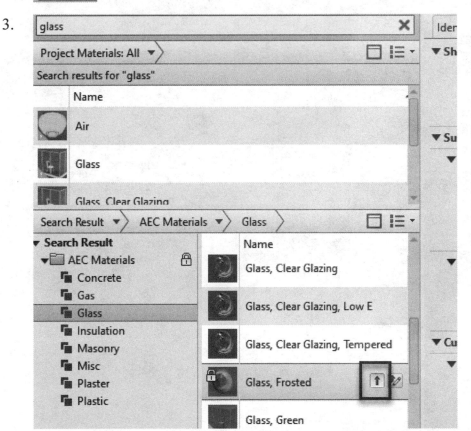   Locate the **Glass, Frosted** material in the Library pane.

Right click and select **Add to→ Document Materials** or use the Up arrow.

4.   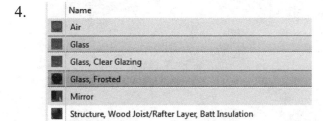   Highlight **Glass, Frosted** in the Documents pane.

5. Activate the Appearance tab.

Select **Replace this asset**.

This will launch the Asset Browser.

6. The Asset Browser will launch.

Type **glass** in the search field.

Highlight **Glass** under Appearance Library and scroll through to see all the different glass colors.

Select **Clear - Amber**.

Select the Replace asset tool on the far right.

7.

On the Identity tab:
Set the fields:

Name: Glass, Frosted Amber
Description: Frosted glass, Amber
Class: Glass
Keywords: amber, frosted, glass
Keynote: 08 81 00.A1

8.

On the Graphics tab:

Enable **Use Render Appearance**.

9.

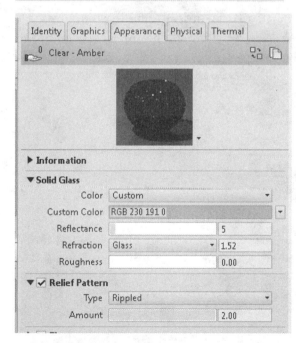

On the Appearance tab:

Under Relief Pattern:

Set the Type to **Rippled**.
Set the Amount to **2.0**.

Press **Apply**.

Apply saves the changes and keeps the dialog box open.
OK saves the changes and closes the dialog box.

10.

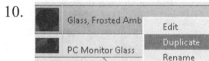

Highlight the **Glass, Frosted Amber** material in the Project Materials pane. Right click and select **Duplicate**.

11. Highlight the duplicated material.
Right click and select **Rename**.

12. Change the name to **Glass, Frosted Green**.

13. On the Identity tab:
Set the fields:

Name: Glass, Frosted Green
Description: Frosted glass, Green
Class: Glass
Keywords: green, frosted, glass
Keynote: 08 81 00.A1

14. In the Asset Browser:
In the Appearance Library

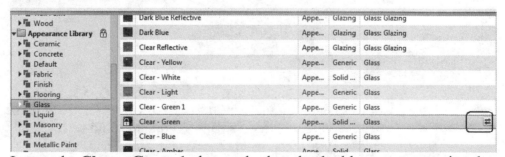

Locate the **Clear - Green 1** glass and select the double arrows to assign that
material appearance to the Glass, Frosted Green material.

15.

| Identity | Graphics | Appearance | Physical | Thermal |

0 Clear - Green 1

▶ Information

▼ Generic

Color   RGB 16 171 74

Image

(no image selected)

Image Fade   0

Glossiness   20

Highlights   Non-Metallic

The Appearance properties will update.

16.

| Identity | Graphics | Appearance | Physical | Thermal |

0 Clear - Green

▶ Information

▼ Solid Glass

Color   Green

Reflectance   5

Refraction   Glass   1.52

Roughness   0.00

▼ Relief Pattern

Type   Rippled

Amount   0.30

▶ Tint

Check the preview image and verify that the material looks good.

Adjust the appearance settings to your preferences.

17.

| Identity | Graphics | Appearance | Physical | Thermal |

▼ Shading

☑ Use Render Appearance

Color   RGB 172 203 187

Transparency   95

Activate the Graphics tab.

Enable **Use Render Appearance**.

Press **Apply** to save the settings.

18. Save as *ex4-8.rvt*.

Exercise 4-9
Defining Wood Materials

Drawing Name: ex4-8.rvt
Estimated Time: 5 minutes

This exercise reinforces the following skills:

- Materials
- Appearance Settings
- Identity Settings
- Copying from Library to Project Materials

1.   Go to **Manage→Materials**.

   Materials

2.   `oak`   Type **oak** in the search field.

3.   Locate the **Oak, Red** in the Project Materials.
   This material was added earlier.

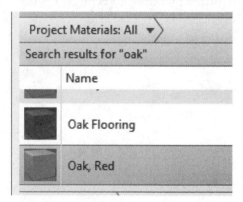

4.   In the Identity tab:

   Name: Oak, Red
   Description: Oak, Red
   Class: Wood
   Keywords: red, oak
   Keynote: 09 64.00

5.

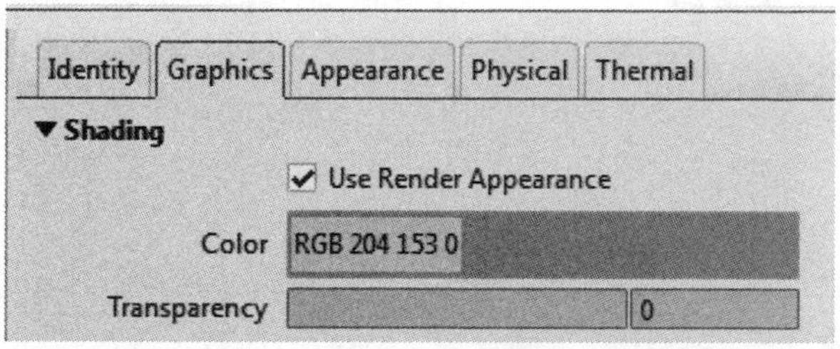

| Identity | Graphics | Appearance | Physical | Thermal |

▼ **Shading**

☑ Use Render Appearance

Color RGB 204 153 0

Transparency 0

On the Graphics tab:

Enable **Use Render Appearance**.

6.

▼ Surface Pattern

▼ Foreground

Pattern Wood 1

Color RGB 120 120 120

Alignment Texture Alignment...

Set the Foreground Surface Pattern to **Wood 1**.

Press **Apply** to save the settings.

7. Save as *ex4-9.rvt*.

Exercise 4-10
Defining Site Materials

Drawing Name: ex4-9.rvt
Estimated Time: 15 minutes

This exercise reinforces the following skills:

- ❏ Copying Materials from the Library to Document
- ❏ Editing Materials
- ❏ Using the Asset Browser
- ❏ Assigning a custom image to a material

1. Go to **Manage→Materials**.

    Materials

2. `grass` Type **grass** in the search field at the top of the Material browser.

3. Grass is listed in the lower library pane.

    Highlight **grass**.
    Right click and select **Add to→**
    Document Materials.

4. In the Identity tab:

    Name: Grass
    Description: Grass
    Class: Earth
    Keywords: grass
    Keynote: 02 81 00.A1

5. On the Graphics tab:

    Enable **Use Render Appearance**.

6. On the Appearance tab:

    Select **Replaces this asset**.

7. Type **grass** in the search field.

8.
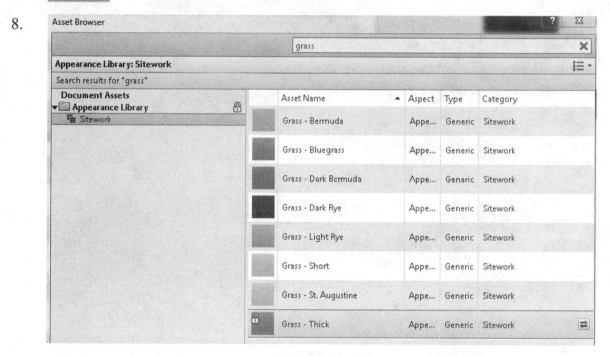

You have a selection of grasses to select from.

9. Select the **St Augustine** grass.

    Right click and select **Replace in Editor**.

    You can select your own choice as well.

    Close the Asset Browser.

10.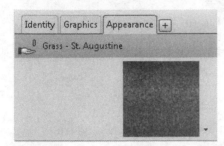

The selected grass name should be listed under Assets.

Press **Apply**.

11. 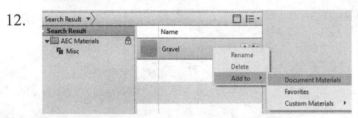 Do a search for **gravel** in the Material Browser.

12.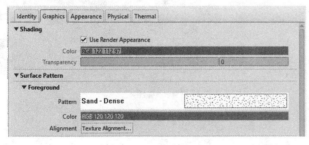

Locate the Gravel in the lower library panel.

Right click and **select Add to → Document Materials**.

13.

In the Identity tab:

Name: Gravel
Description: Crushed stone or gravel
Class: Earth
Keywords: gravel
Keynote: 31 23 00.B1

14.

In the Graphics tab:

Enable **Use Render Appearance**.

Assign the **Sand Dense** hatch pattern to the Foreground Surface Pattern.

15.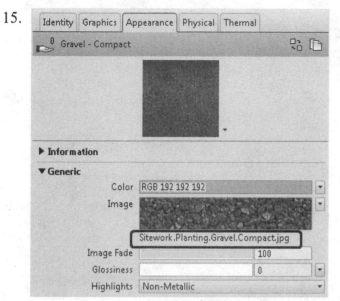

On the Appearance tab:

Change the color to **RGB 192 192 192**.

Left click on the image name.

16.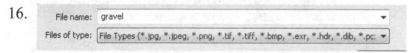

Locate the *gravel* image file in the exercise files.

Press **Open**.

17.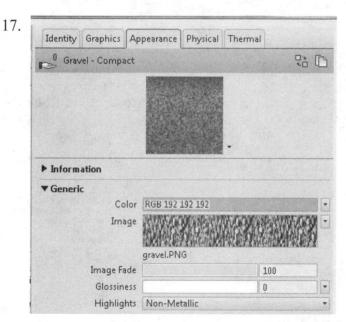

The Appearance will update.

Press **Apply** to save the changes.

18. Save as *ex4-10.rvt*.

Exercise 4-11
Defining Masonry Materials

Drawing Name: ex4-10.rvt
Estimated Time: 15 minutes

This exercise reinforces the following skills:

- ❑ Copying Materials from the Library to the Document Materials
- ❑ Using the Asset Browser

1. Go to **Manage→Materials**.

2. Type **masonry** in the search field.

3.

    Highlight **Stone** in the Libraries pane.

    Right click and select **Add to →
    Document Materials**.

4. In the Identity tab:

    Name: Stone, River Rock
    Description: Stone, River Rock
    Class: Stone
    Keywords: stone, river rock
    Keynote: 32 14 00.D4

32 14 00	Unit Paving
32 14 00.A1	4" x 8" x 1 1/2" Brick Paver
32 14 00.B1	24" x 24" x 4" Concrete Grid Paver
32 14 00.C1	8" x 8" x 2" Concrete Paver
32 14 00.D1	1" Stone Paver
32 14 00.D2	2" Stone Paver
32 14 00.D3	12" x 12" Stone Paver
32 14 00.D4	Cobble Stone

5. In the Graphics tab:

    Enable **Use Render Appearance for Shading**.

6. 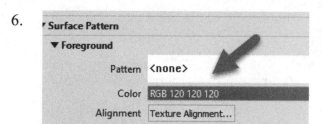 Select the Foreground Pattern button to assign a fill pattern.

7.  Select **New** to create a new fill pattern.

8. 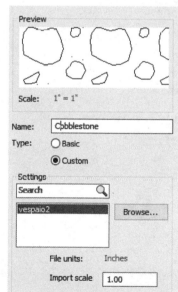 Type **Cobblestone** in the Name field.

Enable **Custom**.

Select **Browse**.

9. Locate the *cobblestone.pat* file in the exercise files downloaded from the publisher's website.

Press **Open**.

10. You will see a preview of the pattern.

Verify that the name is set to **Cobblestone**.

Press **OK**.

11.

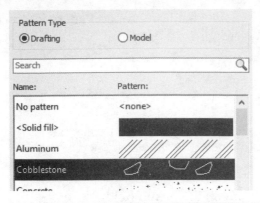

Highlight **Cobblestone**.

Press **OK**.

Revit allows you to use any AutoCAD pattern file as a fill pattern.

12.

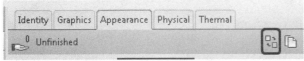

On the Appearance tab:

Select **Replaces this asset**.

13. Type **rock** in the search field.

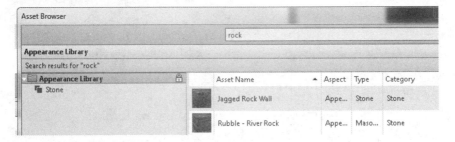

Locate the **Rubble - River Rock**.

14.

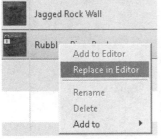

Highlight the **Rubble - River Rock**.

Right click and select **Replace in Editor**.

15.

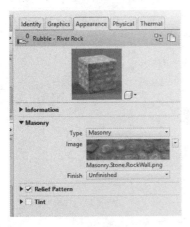

The Material Editor updates.

Press **OK** to save the changes and close the dialog.

16. Save as *ex4-11.rvt*.

Exercise 4-12
Assigning Materials to Stairs

Drawing Name: ex4-11.rvt
Estimated Time: 20 minutes

This exercise reinforces the following skills:

- ❑ Floor
- ❑ Floor Properties

1.   Open *ex4-11.rvt.*

2.
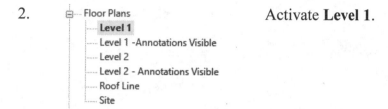
    Activate **Level 1**.

3.   ▼ Edit Type   Select the stairs.

Select **Edit Type** on the Properties pane.

4.   Duplicate...   Select **Duplicate**.

5.   Name: Stairs - Oak Tread with Painted Rise   Type **Stairs - Oak Tread with Painted Riser**.

OK   Cancel   Press **OK**.

6.
Construction	
Run Type	2" Tread 1" Nosing 1/4" Riser
Landing Type	Non Monolithic Landing
Function	Interior

Under Construction:
Locate the Run Type and select the small …
button on the far right.

7.
Parameter	Value
Materials and Finishes	
Tread Material	<By Category>
Riser Material	<By Category>

Select the **Tread Material** column.

8.
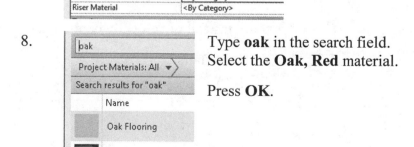

Type **oak** in the search field.
Select the **Oak, Red** material.

Press **OK**.

9.

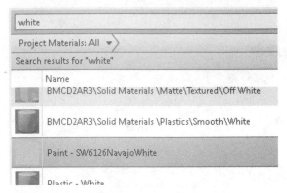

Select the **Riser Material** column.

10.

Type **white** in the search field.

Locate the **Paint - SW6126NavajoWhite** material.

Press **OK**.

11. Press **OK** twice to close the dialog.

12. Switch to a **3D** view.

13.

Change the display to **Realistic**.

Orbit around the model to inspect the stairs.

14.

Observe how the assigned materials affect the stairs' appearance.

Some users may not see the wood pattern on their stairs. This is a graphics card issue. I use a high-end graphics card.

15. If you want the boards to be oriented horizontally instead of vertically on the treads:

16.

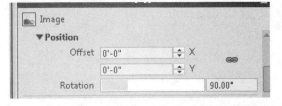

Select the stairs.
Edit Type.

17.

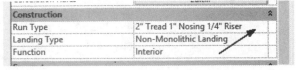

Left click in the Run Type column to open up the Run properties dialog.

18. 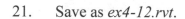 Select the Tread Material.
Select the Appearance tab.

19. 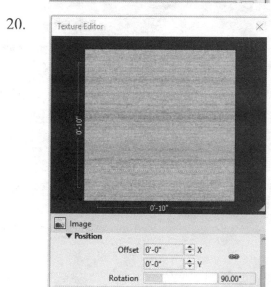 Next to the image picture, select Edit Image from the drop down list.

20. Expand the Transform area.
Set the Rotation to **90 degree**.
Press **Done**.

Press **OK** to close the dialogs.

21. Save as *ex4-12.rvt*.

Extra:

Select the other stairs and assign it to the new stairs type using the type selector.

Add all the custom materials to the Custom Materials library.

Define a custom wallpaper, a custom brick, and a custom wood material using assets from the Asset Browser.

Additional Projects

1) Create a material for the following Sherwin-Williams paints:

SW 6049 Gorgeous White
Interior/Exterior

Color Collection	Soft and Sheer
Color Family	Whites
Color Strip	8
RGB Value	R-232	G-220	B-212
Hexadecimal Value	# E8DCD4
LRV	73

SW 6050 Abalone Shell
Interior/Exterior

Color Family	Reds
Color Strip	8
RGB Value	R-219	G-201	B-190
Hexadecimal Value	# DBC9BE
LRV	60

SW 6052 Sandbank
Interior/Exterior

Color Family	Reds
Color Strip	8
RGB Value	R-196	G-166	B-152
Hexadecimal Value	# C4A698
LRV	41

2) Create a wallpaper material using appearances from the Asset Browser.

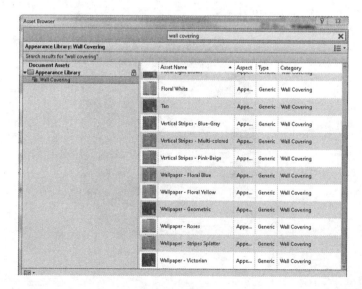

3) Create three different types of carpeting materials using appearances from the Asset Browser.

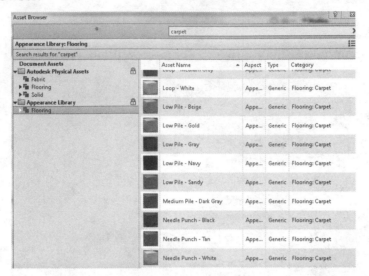

4) On Level 2, in the cafeteria area, apply paint to the walls and modify the floor to use a VCT material.

5) Apply paints to the walls in the conference rooms. Modify the floors to use a carpet material by creating parts.

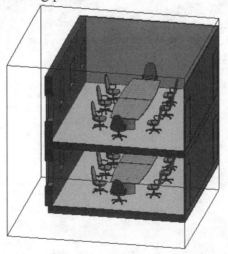

Lesson 4 Quiz

True or False

1. Materials in the default Material Libraries can be modified.
2. Only materials in the current project can be edited.
3. A material can have up to five assets, but only one of each type of asset.

Multiple Choice

4. Which of the following is NOT a Material asset?

    A. Identity
    B. Graphics
    C. Appearance
    D. Structural
    E. Thermal

5. Identify the numbers.

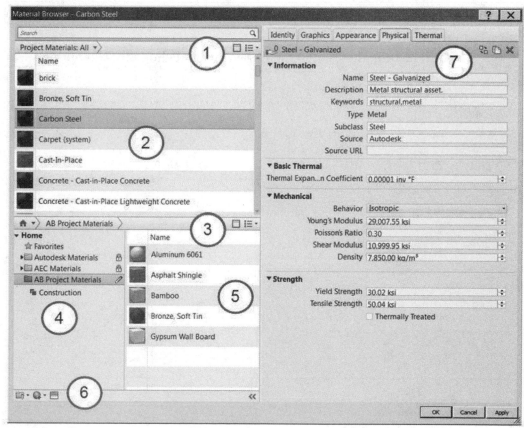

Material Browser

6. The lock icon next to the Material Library indicates the library is:

    A. Secure
    B. Closed
    C. "read-only"
    D. Unable to be moved

7. To access Materials, use this ribbon:

    A. Manage
    B. View
    C. Architecture
    D. Annotate
    E. Materials

8. To add a material to the current project:

    A. Double left click the material in the library list.
    B. Drag and drop the material from the library list into the project materials list.
    C. Right click the material and use the Add to Document materials short-cut menu.
    D. Select the material in the library list and click the Add button located on the far right.
    E. All of the above

Lesson 5
Floors and Ceilings

Ceiling Plans are used to let the contractor know how the ceiling is supposed to look.

When drafting a reflected ceiling plan, imagine that you are looking down on the floor, which is acting as a mirror showing a reflection of the ceiling. You want to locate lighting, vents, sprinkler heads, and soffits.

We start by placing the floors that are going to reflect the ceilings. (Reflected ceiling plans look down, not up. They are the mirror image of what you would see looking up, as if the ceiling were reflected in the floor.)

Exercise 5-1
Creating Floors

Drawing Name: ex4-12.rvt
Estimated Time: 35 minutes

This exercise reinforces the following skills:

- ❑ Floors
- ❑ Floor Properties
- ❑ Materials

1. Open *ex4-12.rvt.*

2.    Activate **Level 1**.

3. In the browser, locate the exterior wall that is used in the project.

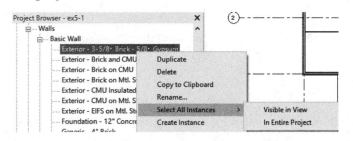

Right click and select
**Select All Instances →
In Entire Project**.

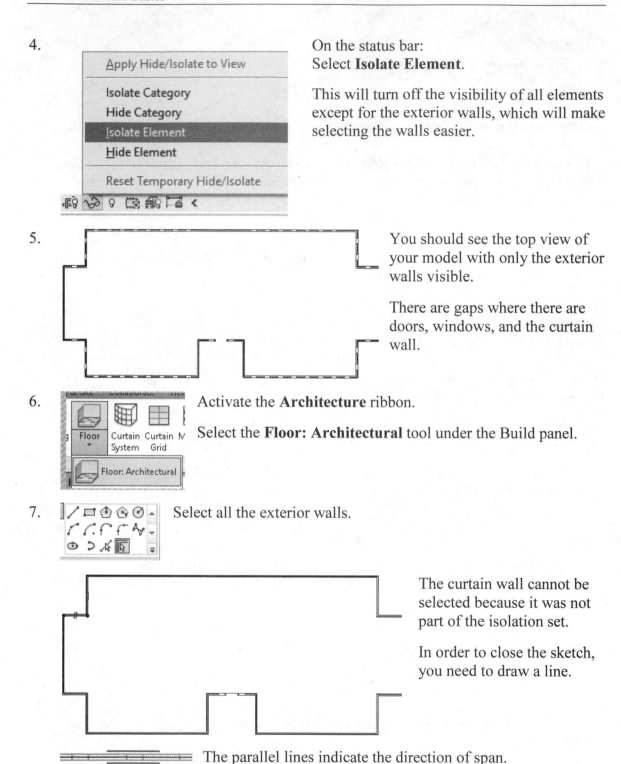

4.	On the status bar:
Select **Isolate Element**.

This will turn off the visibility of all elements except for the exterior walls, which will make selecting the walls easier.

5.	You should see the top view of your model with only the exterior walls visible.

There are gaps where there are doors, windows, and the curtain wall.

6.	Activate the **Architecture** ribbon.

Select the **Floor: Architectural** tool under the Build panel.

7.	Select all the exterior walls.

The curtain wall cannot be selected because it was not part of the isolation set.

In order to close the sketch, you need to draw a line.

The parallel lines indicate the direction of span.

By default, the first line placed/wall selected indicates the direction of span.
To change the span direction, select the Span Direction tool on the ribbon and select a different line.

8. Select the Line tool from the Draw panel.

   Draw a line to close the floor boundary sketch.

9. Verify that your floor boundary has no gaps or intersecting lines.

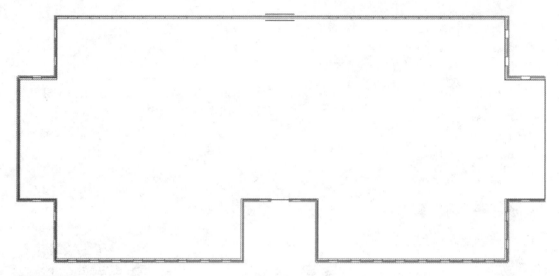

10. Select **Edit Type** from the Properties pane.

11. | | |
    |---|---|
    | Family: | System Family: Floor |
    | Type: | Wood Truss Joist 12" - Carpet Finish |

    Select **Wood Truss Joist 12″ - Carpet Finish** from the Type drop-down list.

12. Duplicate... Select **Duplicate**.

13. Name: Wood Truss Joist 12" -VCT

    Change the Name to **Wood Truss Joist 12″ - VCT**.

    Press **OK**.

14. Type Parameters

Parameter	Value
Construction	
Structure	Edit...

    Select **Edit** under Structure.

15. | | Function | Material |
    |---|---|---|
    | 1 | Finish 1 [4] | Carpet (1) |
    | 2 | Core Boundary | Layers Above |

    For Layer 1: Finish 1 [4]:
    Select the *browse* button in the Material column.

16. 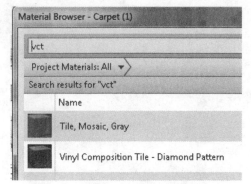   Type **vct** in the search field.

The Tile, Mosaic Gray is listed because you added vct to the keywords.

   Highlight **the Vinyl Composition Tile – Diamond Pattern** to select.

Press **OK**.

17.

	Function	Material	Thickness
1	Finish 1 [4]	Vinyl Composition Tile - Diamond Pattern	0' 0 1/8"
2	**Core Boundary**	**Layers Above Wrap**	**0' 0"**
3	Structure [1]	Plywood, Sheathing	0' 0 3/4"

You should see the new material listed.

Press **OK**.

18. 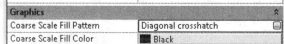   Set the Coarse Scale Fill Pattern to **Diagonal crosshatch**.

Graphics		⌃
Coarse Scale Fill Pattern	Diagonal crosshatch	
Coarse Scale Fill Color	■ Black	

Press **OK** twice to close the dialogs.

19.   Select the **Green Check** under Mode to finish the floor.

    If you get a dialog box indicating an error with the Floor Sketch, press the 'Show' button. Revit will zoom into the area where the lines are not meeting properly. Press 'Continue' and use Trim/Extend to correct the problem.

20. 🏠 Switch to a 3D View.

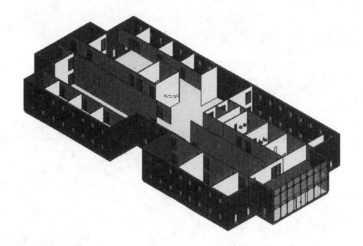

21. Activate **Level 1**.

22. 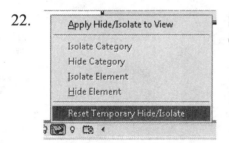 Select **Reset Temporary Hide/Isolate** to restore the elements in the Level 1 floor plan.

23. Save as *ex5-1.rvt*.

Exercise 5-2
Copying Floors

Drawing Name: ex5-1.rvt
Estimated Time: 5 minutes

This exercise reinforces the following skills:

- ❑ Copy
- ❑ Paste Aligned
- ❑ Opening
- ❑ Shaft Opening
- ❑ Opening Properties

1.   Open *ex5-1.rvt.*

2.   Switch to a 3D View.

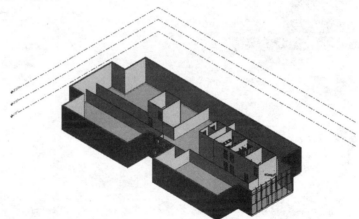

3.   Left click to select the floor.

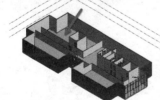

4.   Select **Copy** from the Clipboard panel.

5.   Select **Paste→Aligned to Selected Levels**.

      Paste from Clipboard

      Aligned to Selected Levels

6.   Level 1   Select **Level 2**.
       Level 2
       Roof Line   Press **OK**.

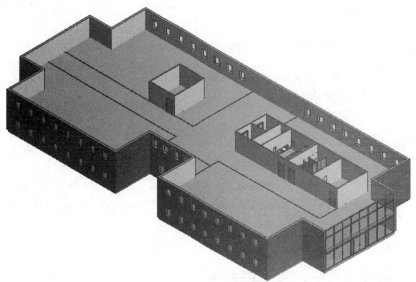

7.   Left click in the window to release the selection.

8.   Save as *ex5-2.rvt*.

Exercise 5-3
Creating a Shaft Opening

Drawing Name: ex5-2.rvt
Estimated Time: 15 minutes

This exercise reinforces the following skills:

- ❑ Shaft Opening
- ❑ Opening Properties
- ❑ Symbolic Lines

1. Open *ex5-2.rvt*.

2. We need to create openings in the Level 2 floor for the elevators.
   Activate **Level 2**.

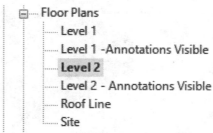

```
⊟ Floor Plans
    Level 1
    Level 1 -Annotations Visible
    Level 2
    Level 2 - Annotations Visible
    Roof Line
    Site
```

3. 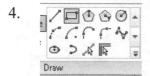 Select the **Shaft Opening** tool under the Opening panel on the Architecture ribbon.

4. 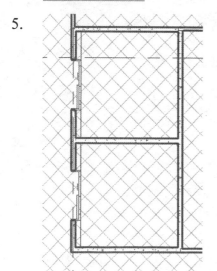 Select the **Rectangle** tool from the Draw panel.

5. Zoom into the elevator area.

   Draw two rectangles for the elevator shafts.

   Align the rectangles to the finish face of the inside walls and behind the elevator doors.

   The shaft will cut through any geometry, so be careful not to overlap any elements you want to keep.

6. 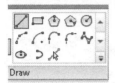 Select the **Symbolic Line** tool.

7. 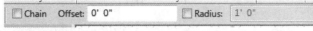 Disable **Chain** on the Options bar.

8. 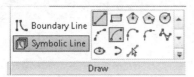 Select the **Line** tool.

9. 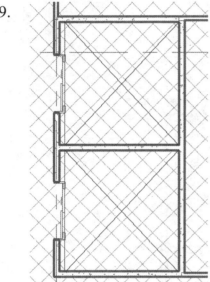 Place an X in each elevator shaft using the corners of the rectangle sketch.

The symbolic lines will appear in the floor plan to designate an opening.

10. 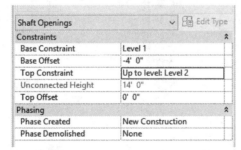 On the Properties pane:

Set the Base Offset at **-4′ 0″ [600 mm]**. This starts the opening below the floor to create an elevator pit.

Set the Base Constraint at **Level 1**.
Set the Top Constraint at **Level 2**.

11. 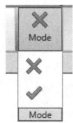 Select the **Green Check** under the Mode panel.

12.  Switch to a 3D view.
Orbit around to inspect the model.

13.

Inspect the elevator shafts.

Note that the shafts go through the bottom floor.

14. Activate **Level 1**.

The symbolic lines are visible.

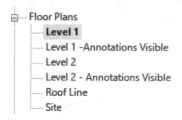

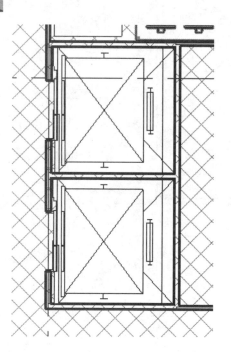

15. Save as *ex5-3.rvt*.

Exercise 5-4
Adding an Opening to a Floor

Drawing Name: ex5-3.rvt
Estimated Time: 20 minutes

This exercise reinforces the following skills:

- ❑ Opening
- ❑ Modify Floor

1.   Open *ex5-3.rvt.*

2.   We need to create openings in the Level 2 floor for the stairs.
Activate **Level 2**.

3.   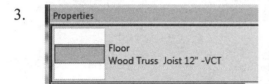   Select the floor.

If it is selected, you should see it listed in the Properties pane.

4.   Select the **Edit Boundary** tool on the ribbon.

5.   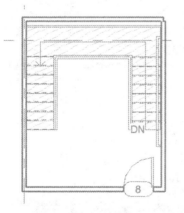   Zoom into the first set of stairs.

6.   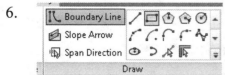   Enable **Boundary Line**.

Select the **Rectangle** tool on the Draw panel.

7.

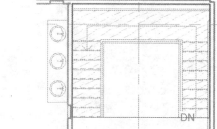

Place the rectangle so it is aligned to the inner walls and the stair risers.

8.

Zoom and pan over to the second set of stairs.

9.

Place the rectangle so it is aligned to the inner walls and the stair risers.

10. Select the **Green Check** to finish modifying the floor.

11. Press **No.**

Would you like walls that go up to this floor's level to attach to its bottom?

Yes   No

12.

> **Revit** ✕
>
> The floor/roof overlaps the highlighted wall(s).
> Would you like to join geometry and cut the
> overlapping volume out of the wall(s)?
>
> [Yes]   [No]

If you see this dialog, press **No**.

13.

Left click to release the selection.

Switch to a 3D view and inspect the openings in the floor.

14. Save as *ex5-4.rvt*.

Challenge Question:

When is it a good idea to use a Shaft Opening to cut a floor and when should you modify the floor boundary?

Shaft Openings can be used to cut through more than one floor at a time. Modifying the floor boundary creates an opening only in the floor selected.

Exercise 5-5
Creating Parts

Drawing Name: ex5-4.rvt
Estimated Time: 30 minutes

This exercise reinforces the following skills:

- ❑ Parts
- ❑ Dividing an Element
- ❑ Assigning materials to parts
- ❑ Modifying materials.

1.   Open *ex5-4.rvt*.

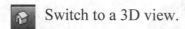 Switch to a 3D view.

2.   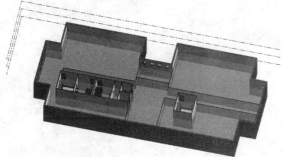   Select the floor on the second level.

3.   Select the **Create Parts** tool under the Create panel.

4.   Select **Divide Parts** from the Part panel.
     Divide
     Parts

5.   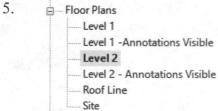   Activate **Level 2**.

6. 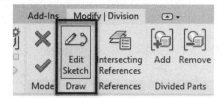 Select **Edit Sketch** on the ribbon.

7. 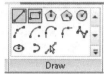 Select the rectangle tool from the Draw panel.

8. 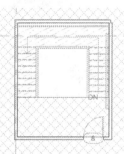 Draw a rectangle to define the landing for the stairwell.

9. 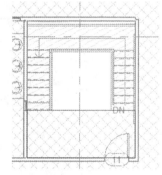 Repeat for the other stairwell.

10. 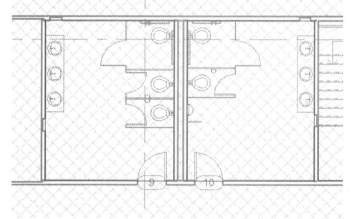 Draw a rectangle using the finish faces of the walls in the lavatories.

11. Select the **Green Check** under Mode twice.

12. 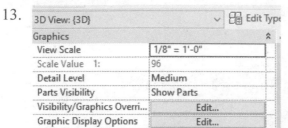 Switch to a 3D View.

13. In the Properties pane for the view:

Under Parts Visibility: Enable **Show Parts**.

In order to see the parts in a view, you need to enable the visibility of the parts.

3D View: {3D}		Edit Type
Graphics		^
View Scale	1/8" = 1'-0"	
Scale Value 1:	96	
Detail Level	Medium	
Parts Visibility	Show Parts	
Visibility/Graphics Overri...	Edit...	
Graphic Display Options	Edit...	

14. Select the landing of the first stairwell.

You will see that it is identified as a part on the Properties pane.

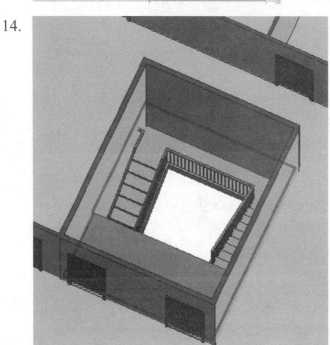

15. In the Properties pane:

Uncheck **Material by Original**.
Left click in the Material field.

Original Type	Wood Truss Joist 12" - Vinyl
Material By Original	☐
Material	VCT - Vinyl Composition Tile Diamond Patter
Construction	Finish

16. Scroll up and select the **Laminate - Ivory, Matte** material.

Press **OK**.

Project Materials: All ▼
Search results for "lamin"

	Name
	Laminate
	Laminate - Ivory, Matte

17. The new material is now listed in the Properties pane.
Left click in the window to release the selection.

Original Type	Wood Truss Joist 12" -VCT
Material By Original	☐
Material	Laminate - Ivory, Matte
Construction	Finish

18. Pan over to the lavatory area.

Hold down the Control key.

Select the lavatory floor in both lavatories.

Note that it is now identified as a part.

19.      Uncheck **Material by Original**. Left click inside the Material field.

20.   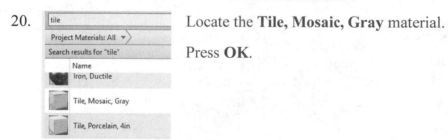   Locate the **Tile, Mosaic, Gray** material.

Press **OK**.

21.      The lavatory parts are now set to the Tile material.
Left click in the window to release the selection.

22.      Pan over to the second stairwell.

Select the landing.

Note that it is now identified as a part.

23.

Uncheck **Material by Original**.
Left click inside the Material field.

24.

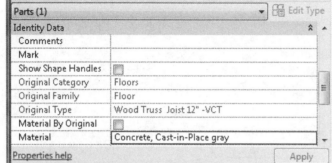

Scroll up in the Material library.
Locate the **Concrete - Cast in-Place gray** material.

Press **OK**.

25.

Verify that the landing has been assigned the correct material in the Properties pane.

26. Left click in the window to release the selection.

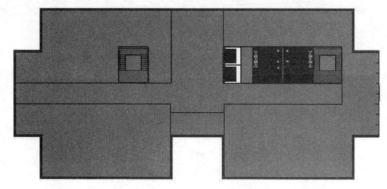

The different materials are displayed in the view.

27. Save as *ex5-5.rvt*.

Understanding View Range

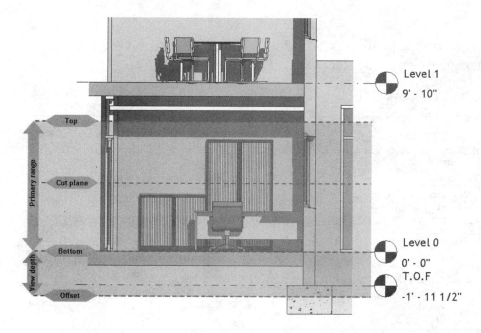

This figure shows an elevation view of a view range setting.

- Elements within the boundaries of the primary range that are not cut are drawn in the element's projection line weight.

- Elements that are cut are drawn in the element's cut line weight.

  Note: Not all elements can display as cut.

- Elements that are within the view depth are drawn in the beyond line style – that is a solid line.

  To modify line styles, go to the Manage ribbon→Additional Settings→Line Styles.

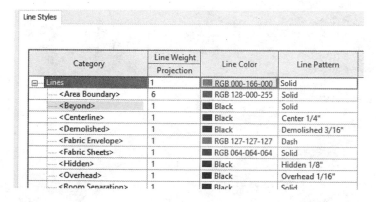

Object
Styles

To determine the line styles and weights set for different elements, go to the
Manage ribbon and select Object Styles.

Filter list: Architecture

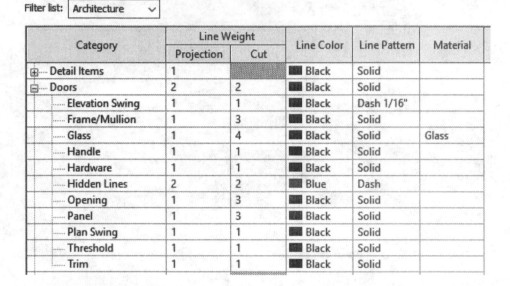

Category	Line Weight		Line Color	Line Pattern	Material
	Projection	Cut			
⊞ Detail Items	1		Black	Solid	
⊟ Doors	2	2	Black	Solid	
Elevation Swing	1	1	Black	Dash 1/16"	
Frame/Mullion	1	3	Black	Solid	
Glass	1	4	Black	Solid	Glass
Handle	1	1	Black	Solid	
Hardware	1	1	Black	Solid	
Hidden Lines	2	2	Blue	Dash	
Opening	1	3	Black	Solid	
Panel	1	3	Black	Solid	
Plan Swing	1	1	Black	Solid	
Threshold	1	1	Black	Solid	
Trim	1	1	Black	Solid	

View Range ✕

Primary Range

Top: Associated Level (Level 2) ⌄ Offset: 7' 6"

Cut plane: Associated Level (Level 2) ⌄ Offset: 4' 0"

Bottom: Associated Level (Level 2) ⌄ Offset: 0' 0"

View Depth

Level: Associated Level (Level 2) ⌄ Offset: 0' 0"

Learn more about view range

<< Show   OK   Apply   Cancel

Select the **Show** button on the View Range dialog.

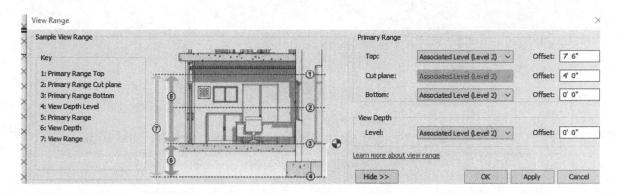

The dialog will expand to show you how the fields correlate with how the view range is defined.

Exercise 5-6
Viewing Parts in a Floor Plan View

Drawing Name: ex5-5.rvt
Estimated Time: 10 minutes

This exercise reinforces the following skills:

- View Range
- Parts

Parts can only be displayed in a view if they are within the view range.

1. Open *ex5-5.rvt*.

2. 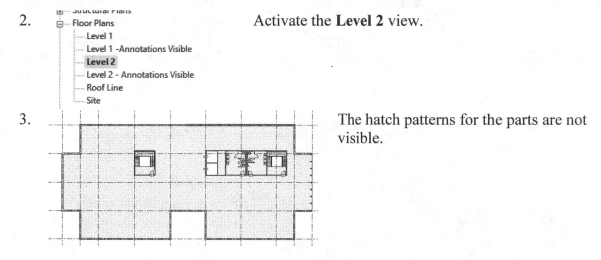 Activate the **Level 2** view.

3. The hatch patterns for the parts are not visible.

4.   In the Properties pane:

Display Model	Normal
Detail Level	Coarse
Parts Visibility	Show Parts
Visibility/Graphics Ov...	Edit
Graphic Display Options	Edit

Set the Parts Visibility to **Show Parts**.

The floor disappears.

5.   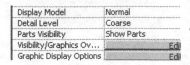   1/8" = 1'-0"

Set the View Display to **Coarse** with **Hidden Lines**.

The view display does not change.

6.

Extents	☆
Crop View	☐
Crop Region Visible	☐
Annotation Crop	☐
View Range	Edit...
Associated Level	Level 2
Scope Box	None

Scroll down the Properties pane.

Press **Edit** next to the View Range field.

7.   << Show

Select the Show button located at the bottom of the dialog.

8.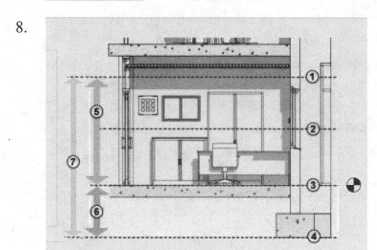

An image helps with understanding how to define the view range.

The numbers indicate the dialog inputs.

1: Top
2: Cut plane
3: Bottom
4. View Depth

9.

Set the Top Offset to **7′ 6″ [228.6]**.
Set the Cut plane Offset to **4′ 0″ [121.92]**.
Set the Bottom Offset to **0′ 0″ [121.92]**.
Set the View Depth Level to Level 1.
Set the View Depth Offset to **-1′ 0″ [-60.0]**.
Press **Apply**.

10. The parts now display the hatch patterns properly.
    Press **OK**.

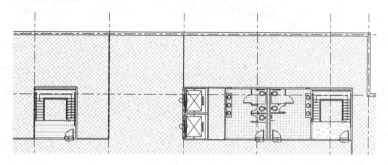

Save as *ex5-6.rvt*.

Exercise 5-7
Adding a Railing

Drawing Name: ex5-6.rvt
Estimated Time: 15 minutes

This exercise reinforces the following skills:

- □ Railing
- □ Railing Properties
- □ Re-Host

1. Open *ex5-6.rvt*.

2. 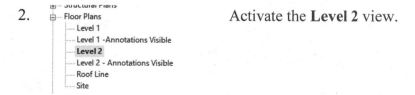 Activate the **Level 2** view.

3. 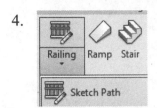   Zoom into the concrete landing next to the lavatories.

4. 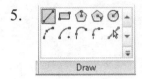   Select the **Railing→Sketch Path** tool under the Circulation panel on the Architecture ribbon.

5. Draw the railing using the **Line** tool from the Draw panel.

   Use the end points of the existing rails to locate it.

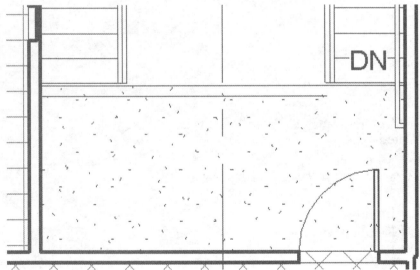

6. Select **Edit Type** from the Properties pane.

7. Select **Duplicate**.

8. Name: Guardrail - Landing   Name the new railing style – **Guardrail-Landing**.

   Press **OK**.

9.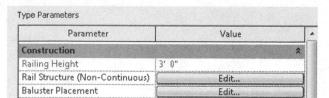
Select the **Edit** button next to Rail Structure.

10. | Insert | Press the **Insert** button two times to add two rails.

11.
	Name
1	Middle Rail
2	Bottom Rail

Name the rails.
Set Rail 1 to **Middle Rail**.
Set Rail 2 to **Bottom Rail**.

12.

| Preview >> |

Use the Preview button located at the bottom of the dialog to see a preview of the railing.

13.
Family: Railing
Type: Guardrail - Landing

	Name	Height	Offset	Profile	Material
1	Middle Rail	1' 6"	0' 0"	Circular Handrail : 1"	Aluminum
2	Bottom Rail	0' 6"	0' 0"	Circular Handrail : 1"	Aluminum

Set the rail heights to **1′-6″** [600], and **6″** [150 mm].
Set the profiles to **Circular Handrail: 1″** [M_Circular Handrail: 30 m].
Set the material to **Aluminum**.
Press **OK**.

14.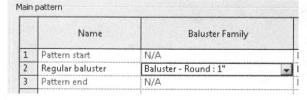
Select **Edit** next to Baluster Placement.

15.
Main pattern

	Name	Baluster Family
1	Pattern start	N/A
2	Regular baluster	Baluster - Round : 1"
3	Pattern end	N/A

Under Main Pattern:

Set the Baluster Family to **Baluster - Round 1″**.

16.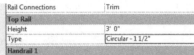

	Name	Baluster Family
1	Start Post	Baluster - Round : 1"
2	Corner Post	Baluster - Round : 1"
3	End Post	Baluster - Round : 1"

Under Posts:

Set the Baluster Family to **Baluster - Round 1″**.

17.

Dist. from previous	
N/A	
0' 4"	
0' 0"	

In the Dist. from previous column, set the distance to **4″ [150]**. Press **OK**.

Close the dialog.

18.

Rail Connections	Trim
Top Rail	
Height	3' 0"
Type	Circular - 1 1/2"
Handrail 1	

Set the Top Rail to **Circular = 1 ½″**.

19.

Click on the **Apply** button to see how the changes you have made affect the preview.

Press **OK** to close the dialog.

20. Select the **Green Check** under Mode to finish the railing.

21. Switch to a 3D view to inspect the railing.

Repeat the exercise to place a railing at the other stair landing.

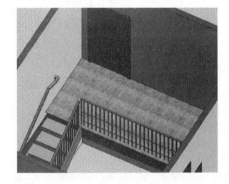

22. Save the file as *ex5-7.rvt*.

Exercise 5-8
Creating Ceilings

Drawing Name: ex5-7.rvt
Estimated Time: 10 minutes

This exercise reinforces the following skills:

- ❑ Ceilings
- ❑ Visibility of Annotation Elements

1. Open *ex5-7.rvt*.

2. 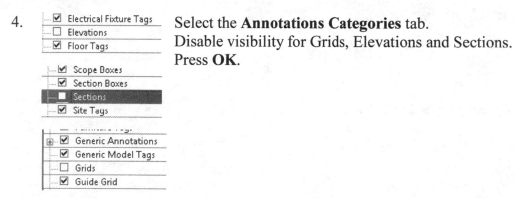 Activate **Level 1** under Ceiling Plans.

3. Type **VV**.

4. Select the **Annotations Categories** tab.
   Disable visibility for Grids, Elevations and Sections.
   Press **OK**.

5. The view display should update.

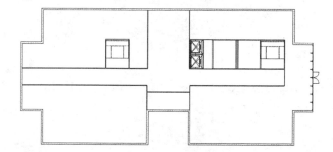

6. Select the **Ceiling** tool from Build panel on the Architecture ribbon.

7. Select the **Ceiling: 2′ × 4′ ACT System [Compound Ceiling: 600 x 1200 mm Grid]** from the Properties pane.

8. 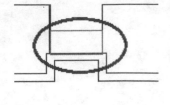 Left click in the building entry to place a ceiling.

The ceiling is placed.

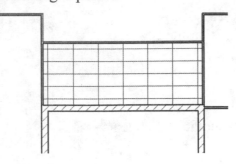

9. Right click and select **Cancel** twice to exit the command.

10. Save as *ex5-8.rvt*.

Exercise 5-9
Adding Lighting Fixtures

Drawing Name: ex5-8.rvt
Estimated Time: 10 minutes

This exercise reinforces the following skills:

❑   Add Component
❑   Load From Library

1. Open or continue working in *ex5-8.rvt*.

2.   Ceiling Plans   Activate **Level 1 Ceiling Plan**.
     └── **Level 1**
     ├── Level 2
     └── Roof Line

3.   Activate the **Architecture** ribbon.

   Component   Select the **Component→Place a Component** tool from the Build panel.

4.   Select **Load Family** from the Mode panel.

   Load
   Family

5.

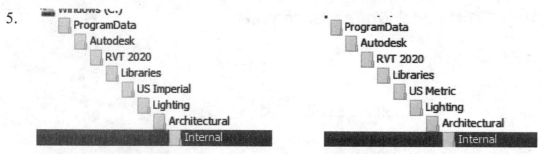

Browse to the *Lighting/Architectural/Internal* folder.

6.

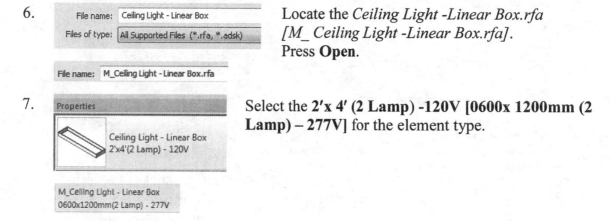

Locate the *Ceiling Light -Linear Box.rfa [M_ Ceiling Light -Linear Box.rfa]*. Press **Open**.

7.

Select the **2′x 4′ (2 Lamp) -120V [0600x 1200mm (2 Lamp) – 277V]** for the element type.

8. Place fixtures on the grid.

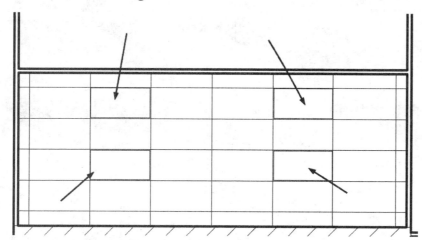

Use the ALIGN tool to position the fixtures in the ceiling grid.

9. Save the file as *ex5-9.rvt.*

Challenge Exercise:

Add ceilings to Level 1 and Level 2.

Define new ceiling types using stucco material.

Exercise 5-10
Applying Paints and Wallpaper to Walls

Drawing Name: ex5-9.rvt
Estimated Time: 30 minutes

This exercise reinforces the following skills:

- Views
- Section Box
- View Properties
- Materials
- Load From Library
- Render Appearance Library

1.   Open or continue working in *ex5-9.rvt*.

2.   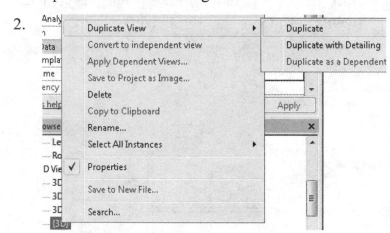 Highlight the {3D} view in the Project Browser.

Right click and select
Duplicate View →
Duplicate.

3.   Right click on the new view.

Select **Rename**.

4.   Type **3D-Lobby**.

Press **OK**.

5.   Set Parts Visibility to **Show Original**.

6. Enable Section Box in the Properties panel.

Extents	
Crop View	☐
Crop Region Visible	☐
Annotation Crop	☐
Far Clip Active	☐
Section Box	☑

7. 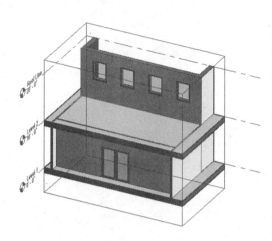 Using the grips on the section box, reduce the view to just the lobby area.

8. Select the floor on Level 2.

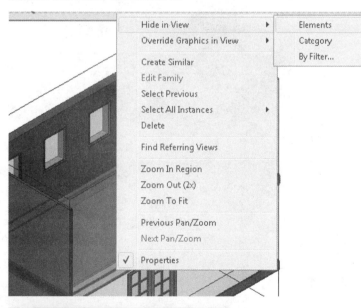

Right click and select **Hide in View→Elements**.

This will hide only the selected element.

9. Select the ceiling. Right click and select **Hide in View → Elements**.

This will hide only the selected element.

10.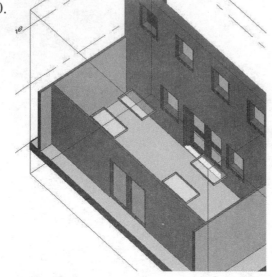

The lighting fixtures in the ceiling are still visible but can be ignored.

Or you can select the lighting fixtures, right click and select Hide In View→Category.

11. Activate the Modify ribbon.

Select the **Split Face** tool on the Geometry panel.

Select one of the side walls. You will see an orange outline to indicate the selection.

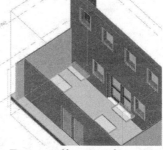

12.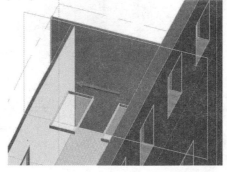

Draw a line on the corner of the room to indicate where the wall should be split.

The end points of the vertical line should be coincident to the orange horizontal lines or you will get an error.

13.

Select the **Green check**.

14. 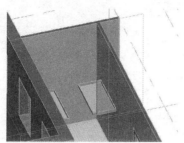 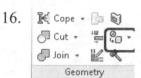 Select the Split Face tool on the Modify ribbon.

Draw a vertical line to divide the wall on the other side of the lobby.

15. Select the **Green check**.

16. Select the **Paint** tool on the Geometry panel.

17. Set the Project Materials filter to **Paint**.

<All>
Carpet
Ceramic
Concrete
Earth
P Gas
Generic
Glass
Laminate
Masonry
Metal
Miscellaneous
Paint
Plastic

18. 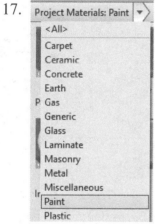 Set the View Type to **List View**.

Document Materials
✓ Show All
Show In Use
Show Unused

View Type
Thumbnail View
✓ List View
Text View

19.

Locate the **Finish – Paint - SW0068 Heron Blue** from the Material Browser list.

20.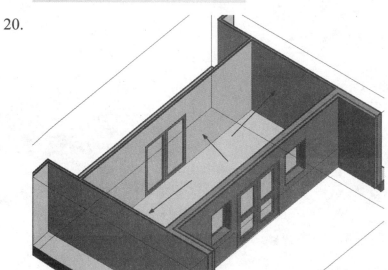

Select the walls indicated.

21.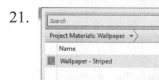

Set the Project Materials to **Wallpaper**.

Locate the striped wallpaper you defined earlier.

22.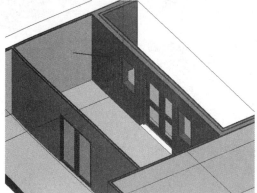

Apply the wallpaper to the inner lobby front wall.

23. 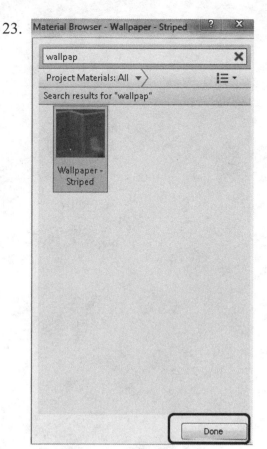 Select the **Done** button at the bottom of the Material Browser.

24. Change the display to **Realistic** so you can see the materials applied.

25. Turn off the visibility of the section box.

26. Save as *ex5-10.rvt*.

Notes:

Additional Projects

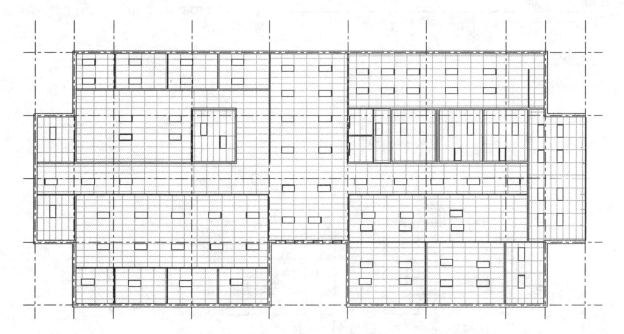

1) Add light fixtures and ceilings to the Second Level Ceiling Plan.
   Create a Ceiling Plan layout/sheet.

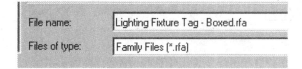

2) Load the Lighting Fixture Tag-Mark – Boxed (this family is in the Library) using
   File→Load Family.
   Use Tag All Not Tagged to tag all the lighting fixtures.
   Create a schedule of lighting fixtures.
   Add to your reflected ceiling plan sheet.

<Lighting Fixture Schedule>			
A	**B**	**C**	**D**
Type	Wattage	Level	Count
2'x4'(2 Lamp) - 120V	80 W	Level 2	93

Sort by Level and then by Type to get a schedule like this. Filter to only display Level 2 fixtures.

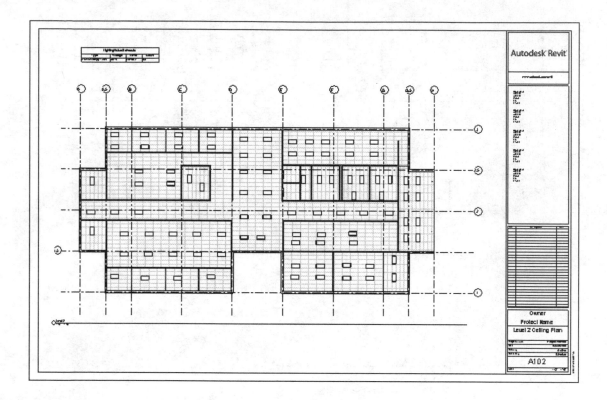

3) This is a department store in Japan. This exterior surface serves as both graphic pattern and structural system and is composed of 300mm-thick concrete and flush-mounted frameless glass. The resulting surface supports floor slabs spanning 10-15 meters without any internal columns.

How would you create a wall that looks like this in Revit?

You might be able to design a curtain wall, or you could modify a concrete wall and create parts of glass. Or you might combine both a curtain wall and a wall using parts.

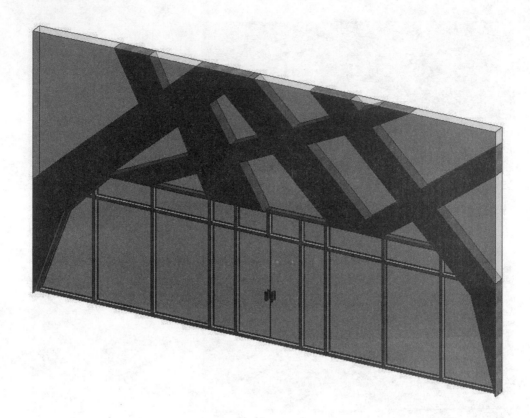

This wall was created by defining a wall using a 300 mm concrete structure, then dividing it into parts. The parts were then designated as either glass or concrete.

Here are the steps.

- Place a concrete wall going from Level 1 to Level 2.
- Edit the profile so that the sides are at an angle.
- Use the type selector to change the concrete wall to a curtain wall.
- Add the curtain wall door.
- Place a concrete wall going from Level 2 to Level 3.
- Edit the profile so that the wall fills in the angles missing from the curtain wall.
- Divide the concrete walls into parts. Sketch the angled sections that resemble tree branches. Assign a glass material to the parts.

See if you can do something similar.

Lesson 5 Quiz

True or False

1. Ceilings are not visible on the Floor Plan view.
2. Ceilings are visible on the Ceiling Plan view.
3. In order to place a ceiling-based light fixture, the model must have a ceiling.
4. To create an opening in a floor, use EDIT BOUNDARY.
5. Floors and ceilings are level-based.
6. Floors and ceilings can be offset from a level.
7. The boundaries for a floor or ceiling must be closed, non-intersecting loops.

Multiple Choice

8. When a floor is placed:

    A. The top of the floor is aligned to the level on which it is placed with the thickness projecting downward.
    B. The bottom of the floor is aligned to the level on which it is placed with the thickness projecting upwards.
    C. The direction of the thickness can be flipped going up or down.
    D. The floor offset can be above or below the level.

9. The structure of a floor is determined by its:

    A. Family type
    B. Placement
    C. Instance
    D. Geometry

10. Floors can be created by these two methods:
    Pick two answers.

    A. Sketching a closed polygon
    B. Picking Walls
    C. Place Component
    D. Drawing boundary lines

11. Ceilings are:
    Pick one answer.

    A. Level-based
    B. Floor-based
    C. Floating
    D. Non-hosted

12. Paint can be applied using the Paint tool which is located on this ribbon:

    A. Architecture
    B. Modify
    C. Rendering
    D. View

13. New materials are created using:

    A. Manage→Materials.
    B. View→Materials
    C. Architecture→Create→Materials
    D. Architecture→Render→Materials

14. This icon:

    A. Adjusts the brightness of a view.
    B. Temporarily isolates selected elements or categories.
    C. Turns off visibility of elements.
    D. Zooms into an element.

15. To place a Shaft Opening, activate the _____ ribbon.

    A. Modify
    B. View
    C. Manage
    D. Architecture

16. In order for parts to be visible in a view:

    A. Parts should be enabled on the Materials ribbon.
    B. Parts should be enabled in the Visibility/Graphics dialog.
    C. Show Parts should be enabled in the Properties pane for the view.
    D. The view display should be set to Shaded.

ANSWERS:

1) T; 2) T; 3) T; 4) T; 5) T; 6) T; 7) T; 8) A; 9) A; 10) A & B; 11) A; 12) B; 13) A; 14) B 15) D; 16) C

Lesson 6
Schedules

Revit proves its power in the way it manages schedules. Schedules are automatically updated whenever the model changes. In this lesson, users learn how to add custom parameters to elements to be used in schedules, how to create and apply keynotes, and how to place schedules on sheets.

Tags are annotations and they are **view-specific**. This means if you add tags in one view, they are not automatically visible in another view.

Tags are Revit families. Tags use parameter data. For example, a Room tag might display the room name and the room area. The room name and room area are both parameters.

A schedule is a tabular view of information, using the parameters of elements in a project. You can organize a schedule using levels, types of elements, etc. You can filter a schedule so it only displays the elements of a specific type or located on a specific level.

Because schedules are views, they are listed in the Project Browser in the view area. Schedules are placed on sheets, just like any other view; simply drag and drop from the Project Browser.

You cannot import schedules from Excel, but you can export schedules to Excel.

Exercise 6-1
Adding Door Tags

Drawing Name: ex5-10.rvt
Estimated Time: 10 minutes

This exercise reinforces the following skills:

    ❑   Door Tags

1.
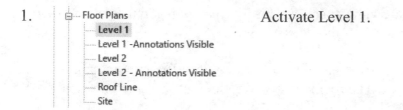
Activate Level 1.

2.   The floor plan has several doors.

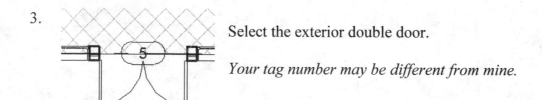

3.
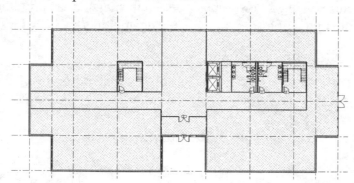
Select the exterior double door.

Your tag number may be different from mine.

4. 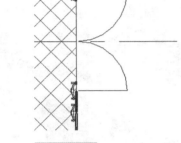 Go to the **Properties** panel.

    Locate the **Mark** parameter under Identity Data.

The Mark or Door Number can be modified in the Properties panel.

You cannot use the same door number twice. You will see an error message if you attempt to assign a door number already in use.

If you delete a door and add a new door, the door will be assigned the next mark number. It will not use the deleted door's mark number. Many users wait until the design is complete and then re-number all the doors on each level.

5.   Zoom into the curtain wall door.

6. Go to the Annotate ribbon.

    Select **Tag by Category**.

    Select the curtain wall door to add a door tag.

7. 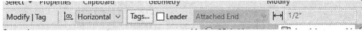 On the Options bar:
    Disable Leader.
    Verify the orientation of the tag is set to **Horizontal**.

8. 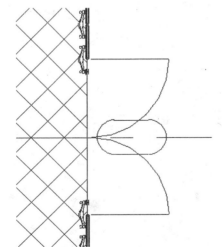 The tag is empty.

    Use the Move function to position the tag over the door.

    Select the door.

9.

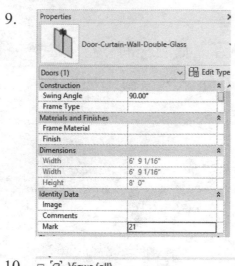
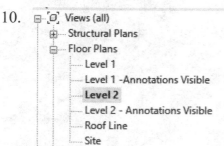

Type **21** in the Mark field.

The tag updates with the new information.

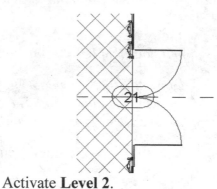

10.

Views (all)
  Structural Plans
  Floor Plans
    Level 1
    Level 1 -Annotations Visible
    Level 2
    Level 2 - Annotations Visible
    Roof Line
    Site

Activate **Level 2**.

11. Door tags have already been added to all the doors.
   This is because Tag All was selected when we tagged the doors. All doors are tagged regardless of whether the view is active or not.

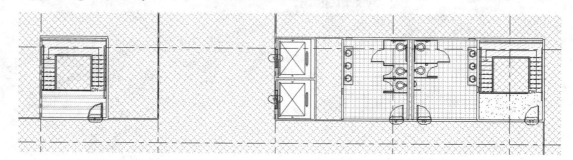

12. Save as *ex6-1.rvt*.

Exercise 6-2
Creating a Door Schedule

Drawing Name: ex6-1.rvt
Estimated Time: 30 minutes

This exercise reinforces the following skills:

- ❑   Door Tags
- ❑   Schedules

1.   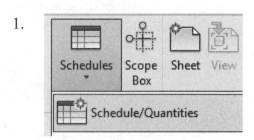   Activate the View ribbon.

   Select **Schedules → Schedule/Quantities**.

2.   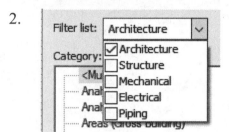   Under Filter list:
   Uncheck all the disciplines except for Architecture.

3.   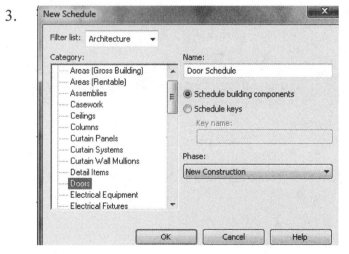   Select **Doors**.

   Press **OK**.

4. The available fields are Parameters assigned to the door families.

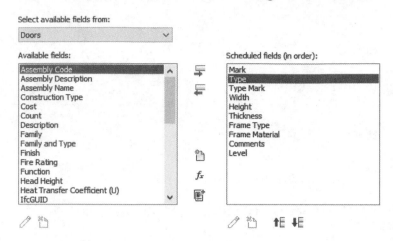

You can add which parameters to use in the schedule by highlighting and selecting the Add button.

The order the fields appear in the right pane is the order the columns will appear in the schedule.

5.  Select the following fields in this order:

Mark	Height	Frame Material
Type	Thickness	Comments
Type Mark	Frame Type	Level
Width		

6.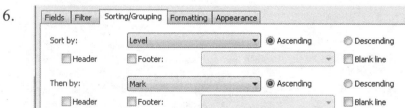

Select the Sorting/ Grouping tab.

Sort by **Level**.
Then by **Mark**.

7.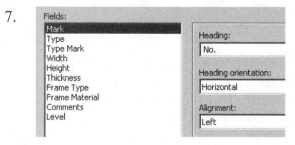

The Formatting tab allows you to rename the Column Headers.

Select the Formatting tab.

Rename Mark to **No**.

8. 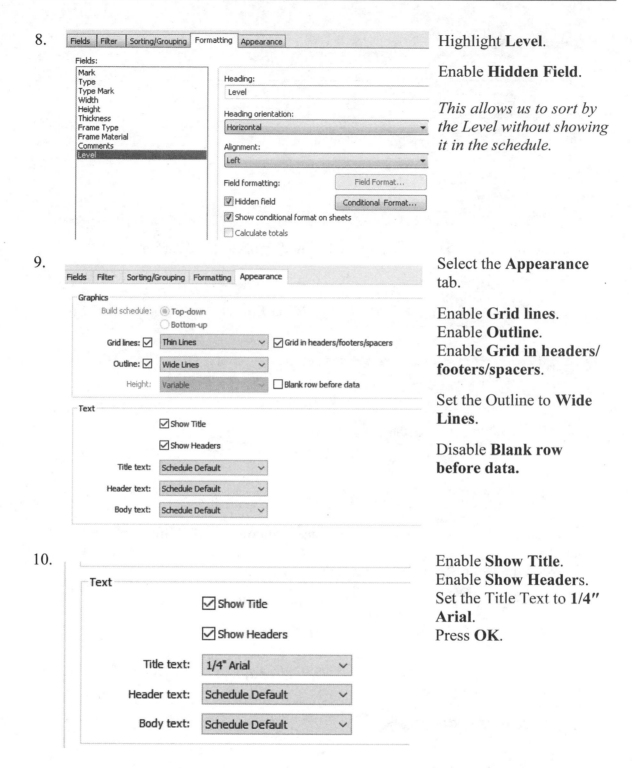 Highlight **Level**.

Enable **Hidden Field**.

This allows us to sort by the Level without showing it in the schedule.

9. Select the **Appearance** tab.

Enable **Grid lines**.
Enable **Outline**.
Enable **Grid in headers/footers/spacers**.

Set the Outline to **Wide Lines**.

Disable **Blank row before data**.

10. Enable **Show Title**.
Enable **Show Headers**.
Set the Title Text to **1/4″ Arial**.
Press **OK**.

11. A view opens with the schedule.

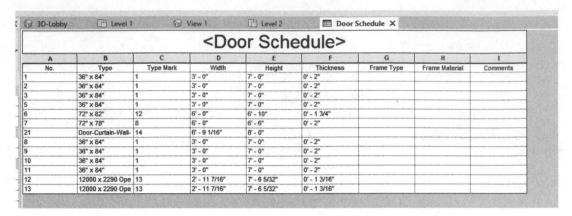

	A	B	C	D	E	F	G	H	I
	No.	Type	Type Mark	Width	Height	Thickness	Frame Type	Frame Material	Comments
	1	36" x 84"	1	3' - 0"	7' - 0"	0' - 2"			
	2	36" x 84"	1	3' - 0"	7' - 0"	0' - 2"			
	3	36" x 84"	1	3' - 0"	7' - 0"	0' - 2"			
	5	36" x 84"	1	3' - 0"	7' - 0"	0' - 2"			
	6	72" x 82"	12	6' - 0"	6' - 10"	0' - 1 3/4"			
	7	72" x 78"	8	6' - 0"	6' - 6"	0' - 2"			
	21	Door-Curtain-Wall-	14	6' - 9 1/16"	8' - 0"				
	8	36" x 84"	1	3' - 0"	7' - 0"	0' - 2"			
	9	36" x 84"	1	3' - 0"	7' - 0"	0' - 2"			
	10	36" x 84"	1	3' - 0"	7' - 0"	0' - 2"			
	11	36" x 84"	1	3' - 0"	7' - 0"	0' - 2"			
	12	12000 x 2290 Ope	13	2' - 11 7/16"	7' - 6 5/32"	0' - 1 3/16"			
	13	12000 x 2290 Ope	13	2' - 11 7/16"	7' - 6 5/32"	0' - 1 3/16"			

Now, you can see where you may need to modify or tweak the schedule.

12.  In the Properties pane:

Select **Edit** next to Sorting/Grouping.

13. 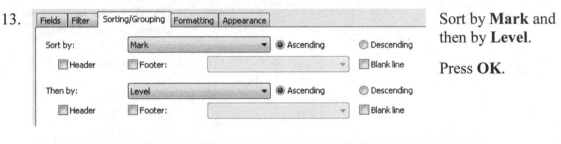 Sort by **Mark** and then by **Level**.

Press **OK**.

14.

You can use the tabs at the top of the display window to switch from one view to another.
Click on the Door Schedule to activate/open the view.

15.  In the Properties Pane:

Select **Edit** next to Fields.

16. 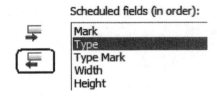 Highlight **Type** in the Scheduled fields list.

Select **Remove** (the red arrow) to remove from the list.

17.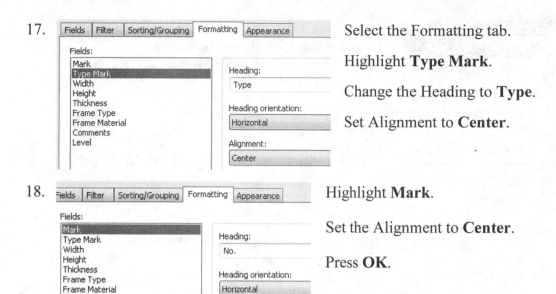

Select the Formatting tab.

Highlight **Type Mark**.

Change the Heading to **Type**.

Set Alignment to **Center**.

18.

Highlight **Mark**.

Set the Alignment to **Center**.

Press **OK**.

19. The schedule updates.

\<Door Schedule\>

A	B	C	D	E	F	F
No.	Type	Width	Height	Thickness	Frame Type	Fi
1	1	3' - 0"	7' - 0"	0' - 2"		
2	1	3' - 0"	7' - 0"	0' - 2"		
3	1	3' - 0"	7' - 0"	0' - 2"		
4	1	3' - 0"	7' - 0"	0' - 2"		
5	8	6' - 0"	6' - 6"	0' - 2"		
6	8	6' - 0"	6' - 6"	0' - 2"		
7	1	3' - 0"	7' - 0"	0' - 2"		
8	1	3' - 0"	7' - 0"	0' - 2"		
9	1	3' - 0"	7' - 0"	0' - 2"		
10	1	3' - 0"	7' - 0"	0' - 2"		
11	13	2' - 11 3/8"	7' - 6 1/8"	0' - 1 1/8"		
12	13	2' - 11 3/8"	7' - 6 1/8"	0' - 1 1/8"		
21	14	6' - 9 1/8"	8' - 0"			

20.

B	C	D	E	F	F
Type	Width	Height	Thickness	Frame Type	Fi
1	3' - 0"	7' - 0"	0' - 2"		
1	3' - 0"	7' - 0"	0' - 2"		

Place the cursor over the **width** column.
Hold down the left mouse button.

Drag the mouse across the Width, Height, and Thickness columns.

21. 🗔 Select **Group** on the Titles & Headers panel on the ribbon.
Group

22.

C	D	E
	Size	
Width	Height	Thickness
3' - 0"	7' - 0"	0' - 2"

Type **Size** as the header in the sub-header.

23. 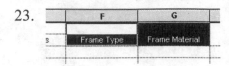 Place the cursor over the **Frame Type** column.
Hold down the left mouse button.
Drag the mouse across the Frame Type and Frame
Material columns.

24. Select **Group** on the Titles & Headers panel on the ribbon.

Group

25. 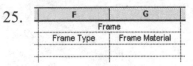 Type **Frame** as the header in the sub-header.

26. Save as *ex6-2.rvt*.

The door schedule is missing information. Many architects prefer to use letter
designations for door types and not numbers. To modify the information and update the
schedule, we need to update the parameters. The values displayed in schedules are the
values stored in family parameters.

Revit uses two types of parameters: *instance* and *type*. Instance parameters can be
modified in the Properties pane. They are unique to each **instance** of an element.
Examples of instance parameters might be level, hardware, or sill height. Type
parameters do not change regardless of where the element is placed. Examples of type
parameters are size, material, and assembly code. To modify a type parameter, you must
edit type.

Exercise 6-3
Modifying Family Parameters

Drawing Name: ex6-2.rvt
Estimated Time: 10 minutes

This exercise reinforces the following skills:

- ❑ Family Parameters
- ❑ Schedules

1.

    Select the Level 1 tab to activate the Level 1 floor plan view.

2. 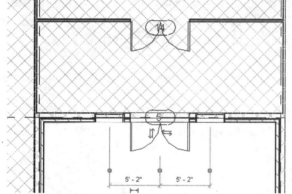 Select the exterior front door.

3.

    Fill in the Frame Type as 1.

    Fill in Frame Material as **WD**.
    WD is the designator for wood.
    Fill in Comments as **BY MFR**.

    Press **Apply** to save the changes.

    These are instance parameters, so they only affect each element selected.

4. Open the Door Schedule.

4	1	3 - 0	7 - 0	0 - 2			
5	8	6' - 0"	6' - 6"	0' - 2"	1	WD	BY MFR
6	8	6' - 0"	6' - 6"	0' - 2"			

   The schedule has updated with the information, but it only affected the one door. That is because the parameters that were changed in the Properties were *instance* parameters.

5.

    Select the Level 1 tab to activate the Level 1 floor plan view.

6.

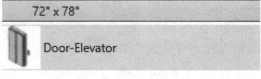

Select the first exterior double door.

Note that this is the **Door-Exterior-Double-Two_Lite 2 [M_ Door-Exterior-Double-Two_Lite 2]**.

7.

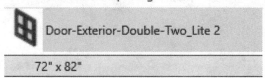

Select the second exterior double door.

Note that this is the **Door-Double-Glass [M_ Door-Double-Glass]**.

8.

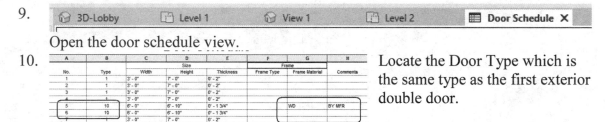

Use the Type Selector to change the second exterior door to be the same type as the outer exterior door.

Note the door tag value for this door.

Your door tag values might be different from mine.

9.

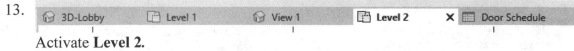

Open the door schedule view.

10.

A	B	C	D	E	F	G	H
			Size		Frame		
No.	Type	Width	Height	Thickness	Frame Type	Frame Material	Comments
1	1	3' - 0"	7' - 0"	0' - 2"			
2	1	3' - 0"	7' - 0"	0' - 2"			
3	1	3' - 0"	7' - 0"	0' - 2"			
4	1	3' - 0"	7' - 0"	0' - 2"			
5	10	6' - 0"	6' - 10"	0' - 1 3/4"		WD	BY MFR
6	10	6' - 0"	6' - 10"	0' - 1 3/4"			
7		3' - 0"	7' - 0"	0' - 2"			
8	1	3' - 0"	7' - 0"	0' - 2"			

Locate the Door Type which is the same type as the first exterior double door.

11. If you click in the Frame Type column, you can select the same values.

12.

5	1	3' - 0"	7' - 0"	0' - 2"			
6	12	6' - 0"	6' - 10"	0' - 1 3/4"	1	WD	BY MFR
7	12	6' - 0"	6' - 10"	0' - 1 3/4"	1	WD	BY MFR
8	1	3' - 0"	7' - 0"	0' - 2"			

Modify the schedule so that the two exterior doors show the same information.

13.

3D-Lobby	Level 1	View 1	Level 2	X	Door Schedule

Activate **Level 2.**

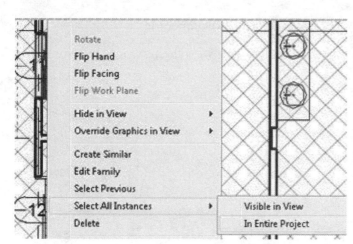

Zoom into the area where the two elevators are located.

Select one of the elevator doors.

Right click and select **Select All Instances → In Entire Project**.

14. 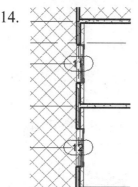   Both elevator doors on both levels are selected.

15. 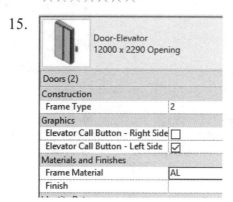   In the Properties pane:

Set the Frame Type to **2**.
Set the Frame Material to **AL**.

AL is the designator for aluminum.

Press **Apply**.

16.

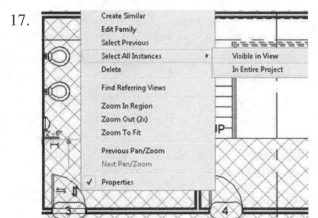

Check the Door Schedule.

It has updated with the new information.

17.

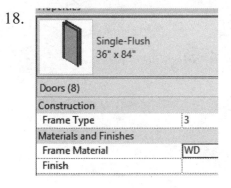

Activate **Level 1**.

Zoom into the area where the lavatories are located.

Select one of the interior single flush doors.

Right click and select **Select All Instances → In Entire Project**.

18.

Single-Flush 36" x 84"

Doors (8)	
Construction	
Frame Type	3
Materials and Finishes	
Frame Material	WD
Finish	

In the Properties pane:

Set the Frame Type to **3**.
Set the Frame Material to **WD**.
WD is the designator for wood.
Press **Apply**.

19. Check the Door Schedule.

A	B	C	D	E	F	G	H
		Size			Frame		
No.	Type	Width	Height	Thickness	Frame Type	Frame Material	Comments
1	1	3' - 0"	7' - 0"	0' - 2"	3	WD	
2	1	3' - 0"	7' - 0"	0' - 2"	3	WD	
3	1	3' - 0"	7' - 0"	0' - 2"	3	WD	
4	1	3' - 0"	7' - 0"	0' - 2"	3	WD	
5	10	6' - 0"	6' - 10"	0' - 1 3/4"		WD	BY MFR
6	10	6' - 0"	6' - 10"	0' - 1 3/4"		WD	BY MFR
7	1	3' - 0"	7' - 0"	0' - 2"	3	WD	
8	1	3' - 0"	7' - 0"	0' - 2"	3	WD	
9	1	3' - 0"	7' - 0"	0' - 2"	3	WD	
10	1	3' - 0"	7' - 0"	0' - 2"	3	WD	
11	11	2' - 11 1/2"	7' - 6 1/4"	0' - 1 1/4"	2	AL	
12	11	2' - 11 1/2"	7' - 6 1/4"	0' - 1 1/4"	2	AL	
21	12	6' - 9"	8' - 0"				

It has updated with the new information.

20.

Activate **Level 1**.

Select one of the lavatory doors.

In the Properties Pane, select **Edit Type**.

21.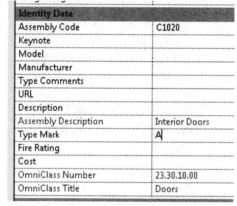

Scroll down to locate the Type Mark.

Change the Type Mark to **A**.

This only changes the parameter value for the doors loaded in the project.

If you would like to change the parameter for the door family so that it will always show a specific value whenever it is loaded into a family, you have to edit the door family.

Press **OK**.

22. Release the door selection by left clicking anywhere in the display window.

23. Check the Door Schedule.

A	B
No.	Type
1	A
2	A
3	A
4	A
5	10
6	10
7	A
8	A
9	A
10	A
11	11
12	11
21	12

It has updated with the new information.

Notice that all the single flush doors updated with the new Type Mark value.

*Because this was a **type** parameter, not an **instance** parameter, all doors that are the same type are affected.*

24. Select the **Edit** tab next to Fields.

Other	⌃
Fields	Edit...
Filter	Edit...
Sorting/Group...	Edit...
Formatting	Edit...
Appearance	Edit...

We want to add a parameter for hardware.

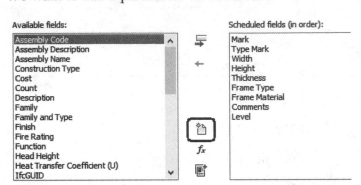

If you look in the available fields, you will see that Hardware is not listed.

So, we will create that parameter.

25. Select the **Add Parameter** button.

26. Enable **Project parameter**.

This means the parameter is only available in this project. If you wanted to create a parameter which could be used in any project or family, you would use a shared parameter.

Under Parameter Data:
Type **Hardware** for the Name.
Select **Text** for the Type of Parameter.

By default, parameters are set to length, which is a dimension parameter. If you do not set the type of parameter to text, you will have to delete the parameter and re-create it.

Group the parameter under **Construction**.
Enable **Instance**.

27. Select **Edit Tooltip**.

28. 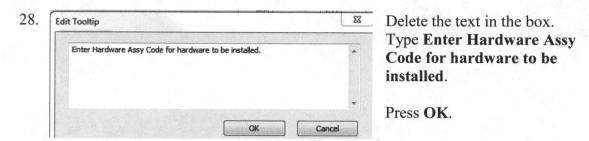 Delete the text in the box.
Type **Enter Hardware Assy Code for hardware to be installed**.

Press **OK**.

You have supplied a helpful tip to remind users what to enter for this parameter.

29. Press **OK**.

30. 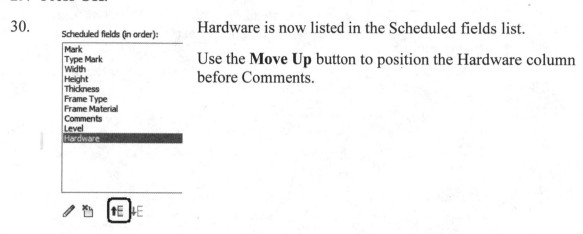 Hardware is now listed in the Scheduled fields list.

Use the **Move Up** button to position the Hardware column before Comments.

31. 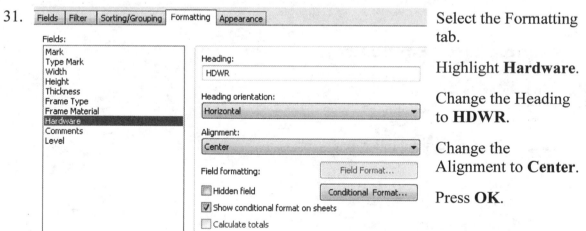 Select the Formatting tab.

Highlight **Hardware**.

Change the Heading to **HDWR**.

Change the Alignment to **Center**.

Press **OK**.

32. The Hardware column is added to the schedule.

A	B	C	D	E	F	G	H	I
			Size		Frame			
No.	Type	Width	Height	Thickness	Frame Type	Frame Material	HDWR	Comments
1	A	3' - 0"	7' - 0"	0' - 2"	3	WD		
2	A	3' - 0"	7' - 0"	0' - 2"	3	WD		
3	A	3' - 0"	7' - 0"	0' - 2"	3	WD		
4	A	3' - 0"	7' - 0"	0' - 2"	3	WD		
5	10	6' - 0"	6' - 10"	0' - 1 3/4"		WD		BY MFR
6	10	6' - 0"	6' - 10"	0' - 1 3/4"		WD		BY MFR
7	A	3' - 0"	7' - 0"	0' - 2"	3	WD		
8	A	3' - 0"	7' - 0"	0' - 2"	3	WD		
9	A	3' - 0"	7' - 0"	0' - 2"	3	WD		
10	A	3' - 0"	7' - 0"	0' - 2"	3	WD		
11	11	2' - 11 1/2"	7' - 6 1/4"	0' - 1 1/4"	2	AL		
12	11	2' - 11 1/2"	7' - 6 1/4"	0' - 1 1/4"	2	AL		
21	12	6' - 9"	8' - 0"					

33. 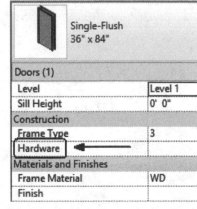 Select a lavatory door on Level 1.

34.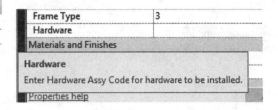

Note that Hardware is now available as a parameter in the Properties pane.

It is available in the Properties palette because it was designated as an **instance** *parameter.*

Hover your mouse over the Hardware column to see your tool tip.

35. Save as *ex6-3.rvt*.

Exercise 6-4
Creating Shared Parameters

Drawing Name: ex6-3.rvt
Estimated Time: 30 minutes

This exercise reinforces the following skills:

- Shared Parameters
- Schedules
- Family Properties

Many architectural firms have a specific format for schedules. The parameters for these schedules may not be included in the pre-defined parameters in Revit. If you are going to be working in multiple projects, you can define a single file to store parameters to be used in any schedule in any project.

1. Activate the **Manage** ribbon.

2. Select the **Shared Parameters** tool from the Settings panel.

3. Press **Create** to create a file where you will store your parameters.

4. Locate the folder where you want to store your file. Set the file name to *custom parameters.txt*. Press **Save**.

Note that this is a txt file.

5. Under Groups, select **New**.

6. Enter **Door**. Press **OK**.

7. Under Parameters, select **New**.

8. Enter **Head Detail** for Name.

In the Type field, we have a drop-down list. Select **Text**.

Press **OK**.

9. Notice that we have a Parameter Group called **Door** now.

There is one parameter listed.

Select **New** under Parameters.

10. Enter **Jamb Detail** for Name. In the Type of Parameter field, select **Text**. Press **OK**.

Select **New** under Parameters.

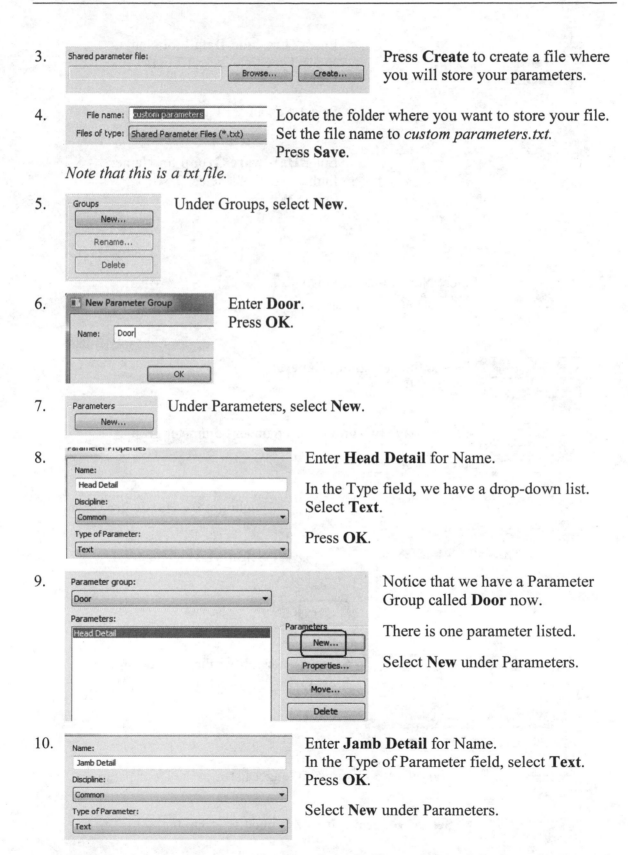

Hint: By default, parameters are set to Length. If you rush and don't set the Type of Parameter, you will have to delete it and re-do that parameter.

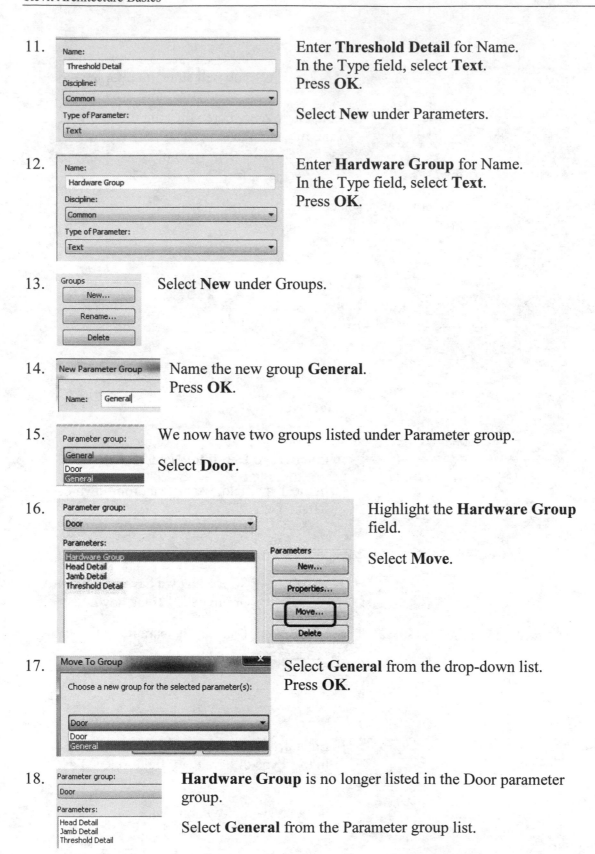

11. Enter **Threshold Detail** for Name.
     In the Type field, select **Text**.
     Press **OK**.

     Select **New** under Parameters.

12. Enter **Hardware Group** for Name.
     In the Type field, select **Text**.
     Press **OK**.

13. Select **New** under Groups.

14. Name the new group **General**.
     Press **OK**.

15. We now have two groups listed under Parameter group.

     Select **Door**.

16. Highlight the **Hardware Group** field.

     Select **Move**.

17. Select **General** from the drop-down list.
     Press **OK**.

18. **Hardware Group** is no longer listed in the Door parameter group.

     Select **General** from the Parameter group list.

19. **Parameter group:**
    General

    Parameters:
    Hardware Group

    We see **Hardware Group** listed.

    Highlight **Hardware Group**.

    Select **Properties**.

20. **Parameter Properties**

    Name:
    Hardware Group

    Discipline:
    Common

    Type of Parameter:
    Text

    We see how **Hardware Group** is defined.
    Press **OK**.

    Notice that the properties cannot be modified.

    Press **OK** to close the dialog.

21. custom parameters.txt
    Dispenser - Towel.rfa

    Locate the *custom parameters.txt* file using Windows Explorer.

22. custom parameters txt
    Dispenser - Towel.r   **Open**
    Elevator_Cab-Tract   Print

    Right click and select **Open**.

    This should open the file using Notepad and assumes that you have associated txt files with Notepad.

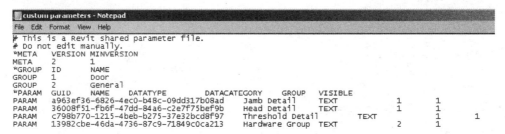

```
custom parameters - Notepad
File  Edit  Format  View  Help
# This is a Revit shared parameter file.
# Do not edit manually.
*META    VERSION MINVERSION
META     2       1
*GROUP   ID      NAME
GROUP    1       Door
GROUP    2       General
*PARAM   GUID         NAME      DATATYPE       DATACATEGORY     GROUP     VISIBLE
PARAM    a963ef36-6826-4ec0-b48c-09dd317b08ad    Jamb Detail      TEXT            1        1
PARAM    36008f51-fb6f-47dd-84a6-c2e7f75bef9b    Head Detail      TEXT            1        1
PARAM    c798b770-1215-4beb-b275-37e32bcd8f97    Threshold Detail       TEXT        1        1
PARAM    13982cbe-46da-4736-87c9-71849c0ca213    Hardware Group   TEXT            2        1
```

We see the format of the parameter file.
Note that we are advised not to edit manually.
However, currently this is the only place you can modify the parameter type from Text to Integer, etc.

23. Change the data type to **INTEGER** for Hardware Group.

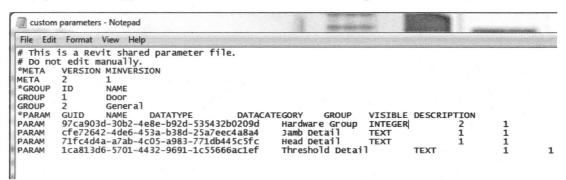

```
custom parameters - Notepad
File  Edit  Format  View  Help
# This is a Revit shared parameter file.
# Do not edit manually.
*META    VERSION MINVERSION
META     2       1
*GROUP   ID      NAME
GROUP    1       Door
GROUP    2       General
*PARAM   GUID         NAME      DATATYPE       DATACATEGORY     GROUP     VISIBLE DESCRIPTION
PARAM    97ca903d-30b2-4e8e-b92d-535432b0209d    Hardware Group   INTEGER        2        1
PARAM    cfe72642-4de6-453a-b38d-25a7eec4a8a4    Jamb Detail      TEXT            1        1
PARAM    71fc4d4a-a7ab-4c05-a983-771db445c5fc    Head Detail      TEXT            1        1
PARAM    1ca813d6-5701-4432-9691-1c55666ac1ef    Threshold Detail       TEXT        1        1
```

Do not delete any of the spaces!

24. Save the text file and close.

25. Select the **Shared Parameters** tool on the Manage ribbon.

If you get an error message when you attempt to open the file, it means that you made an error when you edited the file. Re-open the file and check it.

26.  Set the parameter group to **General**.

Highlight **Hardware Group**.
Select **Properties**.

27. 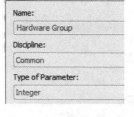 Note that Hardware Group is now defined as an Integer.

Press **OK** twice to exit the Shared Parameters dialog.

28. Save as *ex6-4.rvt.*

Exercise 6-5
Adding Shared Parameters to a Schedule

Drawing Name: ex6-4.rvt
Estimated Time: 30 minutes

This exercise reinforces the following skills:

- Shared Parameters
- Family Properties
- Schedules

1. Open *ex6-4.rvt.*

| View | Activate the **View** ribbon.

2. 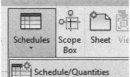 Select **Create→Schedule/ Quantities**.

3.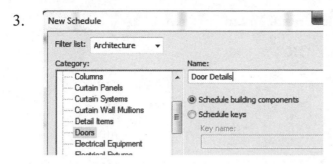

Highlight **Doors**.

Enter **Door Details** for the schedule name.

Press **OK**.

Note that you can create a schedule for each phase of construction.

4. Add **Mark**, **Type**, **Width**, **Height**, and **Thickness**.

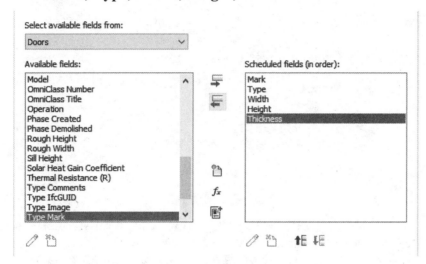

The order of the fields is important. The first field is the first column, etc. You can use the Move Up and Move Down buttons to sort the columns.

5. Select the **Add Parameter** button.

6.

Enable **Shared parameter**.

Press **Select**.

*If you don't see any shared parameters, you need to browse for the shared parameters.txt file to load it. Go to **Shared Parameters** on the Manage ribbon and browse for the file.*

7.

Select **Head Detail**.

Press **OK**.

8.

Group it under **Construction**.

Press **OK**.

9.

☑ Add to all elements in the category

Place a check next to **Add to all elements in the category.**
This will add the parameter to the Type or Instance Properties for all the door elements in the project.

10.

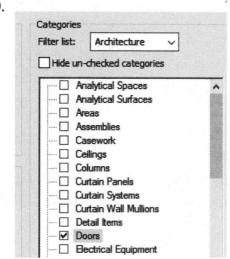

Select **Doors** in the right panel list.

11.

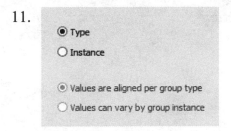

Enable **Type**.

This means all the door elements of the same type will use this parameter.

Press **OK**.

12.

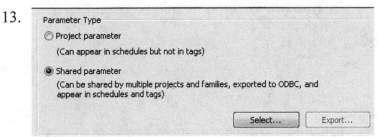

Head Detail is now listed as a column for the schedule.

Select **Add parameter**.

13.

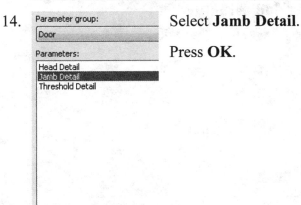

Enable **Shared parameter**.

Press **Select**.

14.

Select **Jamb Detail**.

Press **OK**.

15. **Parameter Data**
Name:
Jamb Detail
Discipline:
Common
Type of Parameter:
Text
Group parameter under:
Construction

Group it under **Construction**.

Press **OK**.

16. ☑ Add to all elements in the category

Place a check next to **Add to all elements in the category.**
This will add the parameter to the Type or Instance Properties for all the door elements in the project.

17. ● Type
○ Instance

● Values are aligned per group type
○ Values can vary by group instance

Enable **Type**.

This means all the door elements of the same type will use this parameter.

Press **OK**.

18. Scheduled fields (in order):
Mark
Type
Width
Height
Thickness
Head Detail
Jamb Detail

fx

Jamb Detail is now listed as a column for the schedule.

Select **Add parameter**.

19. Parameter Type
○ Project parameter
    (Can appear in schedules but not in tags)
● Shared parameter
    (Can be shared by multiple projects and families, exported to ODBC, and appear in schedules and tags)

    Select...   Export...

Enable **Shared parameter**.

Press **Select**.

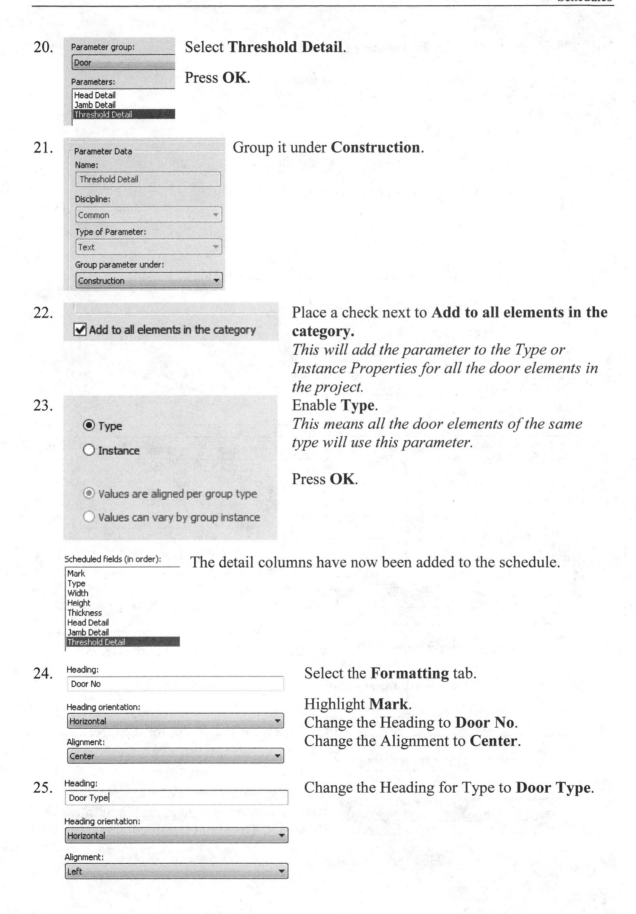

20. Select **Threshold Detail**.

Press **OK**.

21. Group it under **Construction**.

22. Place a check next to **Add to all elements in the category.**
This will add the parameter to the Type or Instance Properties for all the door elements in the project.

23. Enable **Type**.
This means all the door elements of the same type will use this parameter.

Press **OK**.

The detail columns have now been added to the schedule.

24. Select the **Formatting** tab.

Highlight **Mark**.
Change the Heading to **Door No**.
Change the Alignment to **Center**.

25. Change the Heading for Type to **Door Type**.

26. Heading:
W

Heading orientation:
Horizontal

Alignment:
Center

Change the Heading for Width to **W**.

Change the Alignment to **Center**.

27. Heading:
H

Heading orientation:
Horizontal

Alignment:
Center

Change the Heading for Height to **H**.

Change the Alignment to **Center**.

28. Heading:
THK

Heading orientation:
Horizontal

Alignment:
Center

Change the Heading for Thickness to **THK**.

Change the Alignment to **Center**.

Press **OK**.

29. A window will appear with your new schedule.

<Door Details>

A	B	C	D	E	F	G	H
Door No.	Door Type	W	H	THK	Head Detail	Jamb Detail	Threshold Detail
1	36" x 84"	3' - 0"	7' - 0"	0' - 2"			
2	36" x 84"	3' - 0"	7' - 0"	0' - 2"			
3	36" x 84"	3' - 0"	7' - 0"	0' - 2"			
4	36" x 84"	3' - 0"	7' - 0"	0' - 2"			
5	72" x 82"	6' - 0"	6' - 10"	0' - 1 3/4"			
6	72" x 82"	6' - 0"	6' - 10"	0' - 1 3/4"			
7	36" x 84"	3' - 0"	7' - 0"	0' - 2"			
8	36" x 84"	3' - 0"	7' - 0"	0' - 2"			
9	36" x 84"	3' - 0"	7' - 0"	0' - 2"			
10	36" x 84"	3' - 0"	7' - 0"	0' - 2"			
11	12000 x 2290 Ope	2' - 11 1/2"	7' - 6 1/4"	0' - 1 1/4"			
12	12000 x 2290 Ope	2' - 11 1/2"	7' - 6 1/4"	0' - 1 1/4"			
21	Door-Curtain-Wall-	6' - 9"	8' - 0"				

30.
C	D	E
W	H	THK

Select the W column, then drag your mouse to the right to highlight the H and THK columns.

31. Group

Select **Titles & Headers→Group**.

32.
C	D	E
	SIZE	
W	H	THK

Type '**SIZE**' as the header for the three columns.

33.
F	G	H
Head Detail	Jamb Detail	Threshold Detail

Select the Head Detail column, then drag your mouse to the right to highlight the Jamb Detail and Threshold Detail columns.

34. Group

Select **Titles & Headers→Group**.

35.

F	G	H
	DETAILS	
Head Detail	Jamb Detail	Threshold Detail

Type '**DETAILS**' as the header for the three columns.

36. Our schedule now appears in the desired format.

| 3D-Lobby | Level 1 | View 1 | Level 2 | Door Schedule | Door Details ✕ |

<Door Details>

A	B	C	D	E	F	G	H
			SIZE			DETAILS	
Door No	Door Type	W	H	THK	Head Detail	Jamb Detail	Threshold Detail
1	36" x 84"	3' - 0"	7' - 0"	0' - 2"			
2	36" x 84"	3' - 0"	7' - 0"	0' - 2"			
3	36" x 84"	3' - 0"	7' - 0"	0' - 2"			
5	36" x 84"	3' - 0"	7' - 0"	0' - 2"			
6	72" x 82"	6' - 0"	6' - 10"	0' - 1 3/4"			
7	72" x 82"	6' - 0"	6' - 10"	0' - 1 3/4"			
8	36" x 84"	3' - 0"	7' - 0"	0' - 2"			
9	36" x 84"	3' - 0"	7' - 0"	0' - 2"			
10	36" x 84"	3' - 0"	7' - 0"	0' - 2"			
11	36" x 84"	3' - 0"	7' - 0"	0' - 2"			
12	12000 x 2290 Ope	2' - 11 7/16"	7' - 6 5/32"	0' - 1 3/16"			
13	12000 x 2290 Ope	2' - 11 7/16"	7' - 6 5/32"	0' - 1 3/16"			
21	Door-Curtain-Wall-	6' - 9 1/16"	8' - 0"				

37. Save as *ex6-5.rvt*.

Exercise 6-6
Adding Shared Parameters to Families

Drawing Name: ex6-5.rvt
Estimated Time: 25 minutes

In order to add data to the shared parameters, we have to add those shared parameters to the families used in the projects. We have four door families in use.

This exercise reinforces the following skills:

❏ Shared Parameters
❏ Families

1. Open *ex6-5.rvt*.

2. In the Project Browser:

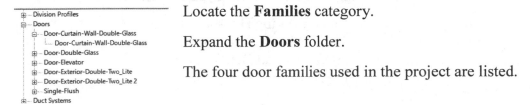

Locate the **Families** category.

Expand the **Doors** folder.

The four door families used in the project are listed.

3. Highlight the **Curtain Wall Double Glass [M_Curtain Wall Double Glass]** door.

Right click and select **Type Properties**.

4. The type properties we added in the schedule are listed under the Construction group.

5. Enter **TD1** for Threshold Detail, **JD1** for Jamb Detail, and **HD1** for Head Detail. Press **OK**.

6. Note that the schedule has updated with the new information.

11	12000 x 2290 O	7' - 6 1/8"	2' - 11 3/8"	0' - 1 1/8"			
12	12000 x 2290 O	7' - 6 1/8"	2' - 11 3/8"	0' - 1 1/8"			
21	Curtain Wall Dbl	8' - 0"	6' - 9 1/8"		HD1	JD1	TD1

7. Highlight the **Exterior-Double-Two_Lite 2** door size under the Door-Double Glass family.

Right click and select **Type Properties**.

8. Enter **TD2** for Threshold Detail, **JD2** for Jamb Detail, and **HD2** for Head Detail. Press **OK**.

9. Highlight the **Door-Elevator** door size under the **Door-Elevator** family.

Right click and select **Type Properties**.

10. Enter **TD3** for Threshold Detail, **JD3** for Jamb Detail, and **HD3** for Head Detail. Press **OK**.

11.

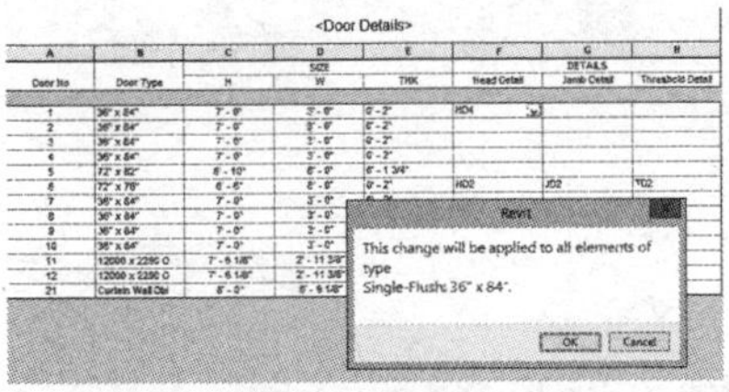

In the first row of the schedule, type HD4 in the Head Detail column.

A dialog will come up saying that all single flush doors will now reference that head detail.

Press OK.

12.

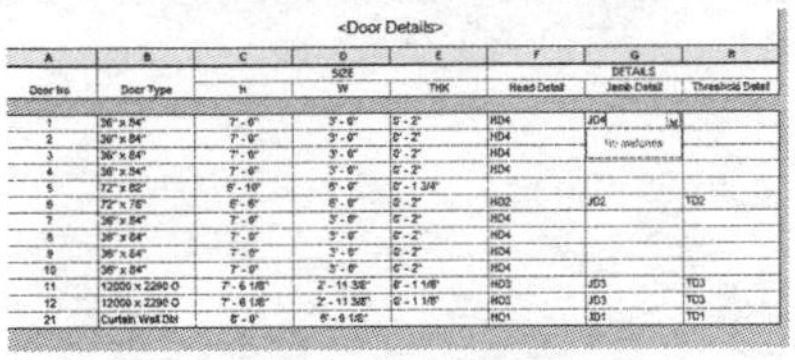

On the first row, in the Jamb Detail column, type **JD4**.

13.

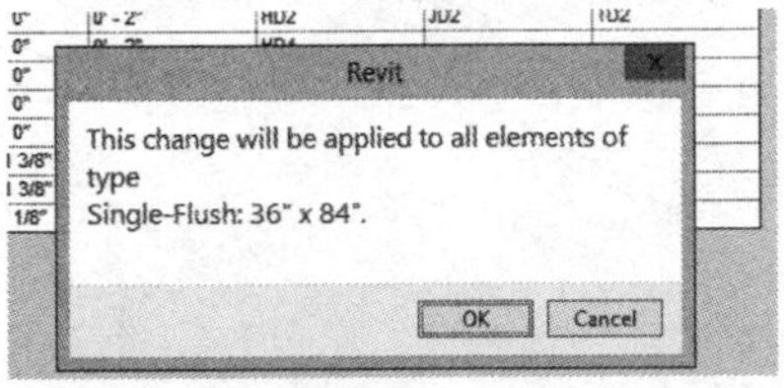

Press OK.

14.

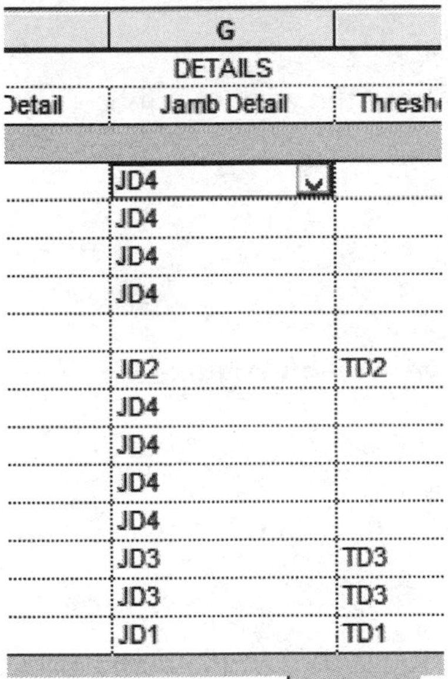

Note that all the doors of the same type fill in with the information.

G	
DETAILS	
Jamb Detail	Thresh(
JD4	
JD4	
JD4	
JD4	
JD2	TD2
JD4	
JD4	
JD4	
JD4	
JD3	TD3
JD3	TD3
JD1	TD1

15.

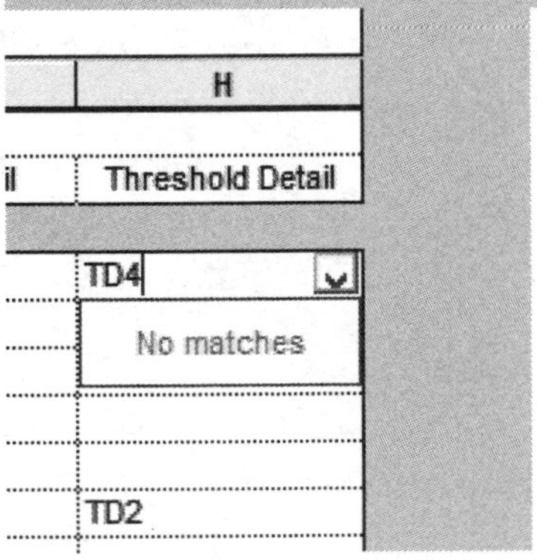

On the first row, in the Threshold Detail column, type **TD4**.

H
Threshold Detail
TD4
No matches
TD2

16.

| This change will be applied to all elements of type Single-Flush: 36" x 84". | Press **OK**. |

17. The detail information is listed for all the doors.

E	F	G	H
		DETAILS	
HK	Head Detail	Jamb Detail	Threshold Detail
- 2"	HD4	JD4	TD4
- 2"	HD4	JD4	TD4
- 2"	HD4	JD4	TD4
- 2"	HD4	JD4	TD4
1 3/4"	HD2	JD2	TD2
1 3/4"	HD2	JD2	TD2
- 2"	HD4	JD4	TD4
- 2"	HD4	JD4	TD4
- 2"	HD4	JD4	TD4
- 2"	HD4	JD4	TD4
1 1/4"	HD3	JD3	TD3
1 1/4"	HD3	JD3	TD3
	HD1	JD1	TD1

18. Activate **Level 1 Floor Plan**.

19.  Activate the **Manage** ribbon.
Select **Settings → Purge Unused** from the Manage ribbon.

20. [Check None] Press **Check None**.

By checking none, you ensure that you don't purge symbols or families you may want to use later.

21.  Expand the Doors category.
Place a checkmark on the door families that are no longer used.

Note only families that are not in use will be listed. If you see families listed that should not be in use, go back and replace those families with the custom families.

Press **OK**.

22.
```
Doors
   Door-Curtain-Wall-Double-Glass
      Door-Curtain-Wall-Double-Glass
   Door-Elevator
      12000 x 2290 Opening
   Door-Exterior-Double-Two_Lite 2
      72" x 82"
   Single-Flush
      36" x 84"
```
The browser now only lists the doors which are used.

23. Activate the Door Details schedule.

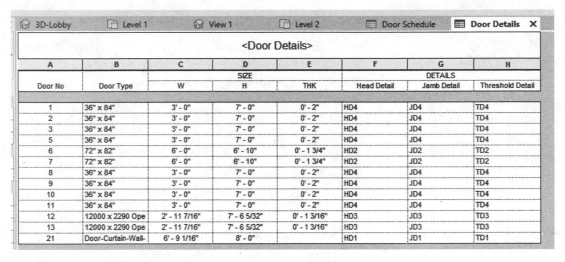

| | 3D-Lobby | Level 1 | View 1 | Level 2 | Door Schedule | Door Details X |

			<Door Details>				
A	B	C	D	E	F	G	H
			SIZE			DETAILS	
Door No	Door Type	W	H	THK	Head Detail	Jamb Detail	Threshold Detail
1	36" x 84"	3' - 0"	7' - 0"	0' - 2"	HD4	JD4	TD4
2	36" x 84"	3' - 0"	7' - 0"	0' - 2"	HD4	JD4	TD4
3	36" x 84"	3' - 0"	7' - 0"	0' - 2"	HD4	JD4	TD4
5	36" x 84"	3' - 0"	7' - 0"	0' - 2"	HD4	JD4	TD4
6	72" x 82"	6' - 0"	6' - 10"	0' - 1 3/4"	HD2	JD2	TD2
7	72" x 82"	6' - 0"	6' - 10"	0' - 1 3/4"	HD2	JD2	TD2
8	36" x 84"	3' - 0"	7' - 0"	0' - 2"	HD4	JD4	TD4
9	36" x 84"	3' - 0"	7' - 0"	0' - 2"	HD4	JD4	TD4
10	36" x 84"	3' - 0"	7' - 0"	0' - 2"	HD4	JD4	TD4
11	36" x 84"	3' - 0"	7' - 0"	0' - 2"	HD4	JD4	TD4
12	12000 x 2290 Ope	2' - 11 7/16"	7' - 6 5/32"	0' - 1 3/16"	HD3	JD3	TD3
13	12000 x 2290 Ope	2' - 11 7/16"	7' - 6 5/32"	0' - 1 3/16"	HD3	JD3	TD3
21	Door-Curtain-Wall-	6' - 9 1/16"	8' - 0"		HD1	JD1	TD1

24.   Save as *ex6-6.rvt*.

Exercise 6-7
Creating a Custom Window Schedule

Drawing Name: ex6-6.rvt
Estimated Time: 25 minutes

This exercise reinforces the following skills:

- ❑   Shared Parameters
- ❑   Schedule/Quantities

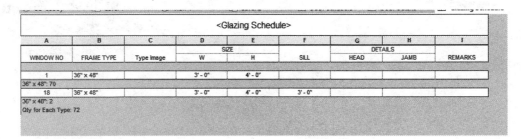

			<Glazing Schedule>					
A	B	C	D	E	F	G	H	I
			SIZE			DETAILS		
WINDOW NO	FRAME TYPE	Type Image	W	H	SILL	HEAD	JAMB	REMARKS
1	36" x 48"		3' - 0"	4' - 0"				
36" x 48": 70								
18	36" x 48"		3' - 0"	4' - 0"	3' - 0"			
36" x 48": 2								
Qty for Each Type: 72								

We want our window schedule to appear as shown.

1.   Open *ex6-6.rvt*.

2.   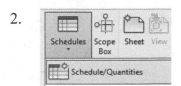   Activate the **View** ribbon.

Select **Create → Schedule/Quantities**.

3.

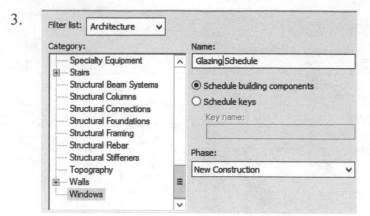

Highlight **Windows**.

Change the Schedule name to **Glazing Schedule**.

Press **OK**.

4. Add the following fields from the Available Fields.

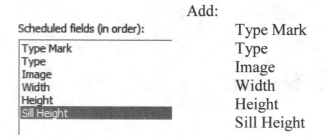

Add:

Type Mark
Type
Image
Width
Height
Sill Height

5. Select **Add Parameter**.

6.

Enable **Shared Parameter**. Press **Select**.

7.

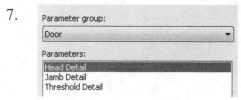

Select **Head Detail**.

Press **OK**.

8. Group it under **Construction**.
Enable **Add to all elements in the category**.
Enable **Type**.

Press **OK**.

9. Head Detail is now listed as a column for the schedule.

Select **Add parameter**.

10. Enable **Shared parameter**.

Press **Select**.

Parameter Type

○ Project parameter

(Can appear in schedules but not in tags)

● Shared parameter

(Can be shared by multiple projects and families, exported to ODBC, and appear in schedules and tags)

Select...   Export...

11. Parameter group:

Door

Parameters:

Head Detail
Jamb Detail
Threshold Detail

Select **Jamb Detail**.

Press **OK**.

12. Group it under **Construction**.
Enable **Add to all elements in the category**.
Enable **Type**.

Press **OK**.

13. Select **Comments** from the available fields.

Add to the selected fields.

14.

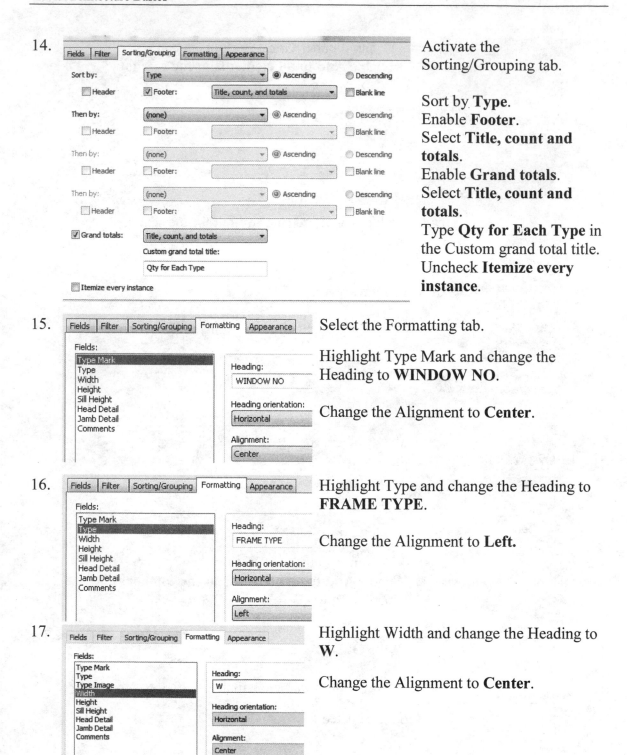

Activate the Sorting/Grouping tab.

Sort by **Type**.
Enable **Footer**.
Select **Title, count and totals**.
Enable **Grand totals**.
Select **Title, count and totals**.
Type **Qty for Each Type** in the Custom grand total title.
Uncheck **Itemize every instance**.

15. Select the Formatting tab.

Highlight Type Mark and change the Heading to **WINDOW NO**.

Change the Alignment to **Center**.

16. Highlight Type and change the Heading to **FRAME TYPE**.

Change the Alignment to **Left.**

17. Highlight Width and change the Heading to **W**.

Change the Alignment to **Center**.

18.

Highlight Height and change the Heading to **H**.

Change the Alignment to **Center**.

19.

Highlight Sill Height and change the Heading to **SILL**.

Change the Alignment to **Center**.

20.

Highlight Head Detail and change the Heading to **HEAD.**

Change the Alignment to **Center**.

21.

Highlight Jamb Detail and change the Heading to **JAMB**.

Change the Alignment to **Center**.

22.

Change the heading for Comments to **REMARKS**.

Change the Alignment to **Left**.

23. Press **OK** to create the schedule.

24.

Select the W and H columns to group.

25. [icon] Select **Titles & Headers→Group**.

    Group

26. | C | D |
    |---|---|
    | SIZE | |
    | W | H |

    Add a header called **SIZE** above the W and H columns.

27. | F | G |
    |---|---|
    | DETAILS | |
    | HEAD | JAMB |

    Group the Head and Jamb columns.

28. Add the header DETAILS over the Head/Jamb group.

<Glazing Schedule>								
A	B	C	D	E	F	G	H	I
			SIZE			DETAILS		
WINDOW NO	FRAME TYPE	Type Image	W	H	SILL	HEAD	JAMB	REMARKS
1	36" x 48"		3' - 0"	4' - 0"				
36" x 48": 70								
18	36" x 48"		3' - 0"	4' - 0"	3' - 0"			
36" x 48": 2								
Qty for Each Type: 72								

We see that only one type of window is being used in the project and there are 54 windows of this type.

Your schedule may appear different depending on the number of windows placed and whether you placed different types.

29. Save as *ex6-7.rvt*.

Exercise 6-8
Adding an Image to a Schedule

Drawing Name: ex6-7.rvt
Estimated Time: 10 minutes

This exercise reinforces the following skills:

❑ Images
❑ Schedule/Quantities

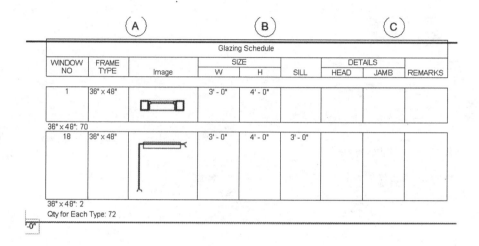

We want our window schedule to appear as shown. We will be adding the images to the schedule.

1. Open *ex6-7.rvt*.

2. Open the **Glazing Schedule** view.

Left click in the Image cell for Window No 1.

You will "wake up" the ... button.

Select the ... button.

3.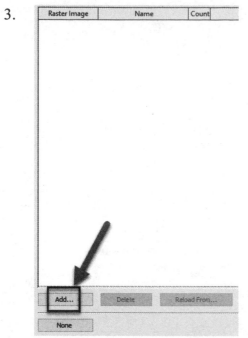

Select the **Add** button at the bottom of the dialog.

4.

File name:	fixed window.PNG
Files of type:	All Image Files (*.bmp, *.jpg, *.jpeg, *.png, *.tif)

Select the fixed window image file located in the exercise files.

I will show you how to create an image file to use in schedules later in this chapter.

5.

Raster Image	Name
	fixed window.PNG

A preview of the image is shown.

Press **OK**.

Autodesk Revit 2019

Error - cannot be ignored

Changes to groups are allowed only in group edit mode. Use the Edit Group command to change to all instances of a group type. You may use the "Ungroup" option to proceed with this change by ungrouping the changed group instances.

Show   More Info   Expand >>

Ungroup   OK   Cancel

If this dialog appears, press **Ungroup**.

This error appears if the windows are part of a grouped array.

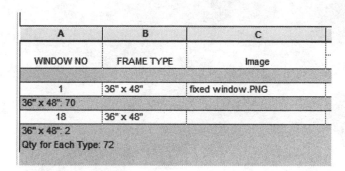

	A	B	C
	WINDOW NO	FRAME TYPE	Image
	1	36" x 48"	fixed window.PNG
36" x 48": 70			
	18	36" x 48"	
36" x 48": 2			
Qty for Each Type: 72			

The image file is now listed in the schedule.

If you placed additional window types, you can add images for them as well if you have created the image files.

6. Open the **A101 – First Level Floor Plan** sheet.

7. Drag and drop the Glazing schedule onto the sheet.

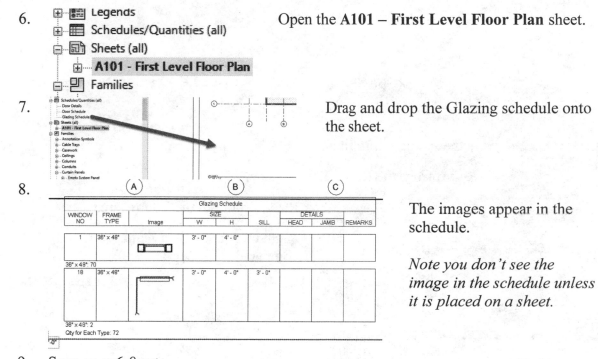

8. The images appear in the schedule.

Note you don't see the image in the schedule unless it is placed on a sheet.

9. Save as *ex6-8.rvt*.

Exercise 6-9
Create a Finish Schedule

Drawing Name: ex6-8.rvt
Estimated Time: 15 minutes

This exercise reinforces the following skills:

- ❑ Schedule/Quantities
- ❑ Rooms
- ❑ Materials

1. Open *ex6-8.rvt*.

2. 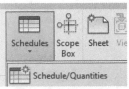 Activate **Level 1** floor plan.

3. Activate the Architecture ribbon.

    Select **Room** from the Room & Area panel.

4.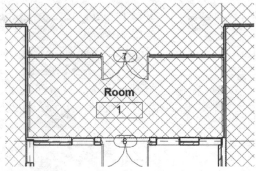

    Place a room in the Lobby Area.

    Right click and select **Cancel** to exit the command.

Materials were applied in Lesson 4. If you skipped those exercises, you should do Lesson 4 before you do this exercise or use the exercise file included with the Class Files available for download from the publisher's website.

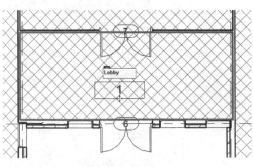

To rename the room, simply left click on the label and type.

5. Activate the View ribbon.

    Select **Create → Schedule/Quantities**.

6.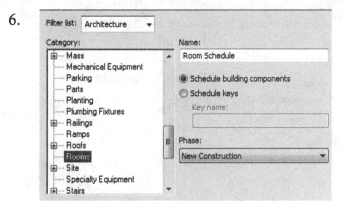

Highlight **Rooms** in the left pane.

Enable **Schedule building components**.

Press **OK**.

7.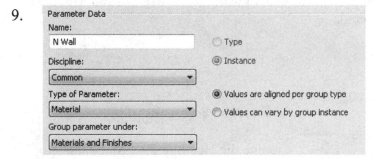

Add the following Room parameters:
- Name
- Base Finish
- Ceiling Finish
- Floor Finish

8. Select **Add Parameter**.

9.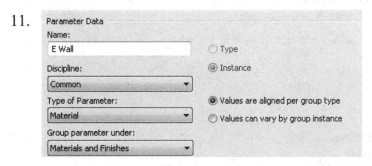

Type **N Wall** for Name.
Set Type of Parameter to **Material**.
Group parameter under **Materials and Finishes**.

Press **OK**.

10. Select **Add Parameter**.

11.

Parameter Data	
Name:	
E Wall	○ Type
Discipline:	◉ Instance
Common	
Type of Parameter:	◉ Values are aligned per group type
Material	○ Values can vary by group instance
Group parameter under:	
Materials and Finishes	

Type **E Wall** for Name.
Set Type of Parameter to **Material**.
Group parameter under **Materials and Finishes**.

Press **OK**.

12. Select **Add Parameter**.

13.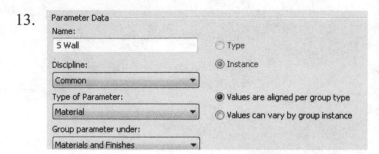

Type **S Wall** for Name.
Set Type of Parameter to **Material**.
Group parameter under **Materials and Finishes**.

Press **OK**.

14. 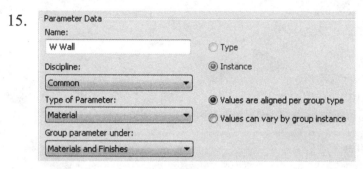 Select **Add Parameter**.

15.

Type **W Wall** for Name.
Set Type of Parameter to **Material**.
Group parameter under **Materials and Finishes**.

Press **OK**.

16.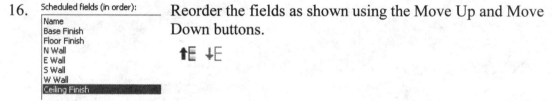

Reorder the fields as shown using the Move Up and Move Down buttons.

17.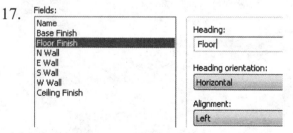

Activate the Formatting tab.

Change the Heading for Floor Finish to **Floor**.

18.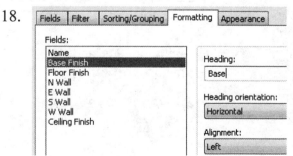

Change the Heading for Base Finish to **Base**.

19.

| Fields | Filter | Sorting/Grouping | Formatting | Appearance |

Fields:

Name
Base Finish
Floor Finish
N Wall
E Wall
S Wall
W Wall
Ceiling Finish

Heading:

Clg

Heading orientation:

Horizontal

Alignment:

Left

Change the Heading for Ceiling Finish to **Clg**.

20.

Graphics

Build schedule: ○ Top-down
○ Bottom-up

Grid lines: ☑ Thin Lines   □ Grid in headers/footers/spacers

Outline: □ Thin Lines

Height: Variable   □ Blank row before data

Activate the Appearance tab.

Disable **Blank row before data**.

Press **OK**.

21. The schedule is displayed.

<Room Schedule>							
A	B	C	D	E	F	G	H
Name	Base	Floor	N Wall	E Wall	S Wall	W Wall	Clg
Lobby							

22.

Activate **Level 1**.

23.

Repeat [Schedule/Quantities]

Recent Commands

Select Previous

Find Referring Views

Right click in the window and **Select Previous** from the shortcut menu.

24.

Rooms (1)   Edit Type

Constraints		
Level	Level 1	
Upper Limit	Level 1	
Limit Offset	10' 0"	
Base Offset	0' 0"	
Materials and Finishes		
N Wall		
E Wall		

The room should re-appear in the Properties pane as selected.

25. Note the parameters added now appear in the Properties panel.

26. Select in the material column and assign the appropriate material for each wall.

27. Under Identity Data:

    Assign letters to each of the remaining finish elements.

    Press **Apply.**

28. In the Project Browser, select the **Room Schedule**.

29. Change the View Name to **FINISH SCHEDULE** in the Properties pane.

The title of the schedule will update.

			<FINISH SCHEDULE>				
A	B	C	D	E	F	G	H
Name	Base	Floor	N Wall	E Wall	S Wall	W Wall	Clg
Lobby	A	D	Paint - Interior-S	Wallpaper-Stripe	Paint - Interior-S	Paint - Interior-S	B

30. Save as *ex6-9.rvt*.

Exercise 6-10
Adding Schedules and Tables to Sheets

Drawing Name: ex6-9.rvt
Estimated Time: 15 minutes

This exercise reinforces the following skills:

- ❑   Shared Parameters
- ❑   Schedule/Quantities
- ❑   Sheets

1.   Open *ex6-9.rvt*.

2.   Right click on Sheets and select **New Sheet**.

3.   Highlight the D 22 x 34 Horizontal title block.

   Press **OK**.

4.   Drag and drop the door schedule from the browser onto the sheet.

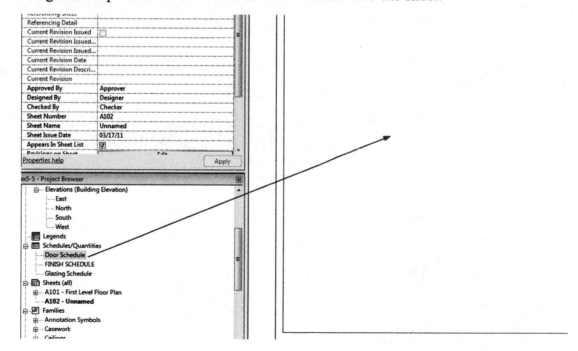

5. Use the blue grips to widen or shorten the columns.

Door Schedule									
	Door				Frame				
No.	Type	Width	Height	Thickness	Frame Type	Frame Material	HDWR	Remarks	
1	A	3' - 0"	7' - 0"	0' - 2"	3	WD			
2	A	3' - 0"	7' - 0"	0' - 2"	3	WD			
3	A	3' - 0"	7' - 0"	0' - 2"	3	WD			
4	A	3' - 0"	7' - 0"	0' - 2"	3	WD			
5	B	6' - 0"	6' - 6"	0' - 2"	1	WD		BY MFR	
7	A	3' - 0"	7' - 0"	0' - 2"	3	WD			
8	A	3' - 0"	7' - 0"	0' - 2"	3	WD			
9	A	3' - 0"	7' - 0"	0' - 2"	3	WD			
10	A	3' - 0"	7' - 0"	0' - 2"	3	WD			
11	13	2' - 11 7/16"	7' - 6 5/32"	0' - 1 3/16"	2	AL			
12	13	2' - 11 7/16"	7' - 6 5/32"	0' - 1 3/16"	2	AL			
13	14	6' - 6 9/16"	7' - 8 1/4"						
14	8	6' - 0"	6' - 6"	0' - 2"	1	WD		BY MFR	

6.

Current Revision	
Approved By	M. Instructor
Designed By	Designer
Checked By	J. Student
Drawn By	J. Student
Sheet Number	A601
Sheet Name	Door Schedule
Sheet Issue Date	05/31/18
Appears In Sheet List	☑
Revisions on Sheet	Edit...
Other	⌃

On the Properties pane:

Change the Sheet Number to **A601**.
Change the Sheet Name to **Door Schedule**.
Add your name to the Designed by and Drawn by fields.
Add your instructor's name to the Approved by and Checked by fields.

7. View   Activate the **View** ribbon.

8. [Sheet icon] Select the **New Sheet** tool from the Sheet Composition panel.

Highlight the D 22 x 34 Horizontal title block.

Press **OK**.

9.

Approved By	M. Instructor
Designed By	J. Student
Checked By	M. Instructor
Drawn By	J. Student
Sheet Number	A602
Sheet Name	Glazing Schedule
Sheet Issue Date	05/31/18

In the Properties pane:

Change the Sheet Number to **A602**.
Change the Sheet Name to **Glazing Schedule**.
Change the Drawn By to your name.
Change the Checked By to your instructor's name.

Press **Apply** to accept the changes.

10.

Owner	
Project Name	
Glazing Schedule	
Project number	Project Number
Date	Issue Date
Drawn by	J. Student
Checked by	M. Instructor
A602	
Scale	

Zoom into the title block and you see that some of the fields have updated.

11. Manage   Activate the **Manage** ribbon.
12. [Project Information icon]   Select **Project Information** from the Settings panel.

13. 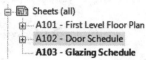 Fill in the Project Data.
Press **OK**.

14.  Note that the title block updates.

Sheets (all)
- A101 - First Level Floor Plan
- A102 - Door Schedule
- **A103 - Glazing Schedule**

Activate the other sheets and note that the project information has updated on all sheets in the project.

15. 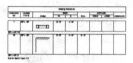 Drag and drop the **Glazing Schedule** onto the A602 sheet.

16. Save the file as *ex6-10.rvt*.

Exercise 6-11
Using Keynotes

Drawing Name: ex6-10.rvt
Estimated Time: 30 minutes

This exercise reinforces the following skills:

- Keynotes
- Duplicate Views
- Cropping a View

1. Open *ex6-10.rvt*.

2. [View] Activate the **View** ribbon.

3. Select the **New Sheet** tool from the Sheet Composition panel.

4. Highlight the D 22 x 34 Horizontal titleblock.

    Select titleblocks:

    D 22 x 34 Horizontal
    E1 30 x 42 Horizontal : E1 30x42 Horizontal
    None

    Press **OK**.

5. | Current Revision | |
    |---|---|
    | Approved By | M. Instructor |
    | Designed By | J. Student |
    | Checked By | M. Instructor |
    | Drawn By | J. Student |
    | Sheet Number | A603 |
    | Sheet Name | LOBBY KEYNOTES |
    | Sheet Issue Date | 05/31/18 |

    In the Properties pane:

    Change the Sheet Name to **LOBBY KEYNOTES**.
    Change the Drawn By to your name.
    Change the Checked By to your instructor's name.

6. ⊞· Structural Plans
    ⊟· Floor Plans
    ····· **Level 1**
    ····· Level 1 -Annotations Visible
    ····· Level 2
    ····· Level 2 - Annotations Visible
    ····· Roof Line
    ····· Site

    Activate the **Level 1** floor plan.

7.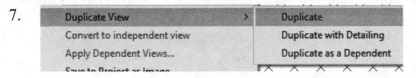

    Highlight **Level 1**.
    Right click and select
    Duplicate View →
    Duplicate.

    This creates a duplicate
    view without any
    annotations, such as
    dimensions and tags.

8. ⊟· Floor Plans
    ····· Level 1
    ····· **Level 1 - Lobby Detail**
    ····· Level 1 -Annotations Visible
    ····· Level 2
    ····· Level 2 - Annotations Visible
    ····· Roof Line
    ····· Site

    Rename the duplicate view **Level 1 - Lobby Detail**.
    Press **OK**.

9.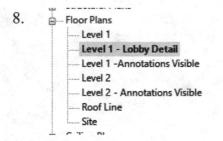

    In the Properties palette:

    Enable **Crop View**.
    Enable **Crop Region Visible**.

10.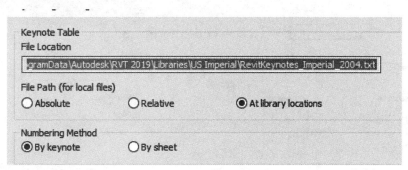

Zoom out to see the crop region.

Select the rectangle surrounding the view. This is the crop region.

11.

Use the blue circle grips to crop the view so that only the lobby area is displayed.

12.

Activate the **Annotate** ribbon.

Select **Keynote → Keynoting Settings**.

Duct Leger
Pipe Leger
Color Fill L
Keynote

Element Keynote
Material Keynote
User Keynote
Keynoting Settings

13. *This shows the path for the keynote database.*

Keynote Table
File Location

:gramData\Autodesk\RVT 2019\Libraries\US Imperial\RevitKeynotes_Imperial_2004.txt

File Path (for local files)
○ Absolute ○ Relative ● At library locations

Numbering Method
● By keynote ○ By sheet

This is a txt file that can be edited using Notepad. You can modify the txt file or select a different txt file for keynote use.

The file can be placed on a server in team environments.

Press **OK**.

14. 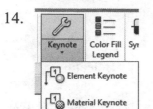 Select **Keynote→ User Keynote**.

15. 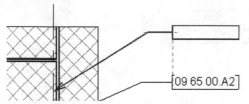 Select the east wall.

16.  Select **Semi-Gloss Paint Finish** from the list.

Press **OK**.

17. Select the tag.

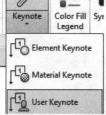

In the Properties palette:
Verify that the material displayed is **Semi-Gloss Paint Finish**.

18. 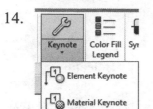 Select **Keynote → User Keynote**.

19.  Select the north wall.

20.  Select **Vinyl Wallcovering** from the list.
Press **OK**.

21. 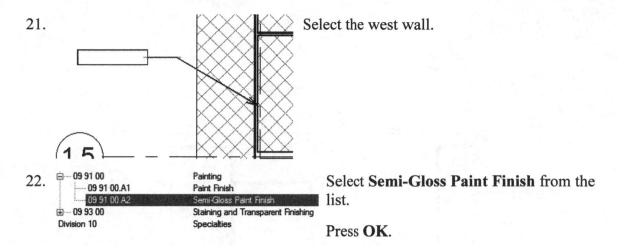 Select the west wall.

22. Select **Semi-Gloss Paint Finish** from the list.

Press **OK**.

23. Select the South wall to add a tag.

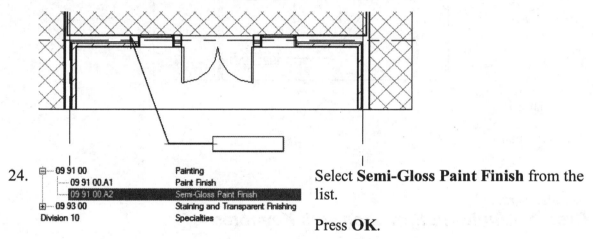

24. Select **Semi-Gloss Paint Finish** from the list.

Press **OK**.

25. The view should appear as shown.

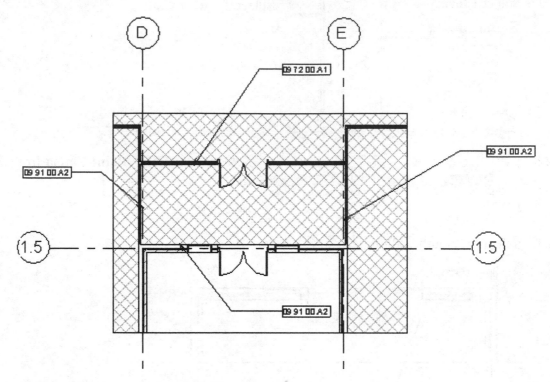

Save as *ex6-11.rvt*.

Exercise 6-12
Create a Building Elevation with Keynotes

Drawing Name: ex6-11.rvt
Estimated Time: 30 minutes

This exercise reinforces the following skills:

- ❏ Elevation View
- ❏ Cropping a View
- ❏ Keynote Tags

1.   Open *ex6-11.rvt*.

2.   **Level 1- Lobby Detail**   Activate the **Level 1 - Lobby Detail** view.

3.   Type **VV** to bring up the Visibility/Graphics dialog.

     Activate the **Annotation Categories** tab.

     Enable the visibility of Elevations.

4.   Activate the View ribbon.

Select **Elevation→ Elevation**. 🏠 Elevation ▾

5. 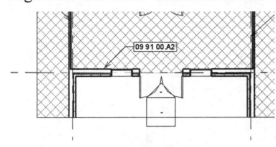 Check in the Properties palette.

You should see that you are placing a Building Elevation.

6. Place an elevation marker as shown in front of the exterior front door.
Right click and select **Cancel** to exit the command.

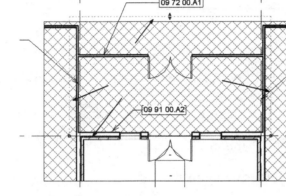

7. Click on the triangle part of the elevation marker to adjust the depth of the elevation view.

Adjust the boundaries of the elevation view to contain the lobby room.

8. ⎯⎯ North
⎯⎯ South
⎯⎯ South - Lobby

Locate the elevation in the Project Browser.

Rename to **South - Lobby**.

9. Elevations (Building Elevation)
⎯⎯ East
⎯⎯ North
⎯⎯ South
⎯⎯ **South Lobby**
⎯⎯ West

Activate the **South - Lobby** view.

10. Adjust the crop region to show the entire lobby including floor and ceiling.

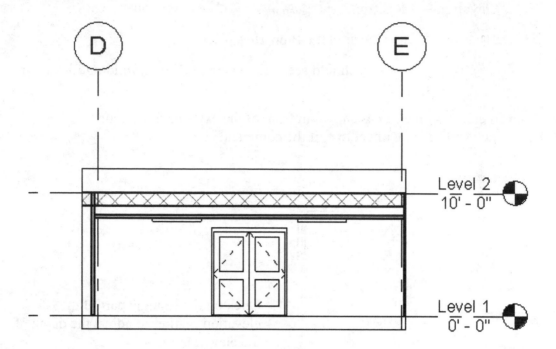

You may need to return to the Lobby Detail floor plan to adjust the location of the elevation view until it appears like the view shown.

This view is set to **Coarse Level Detail, Wireframe**.

11. Activate the Annotation ribbon.

Select **Keynote→Element Keynote**.

12. From the Properties palette:
Set the Keynote Tag to **Keynote Number**.

13. Select the ceiling.

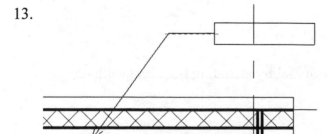

14.

⊟— 09 51 00	Acoustical Ceilings
09 51 00.A1	Square Edge (3/4 x 12 x 12)
09 51 00.A2	Square Edge (3/4 x 12 x 18)
09 51 00.A3	Square Edge (3/4 x 18 x 18)
09 51 00.A4	Square Edge (3/4 x 24 x 12)
09 51 00.A5	Square Edge (3/4 x 24 x 24)
09 51 00.A6	Square Edge (3/4 x 24 x 48)
09 51 00.A7	Tegular Edge (3/4 x 12 x 12)
09 51 00.A8	Tegular Edge (3/4 x 12 x 18)

Select the keynote under **Acoustical Ceilings: Square Edge (3/4 x 24 x 48)**.

Press **OK**.

15.   **?**   Select the Lighting Fixture.

A ? indicates that Revit needs you to assign the keynote value.

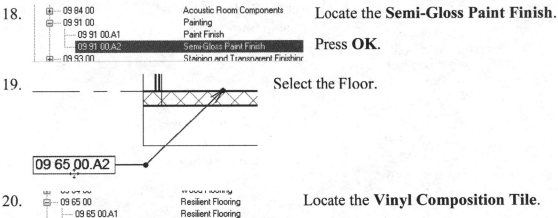

16.

⊟— 26 51 00	Interior Lighting
26 51 00.A1	12" Recessed Fluorescent Light Fixture
26 51 00.A2	Wall Mounted Fluorescent Fixture
26 51 00.A3	1' X 4' Surface Mounted Modular Fluorescent Fixture
26 51 00.A4	2' X 2' Surface Mounted Modular Fluorescent Fixture
26 51 00.A5	2' X 4' Surface Mounted Modular Fluorescent Fixture
26 51 00.A6	Surface Mounted Compact Fluorescent Fixture
26 51 00.A7	Wall Mounted Decorative Fluorescent Fixture

Locate the **2′ x 4′ Surface Mounted Modular Fluorescent Fixture**.

Press **OK**.

17.   Select the west wall.

18.

⊞— 09 84 00	Acoustic Room Components
⊟— 09 91 00	Painting
09 91 00.A1	Paint Finish
09 91 00.A2	Semi-Gloss Paint Finish
⊞— 09 93 00	Staining and Transparent Finishing

Locate the **Semi-Gloss Paint Finish**.

Press **OK**.

19.   Select the Floor.

09 65 00.A2

20.

⊞— 09 64 00	Wood Flooring
⊟— 09 65 00	Resilient Flooring
09 65 00.A1	Resilient Flooring
09 65 00.A2	Vinyl Composition Tile
09 65 00.A3	Rubber Flooring

Locate the **Vinyl Composition Tile**.

Press **OK**.

21. 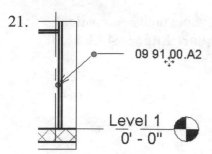 Select the East Wall.

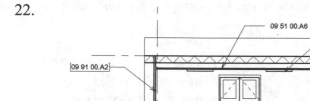 Because you have already assigned a keynote to a similar wall, Revit remembers what was assigned.

22.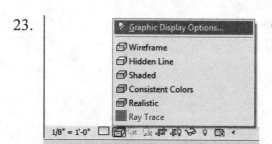

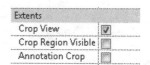

Disable **Crop Region Visible** in the Properties pane.

Your view should look similar to the one shown.

23.

On the Graphics Display bar:
Select **Graphic Display Options**.

24.

Set the Style to **Shaded**.
Enable **Show Edges**.

25.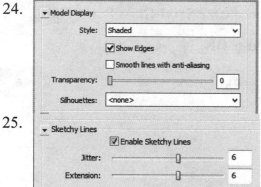

Enable **Sketchy Lines**.
Set the Jitter to **6**.
Set the Extension to **6**.

Press **Apply** to preview.

26.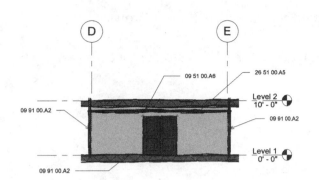

If you are happy with the result, press **OK**.

27.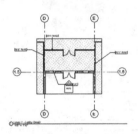

Activate the **A603 - LOBBY KEYNOTES** sheet.

28.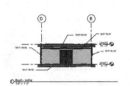

Add the **Level 1 - Lobby Detail** view to the sheet.

Add the **South Lobby Elevation** view to the sheet.

29. Add the Finish Schedule to the sheet.

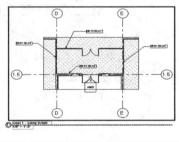

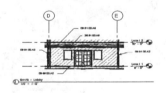

30.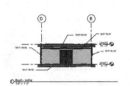

Activate the View ribbon.

Select **Create→Keynote Legend**.

31. Type **FINISH SCHEDULE KEYS**.

Press **OK**.

32. Press **OK** to accept the default legend created.

33.   Legends   The legend appears in the Project Browser.
       FINISH SCHEDULE KEYS

34. Drag and drop the FINISH SCHEDULE KEYS on to the sheet.

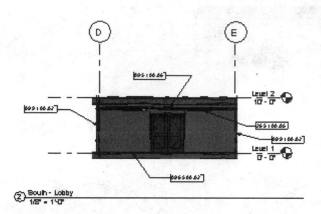

FINISH SCHEDULE KEYS	
Key Value	Keynote Text
09 51 00.A6	Square Edge (3/4 x 24 x 48)
09 65 00.A2	Vinyl Composition Tile
09 72 00.A1	Vinyl Wallcovering
09 91 00.A2	Semi-Gloss Paint Finish
26 51 00.A5	2' X 4' Surface Mounted Modular Fluorescent Fixture

35. 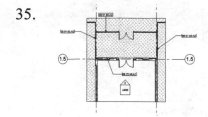   Activate the **Level 1 – Lobby Detail** floor plan. Zoom in to the view and you will see the elevation marker now indicates the Sheet Number and View Number.

36.

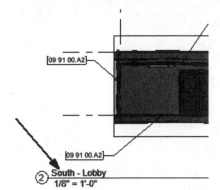

Return to the A603 sheet.

Select the South – Lobby title bar under the view.

37.

Associated Datum	None
Identity Data	
View Template	\<None\>
View Name	South - Lobby
Dependency	Independent
Title on Sheet	Lobby - Finishes
Sheet Number	A603
Sheet Name	Lobby Keynotes
Referencing Sheet	A603

On the Properties pane:
Change the Title on the Sheet to **Lobby – Finishes**.

38. The view title updates.

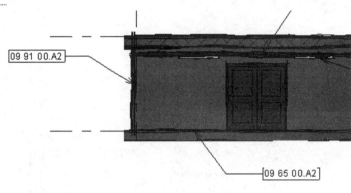

39. Save as *ex6-11.rvt.*

Exercise 6-13
Find and Replace Families

Drawing Name: ex6-12.rvt
Estimated Time: 5 minutes

This exercise reinforces the following skills:

- ❑ Families
- ❑ Project Browser

1.   Open *ex6-12.rvt*.

2.   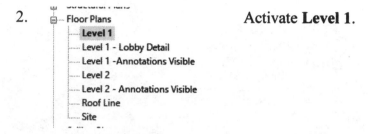   Activate **Level 1**.

3.   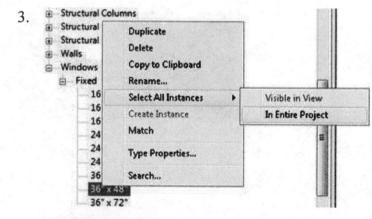   In the Project Browser, locate the *Windows* category under Families.

Highlight **36″ x 48″ [0915 x 1220mm]**.

Right click and select **Select All Instances → In Entire Project**.

You should see the windows highlight.

4.   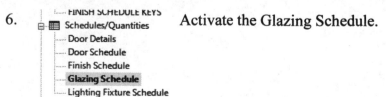   Select **Fixed: 36″ x 24″ [M_Fixed: 0915 x 0610mm]** from the Properties pane Type selector.

Left click in the display window to release the selection.

5.   All the selected windows have been replaced with the new type.

6.   Activate the Glazing Schedule.

7.

A	B	C
WINDOW NO	FRAME TYPE	Image
8	36" x 24"	fixed window.PNG
36" x 24": 70		

Zoom into the window schedule and you will see that it has automatically updated.

8.   Save as *ex6-13.rvt*.

Exercise 6-14
Modifying Family Types in a Schedule

Drawing Name: ex6-13.rvt
Estimated Time: 10 minutes

This exercise reinforces the following skills:

❑   Families
❑   Schedules

1.   Open *ex6-13.rvt*.

2.   Activate the Insert ribbon.

     Select **Load Family**.

     Load Family

3. | File name: | Door-Interior-Single-Flush_Panel-Wood |
     |---|---|
     | Files of type: | All Supported Files (*.rfa, *.adsk) |

     Locate the *Door-Interior-Single-Flush_Panel-Wood [M_Door-Interior-Single-Flush_Panel-Wood]* family in the Residential folder.

4. | 30" x 84" | 2' 6" | 7' 0" |
     |---|---|---|
     | 32" x 84" | 2' 8" | 7' 0" |
     | 34" x 84" | 2' 10" | 7' 0" |
     | 36" x 84" | 3' 0" | 7' 0" |
     | 12" x 96" | 1' 0" | 8' 0" |

     Locate the **34" x 84"** [850 x 2100] size and press **OK**.

5.    Activate **Level 1**.

6.   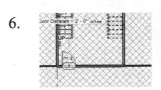 Select the door in the first stairwell.

     Note the door tag value.

7. Use the Type Selector to change the door to the 34" x 84" [850 x 2100] door size.

Left click in the display window to release the selection.

8. Activate the **Door Details** schedule in the Project Browser.

9. The schedule has updated with the new door size.

10. Activate the **View** ribbon.

Select **Close Inactive** on the Windows panel.

This closes any open views/files other than the active view.

Note that only the active tab remains available.

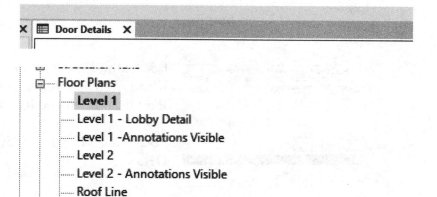

11. Activate **Level 1**.

12. Activate the **View** ribbon.

Select **Tile** on the Windows panel.

13. There should be two windows: The Level 1 floor plan and the Door Details schedule.

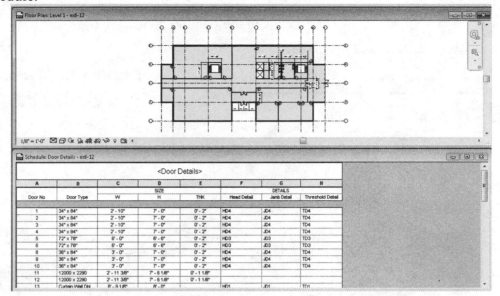

14.

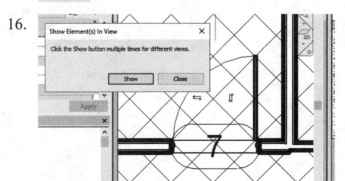

Place your cursor in the cell for Door No 1.

15. Select **Highlight in Model** from the ribbon.

16. The floor plan view will zoom into the door. The door is automatically selected.

Press **Close**.

17. In the Properties palette:

Note that the door is 34″ x 84″.

18. 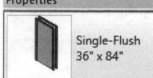 Use the Type Selector to change the door back to **36″ x 84″**.

Left click in the window to release the selection.

19.

Note that the door details schedule has updated.

A	B
Door No.	Door Type
1	36" x 84"
2	36" x 84"
3	36" x 84"
4	36" x 84"
5	72" x 82"
6	72" x 82"
7	36" x 84"
8	36" x 84"
9	36" x 84"
10	36" x 84"
11	12000 x 2290 Opening
12	12000 x 2290 Opening
21	Door-Curtain-Wall-Double-Glass

20. Save as *ex6-14.rvt*.

Exercise 6-15
Export a Schedule

Drawing Name: ex6-14.rvt
Estimated Time: 5 minutes

This exercise reinforces the following skills:

❑   Schedules

1.   Open *ex6-14.rvt*.

2.      Activate the **Glazing Schedule**.

3.   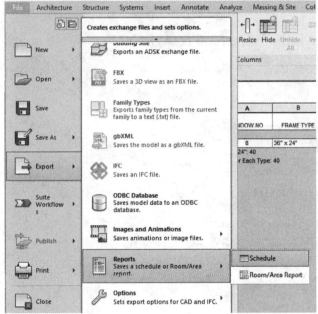   Go to the Applications Menu.

Select **File → Export → Reports → Schedule**.

You will need to scroll down to see the Reports option.

4.
| File name: | Glazing Schedule.txt |
| Files of type: | Delimited text (*.txt) |

The schedule will be saved as a comma delimited text file.

Browse to your exercise folder.
Press **Save**.

5.   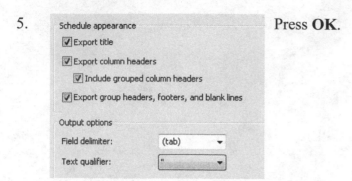   Press **OK**.

6.   Launch **Excel**.

7.    ⌂ Open   Select **Open**.

8.   Text Files   Set the file types to **Text Files**.

9.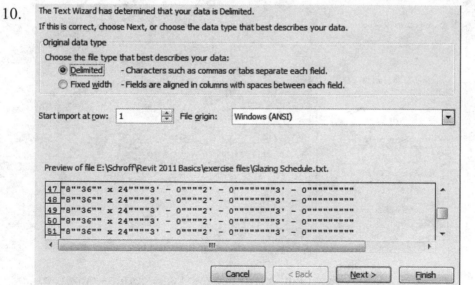

Browse to where you saved the file and select it. Press **Open**.

10.   Press **Next**.

11.

This screen lets you set the delimiters your data contains. You can see how your text is affected in the preview below.

Delimiters
- ☑ Tab
- ☐ Semicolon   ☐ Treat consecutive delimiters as one
- ☐ Comma   Text qualifier: [" ▼]
- ☐ Space
- ☐ Other: [   ]

Data preview

Glazing Schedule							
WINDOW NO	FRAME TYPE	SIZE W	H	DETAILS HEAD	JAMB	SILL	MULLIC
8	36" x 24"	3' - 0"	2' - 0"	this is a line			

Press **Next**.

12.

This screen lets you select each column and set the Data Format.

Column data format
- ◉ General
- ○ Text
- ○ Date: [MDY ▼]
- ○ Do not import column (skip)

'General' converts numeric values to numbers, date values to dates, and all remaining values to text.

[Advanced...]

Data preview

General	General	General	General	General	Gener	Gener	Gener
Glazing Schedule							
WINDOW NO	FRAME TYPE	SIZE W	H	DETAILS HEAD	JAMB	SILL	MULLIC
8	36" x 24"	3' - 0"	2' - 0"	this is a line			

[Cancel] [< Back] [Next >] [Finish]

Press **Finish**.

13.

	A	B	C	D	E	F	G	H	I
1	Glazing Schedule								
2	WINDOW	FRAME TY	SIZE		SILL	DETAILS		REMARKS	
3			W	H		HEAD	JAMB		
4									
5	8	36" x 24"	3' - 0"	2' - 0"					
6	36" x 24": 40								
7	Qty for Each Type: 40								
8									
9									
0									

The spreadsheet opens.

14. Close the Revit and Excel files without saving.

Exercise 6-16
Assigning Fonts to a Schedule

Drawing Name: ex6-15.rvt
Estimated Time: 5 minutes

This exercise reinforces the following skills:

- ❑ Schedules
- ❑ Text Styles

1. Open *ex6-15.rvt*.

2. Activate the Annotate ribbon.

3.  Select the small arrow located in the bottom corner of the Text panel.

4. Duplicate...   Select **Duplicate**.

5. Name: Schedule Title   Enter **Schedule Title**.
   Press **OK**.

   OK   Cancel

6. Type: Schedule Title   Duplicate...
   Rename...

   Set the Text Font to **Technic**.
   Set the Text Size to ¼″.
   Enable **Italic**.

   Any Windows font can be used to create a Revit text style.

   Type Parameters

Parameter	Value
Graphics	
Color	Black
Line Weight	1
Background	Opaque
Show Border	☐
Leader/Border Offset	5/64"
Leader Arrowhead	Arrow 30 Degree
Text	
Text Font	Technic
Text Size	1/4"
Tab Size	1/2"
Bold	☐
Italic	☑
Underline	☐
Width Factor	1.000000

7. Duplicate...   Select **Duplicate**.

8. Name: Schedule Header   Enter **Schedule Header**.
   Press **OK**.

9.

| Type: | Schedule Header | ▼ | Duplicate... |
| | | | Rename... |

Type Parameters

Parameter	Value
Graphics	⌃
Color	■ Black
Line Weight	1
Background	Opaque
Show Border	☐
Leader/Border Offset	5/64"
Leader Arrowhead	Arrow 30 Degree
Text	⌃
Text Font	Technic
Text Size	51/256"
Tab Size	1/2"
Bold	☐
Italic	☐
Underline	☐
Width Factor	1.000000

Set the Text Font to **Technic**.
Set the Text Size to **0.2"**.
Disable **Italic**.

10. ` Duplicate... `   Select **Duplicate**.

11. | Name: | Schedule Cell | |

Enter **Schedule Cell**.
Press **OK**.

12.

| Type: | Schedule Cell | ▼ | Duplicate... |
| | | | Rename... |

Type Parameters

Parameter	Value
Graphics	⌃
Color	■ Black
Line Weight	1
Background	Opaque
Show Border	☐
Leader/Border Offset	5/64"
Leader Arrowhead	Arrow 30 Degree
Text	⌃
Text Font	Technic
Text Size	1/8"
Tab Size	1/2"
Bold	☐
Italic	☐
Underline	☐
Width Factor	1.000000

Set the Text Font to **Technic**.
Set the Text Size to **1/8"**.
Disable **Italic**.

Press **OK**.

13.

```
      FINISH SCHEDULE KEYS
  Schedules/Quantities
      Door Details
      Door Schedule
      FINISH SCHEDULE
      Glazing Schedule
  Sheets (all)
```

Activate the **Glazing Schedule**.

14. Select **Edit** next to Appearance.

Other		⊗
Fields	Edit...	
Filter	Edit...	
Sorting/Grouping	Edit...	
Formatting	Edit...	
Appearance	Edit...	

15. Set the Title text to **Schedule Title**.
Set the Header text to **Schedule Header**.
Set the Body text to **Schedule Cell**.

Press **OK**.

Text
☑ Show Title
☑ Show Headers
Title text: Schedule Title
Header text: Schedule Header
Body text: Schedule Cell

16. The schedule updates with the new fonts assigned.

<GLAZING SCHEDULE>

A	B	C	D	E	F	G	H
		SIZE			DETAILS		
WINDOW NO.	FRAME TYPE	W	H	SILL	HEAD	JAMB	REMARKS
8	36" x 24"	3' - 0"	2' - 0"				
36" x 24": 54							
QTY FOR EACH TYPE: 54							

17. Save as *ex6-16.rvt*.

Exercise 6-17
Using a View Template for a Schedule

Drawing Name: ex6-16.rvt
Estimated Time: 5 minutes

This exercise reinforces the following skills:

❑ Schedules
❑ Text Styles

1. Open *ex6-16.rvt*.

2. Activate the **Glazing Schedule**.

⊟ ▦ Schedules/Quantities
 ⌐ Door Details
 ⌐ Door Schedule
 ⌐ Finish Schedule
 ⌐ **Glazing Schedule**
 ⌐ Lighting Fixture Schedule

3. Activate the **View** ribbon.

Select **Create Template from Current View**.

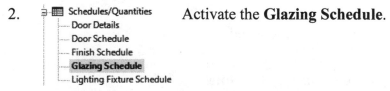

4.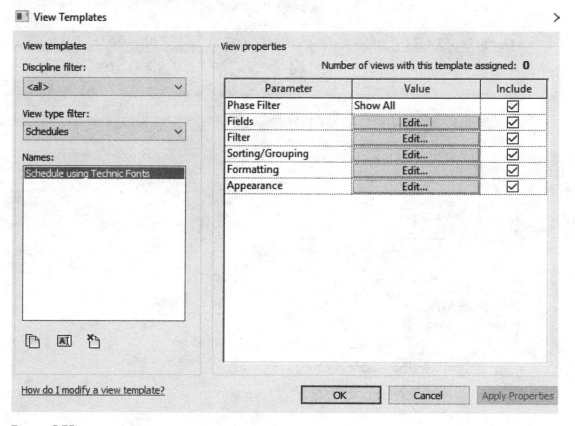

   Type **Schedule using Technic Fonts**.

   Press **OK**.

5. Note that you can use a View Template to control not only the appearance of a schedule, but for selecting fields, formatting, etc.

   Press **OK**.

6. 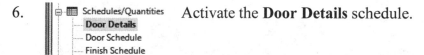 Activate the **Door Details** schedule.

7. On the Properties palette:

   Select the **<None>** button next to View Template.

8. 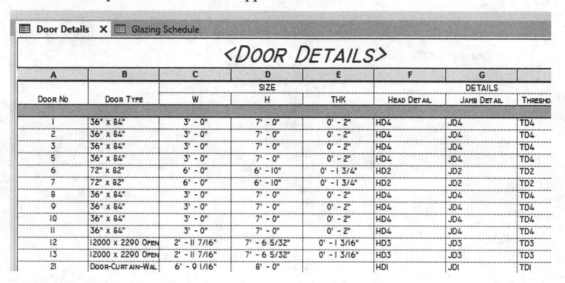 Highlight the **Schedule using Technic Fonts** template.

Press **OK**.

9. The schedule updates with the new appearance.

| Door Details X | Glazing Schedule | | | | | | |

⟨DOOR DETAILS⟩

A	B	C	D	E	F	G	
		SIZE			DETAILS		
Door No	Door Type	W	H	THK	Head Detail	Jamb Detail	Thresho
1	36" x 84"	3' - 0"	7' - 0"	0' - 2"	HD4	JD4	TD4
2	36" x 84"	3' - 0"	7' - 0"	0' - 2"	HD4	JD4	TD4
3	36" x 84"	3' - 0"	7' - 0"	0' - 2"	HD4	JD4	TD4
5	36" x 84"	3' - 0"	7' - 0"	0' - 2"	HD4	JD4	TD4
6	72" x 82"	6' - 0"	6' - 10"	0' - 1 3/4"	HD2	JD2	TD2
7	72" x 82"	6' - 0"	6' - 10"	0' - 1 3/4"	HD2	JD2	TD2
8	36" x 84"	3' - 0"	7' - 0"	0' - 2"	HD4	JD4	TD4
9	36" x 84"	3' - 0"	7' - 0"	0' - 2"	HD4	JD4	TD4
10	36" x 84"	3' - 0"	7' - 0"	0' - 2"	HD4	JD4	TD4
11	36" x 84"	3' - 0"	7' - 0"	0' - 2"	HD4	JD4	TD4
12	12000 x 2290 Open	2' - 11 7/16"	7' - 6 5/32"	0' - 1 3/16"	HD3	JD3	TD3
13	12000 x 2290 Open	2' - 11 7/16"	7' - 6 5/32"	0' - 1 3/16"	HD3	JD3	TD3
21	Door-Curtain-Wal	6' - 9 1/16"	8' - 0"		HD1	JD1	TD1

10. Save as *ex6-16.rvt*.

Exercise 6-18
Exporting a View Template to another Project

Drawing Name: ex6-17.rvt, Office_2.rvt
Estimated Time: 55 minutes

This exercise reinforces the following skills:

- Schedules
- View Templates
- Transfer Project Standards

1.   Open *ex6-17.rvt*.

2.   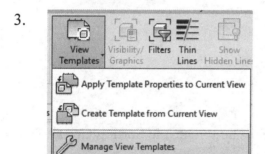   Activate the **Glazing Schedule**.

3.   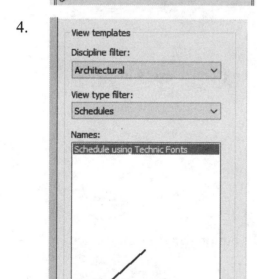

Activate the **View** ribbon.

Select **Manage View Templates**.

4.

In the Discipline filter: Select **Architectural** from the drop-down list.

In the View type filter: Select **Schedules** from the drop-down list.

Highlight the **Schedule using Technic Fonts** template.

Select the **Duplicate** icon at the bottom of the dialog.

5.   Type **Schedule with Shared Parameters**.

Press **OK**.

6.   Verify that the new schedule template is highlighted.

7.   Select the **Edit** button next to Fields.

8.   Note the fields listed include the shared parameters.

Press **OK**.

9.   Select the **Edit** button next to Sorting/Grouping.

10.   Place a check on **Itemize every instance**.

Press **OK**.

11.   Select the **Edit** button next to Appearance.

12. 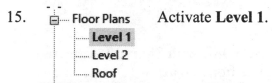 Place a check next to **Grid in headers/footers/spacers**.
Place a check next to **Outline** and set the line style to **Wide Lines**.
Uncheck **Blank row before data**.

Press **OK**.

13.
Parameter	Value	Include
Phase Filter	Show All	☑
Fields	Edit...	☑
Filter	Edit...	☑
Sorting/Grouping	Edit...	☑
Formatting	Edit...	☑
Appearance	Edit...	☑

Note that there is a check to Include all the properties listed.

Press **OK** to close the dialog.

14. Open *Office_2.rvt*.

15. Floor Plans — **Level 1** — Level 2 — Roof   Activate **Level 1**.

16. Activate the **Manage** ribbon.

Select **Transfer Project Standards**.

17. Select **Check None** to uncheck all the items in the list.

Scroll down and select **View Templates**.

Press **OK**.

18. Select **New Only** to copy over only the new view templates.

19. 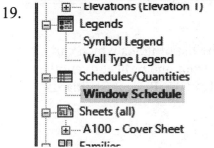 Activate the **Window Schedule** in the Project Browser.

- Elevations (Elevation 1)
- Legends
  - Symbol Legend
  - Wall Type Legend
- Schedules/Quantities
  - **Window Schedule**
- Sheets (all)
  - A100 - Cover Sheet
- Families

20.

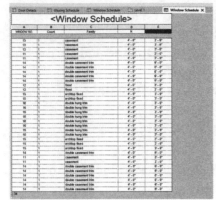

Note the fields included and the appearance of the schedule.

21.

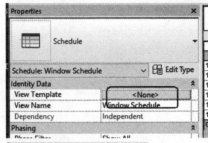

In the Properties pane, select the **<None>** button next to View Template.

22.

Select the **Schedule with Shared Parameters** view template from the list.

Note that the two schedule view templates created were copied into the project.

Press **OK**.

23.

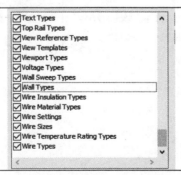

Note how the schedule updates.

Save as *Office_2_A.rvt*.

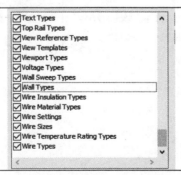

Note the list of items which can be copied from one project to another using Transfer Project Standards.

Additional Projects

		FINISH SCHEDULE														
NO	ROOM NAME	FLOOR		BASE		NORTH WALL		EAST WALL		SOUTH WALL		WEST WALL		CLG		REMARKS
		MAT	FIN	MAT	FIN	MAT	FIN	MAT	FIN	MAT	FIN	MAT	FIN	MAT	FIN	

1) Using Shared Parameters, create a Finish Schedule for the lavatories.

Door Schedule - Ver 2			
Mark	Type	Count	Remarks
1	36" x 84"	1	Single-Flush
2	36" x 84"	1	Single-Flush
3	36" x 84"	1	Single-Flush
4	36" x 84"	1	Single-Flush
5	72" x 78"	1	Double-Exterior
6	72" x 78"	1	Double-Exterior
7	36" x 84"	1	Single-Flush
8	36" x 84"	1	Single-Flush
9	36" x 84"	1	Single-Flush
10	36" x 84"	1	Single-Flush
	Curtain Wall Dbl Glass	1	Double Glazed

2) Create a door schedule.

Door Schedule 2							
			Rough Opening				
Mark	Type	Level	Rough Height	Rough Width	Material	Lock Jamb	Swing
1	36" x 84"	Level 1			ULTEX/WOOD	CENTER	RIGHT
2	36" x 84"	Level 1			ULTEX/WOOD	THROW	RIGHT
3	36" x 84"	Level 1			ULTEX/WOOD	THROW	LEFT
4	36" x 84"	Level 1			ULTEX/WOOD	THROW	LEFT
5	72" x 78"	Level 1			ULTEX/WOOD	THROW	CENTER
6	72" x 78"	Level 1			ULTEX/WOOD	THROW	CENTER
	Curtain Wall Dbl	Level 1			ULTEX/WOOD	THROW	CENTER
7	36" x 84"	Level 2			ULTEX/WOOD	THROW	LEFT
8	36" x 84"	Level 2			ULTEX/WOOD	THROW	RIGHT
9	36" x 84"	Level 2			ULTEX/WOOD	THROW	RIGHT
10	36" x 84"	Level 2			ULTEX/WOOD	THROW	RIGHT

3) Create this door schedule.

4) Create three new fonts for Schedule Title, Schedule Header, and Cell text. Apply them to a schedule.

5) Create a view template using the schedule with your custom fonts.

Lesson 6 Quiz

True or False

1. A schedule displays information about elements in a building project in a tabular form.
2. Each property of an element is represented as a field in a schedule.
3. If you replace a model element, the schedule for that element will automatically update.
4. Shared parameters are saved in an Excel spreadsheet.
5. A component schedule is a live view of a model.

Multiple Choice

6. Select the three types of schedules you can create in Revit:

    A. Component
    B. Multi-Category
    C. Key
    D. Symbol

7. Keynotes can be attached to the following:
   Select THREE answers.

    A. Model elements
    B. Detail Components
    C. Materials
    D. Datum

8. Select the tab that is not available on the Schedule Properties dialog:

    A. Fields
    B. Filter
    C. Sorting/Grouping
    D. Formatting
    E. Parameters

9. Schedules are exported in this file format:

    A. Excel
    B. Comma-delimited text
    C. ASCII
    D. DXF

10. To export a schedule:

    A. Right click on a schedule and select Export.
    B. Place a schedule on a sheet, select the sheet in the Project Browser, right click and select Export.
    C. Go to the Manage ribbon and select the Export Schedule tool.
    D. On the Application Menu, go to File→Export→Reports→Schedule.

11. To create a custom grand total title in a schedule:

    A. Enter the title in the custom grand total title text box on the Sorting/ Grouping tab.
    B. Click in the schedule cell for the grand total title and modify the text.
    C. Select the schedule and type the desired text in the custom grand total title field on the Properties pane.
    D. All of the above.

12. To create a graphic display that looks like a view is sketched:

    A. Select Sketchy from the Graphics Display list.
    B. Enable Sketchy Lines in the Graphic Display Options dialog.
    C. Set the Graphic Display to Sketch.
    D. Apply a sketch view template.

ANSWERS:
  1) T; 2) T; 3) T; 4) F; 5) T; 6) A, B, and C; 7) A, B, and C; 8) E; 9) B; 10) D; 11) A; 12) B

Lesson 7
Roofs

Revit provides three methods for creating roofs: extrusion, footprint or by face. The extrusion method requires you to sketch an open outline. The footprint method uses exterior walls or a sketch. The face method requires a mass.

A roof cannot cut through windows or doors.

Exercise 7-1
Creating a Roof Using Footprint

Drawing Name: ex6-16.rvt
Estimated Time: 15 minutes

This exercise reinforces the following skills:

- Level
- Roof
- Roof Properties
- Roof Options
- Isolate Element
- Select All Instances
- 3D View

1. Open *ex6-16.rvt*.

2. Activate the roof line view under Floor Plans by double clicking on it.

Floor Plans
  Level 1
  Level 1 - Lobby Detail
  Level 1 -Annotations Visible
  Level 2
  Level 2 - Annotations Visible
  Roof Line
  Site

3.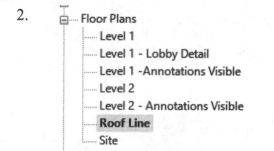

Underlay	
Range: Base Level	Level 1
Range: Top Level	Roof Line
Underlay Orientation	Look down

In the Properties pane:

Scroll down to the Underlay heading.
Set the Base Level Range to **Level 1**.
Set the Top Level Range to **Roof Line**.

4.

In the browser, locate the exterior wall that is used in the project.

Right click and select **Select All Instances → In Entire Project**.

5.

Isolate Category
Hide Category
Isolate Element
Hide Element

Reset Temporary Hide

On the Display bar:

Left click on the icon that looks like sunglasses.

Select **Isolate element**.

This will turn off the visibility of all elements except for the exterior walls.
This will make selecting the walls easier.

You should see the top view of your model with only the exterior walls visible.

You won't see the curtain wall because it is a different wall style.

6. Left click in the display window to release the selection set.

7.

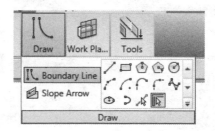

Select **Build → Roof → Roof by Footprint** from the Architecture ribbon.

Notice that **Pick Walls** is the active mode.

8.

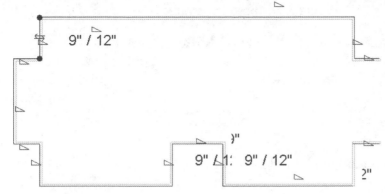

9" / 12"

9" / 1: 9" / 12"

You can zoom in and select the exterior walls.

You can pan or scroll to move around the screen to select the walls.

If you miss a wall, you will get a dialog box advising you of the error.

You can use the ALIGN tool to align the roof edge with the exterior side of the wall.

9.

☑ Defines slope | Overhang: 0' 6" | ☑ Extend to wall core

☑ Defines slope | Overhang: 60.00 | ☑ Extend to wall core

Hold down the CTL key and select all the roof lines.
On the options bar, enable **Defines slope**.
Set the Overhang to **6″ [60]**.
Enable **Extend to wall core**.
It is very important to enable the **Extend into wall (to core)** if you want your roof overhang to reference the structure of your wall.

For example, if you want a 2′-0″ overhang from the face of the stud, not face of the finish, you want this box checked. The overhang is located based on your pick point. If your pick point is the inside face, the overhang will be calculated from the interior finish face.

The angle symbols indicate the roof slope.

10.

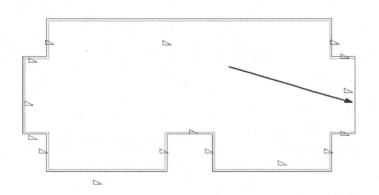

Draw

Select the **Line** tool from the Draw panel.

11.

Draw a line to close the sketch.

Check the sketch to make sure there are no gaps or intersecting lines.

12. 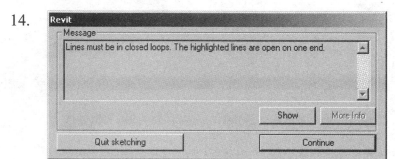 On the Properties palette:

Select the **Wood Rafter 8″ – Asphalt Shingle – Insulated** [**Warm Roof - Timber**] from the Type Selector.

13. 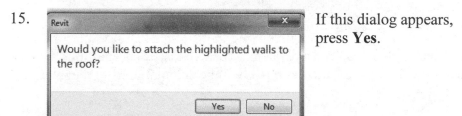 Select **Green Check** under Mode.

14. 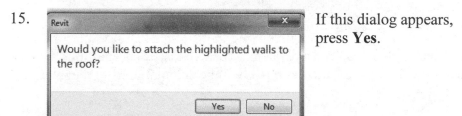 This dialog will appear if you missed a wall.

You can press the **Show** button and Revit will highlight the wall you missed.

15. 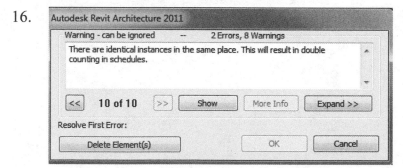 If this dialog appears, press **Yes**.

16. 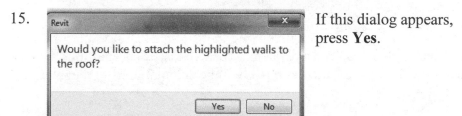 If you get an error message like this, press **Delete Elements** to delete any conflicting elements.

17. 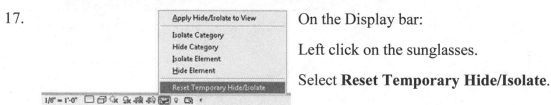 On the Display bar:

Left click on the sunglasses.

Select **Reset Temporary Hide/Isolate**.

18. Go to **3D View**.

19.

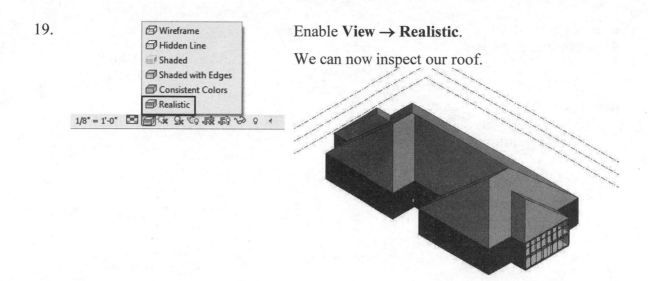

Enable **View → Realistic**.

We can now inspect our roof.

20. Save as *ex7-1.rvt*.

The Cutoff Level Property defines the distance above or below the level at which the roof is cut off.

Roofs (1)	
Constraints	
Base Level	Roof Line
Room Bounding	☑
Related to Mass	☐
Base Offset From Level	0' 0"
Cutoff Level	None
Cutoff Offset	0' 0"

Exercise 7-2
Modifying a Roof

Drawing Name: ex7-1.rvt
Estimated Time: 10 minutes

This exercise reinforces the following skills:

- ❑ Modifying Roofs
- ❑ Edit Sketch
- ❑ Align Eaves
- ❑ Roof
- ❑ Work Plane

1.   Open or continue working in *ex7-1.rvt*.

2.   ⊟ Elevations (Building Elevation)   Activate the East elevation.
               East
               North
               South
               South Lobby
               West

3.   Select the **Level** tool from the Datum panel on the Architecture ribbon.

4.   Add a level 10′ 6″ [270mm] above the Roof Line level.

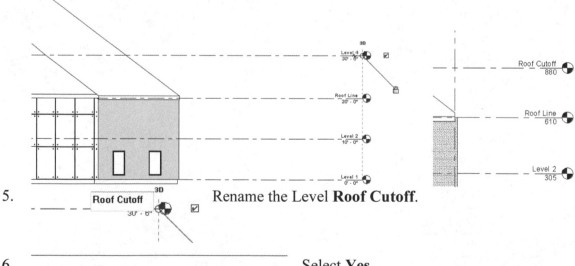

5.      Rename the Level **Roof Cutoff**.

6.   Would you like to rename corresponding views?   Select **Yes**.

                                         [Yes] [No]

7.   Select the roof so it is highlighted.

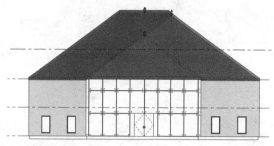

8.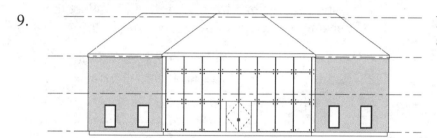

On the Properties pane:

Set the Cutoff Level to **Roof Cutoff**.

9.

Note how the roof adjusts.

10. Level 2 - No Dime
   Roof Cutoff
   Roof Line
   Site

   Activate the **Roof Cutoff** view under Floor Plans.

11. Roof Ceiling Floor

   Roof by Footprint

   Select **Roof → Roof by Footprint** from the Architecture ribbon.

12. Draw

   Select **Pick Lines** from the Draw panel.

13. ☐ Defines slope   Offset: 0' 0"   ☐ Lock

   Disable **Defines slope** on the Options bar.

   Set the Offset to **0' 0"**.

14.

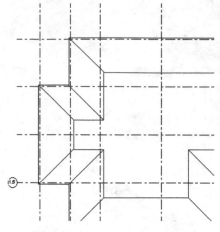

Pick an edge.

You won't see a slope symbol because slope has been disabled.

15.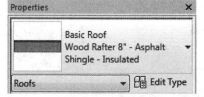

Select the **Modify** button.
To remove a slope from a line, simply select the line and uncheck Defines slope on the Options bar.

16.

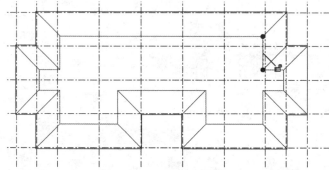

Select the edges to form a closed sketch.

Note that none of the lines have a slope defined.

17.

Select **Edit Type** from the Properties pane.

18.

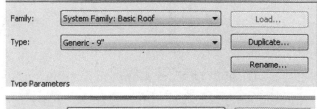

Select the **Generic - 9″ [Generic 400mm]** roof from the drop-down list.

Select **Duplicate**.

19. 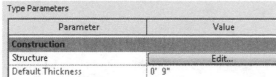 Enter **Tar and Gravel**.

Press **OK**.

20. 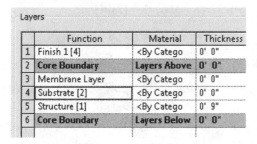 Select **Edit** next to Structure.

21. Insert three layers.

Arrange the layers so:
Layer 1 is set to Finish 1 [4].
Layer 2 is Core Boundary.
Layer 3 is Membrane Layer.
Layer 4 is Substrate [2]
Layer 5 is Structure [1]
Layer 6 is Core Boundary.

	Function	Material	Thickness
1	Finish 1 [4]	<By Catego	0' 0"
2	Core Boundary	Layers Above	0' 0"
3	Membrane Layer	<By Catego	0' 0"
4	Substrate [2]	<By Catego	0' 0"
5	Structure [1]	<By Catego	0' 9"
6	Core Boundary	Layers Below	0' 0"

22. Select the Material column for **Finish 1 [4]**.

23. 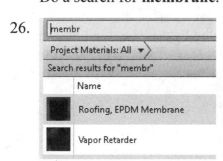 Locate the **gravel** material in the Document materials.

24. Set the thickness to **1/4″** [6].

	Function	Material	Thickness
1	Finish 1 [4]	Gravel	0' 1/4"

25. Select the Material column for **Membrane Layer**.

Do a search for **membrane**.

26. 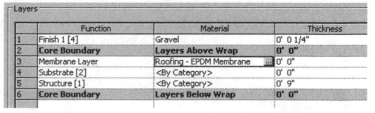 Set the Material to **Roofing - EPDM Membrane**.

Set the Membrane thickness to **0″**.

	Function	Material	Thickness
1	Finish 1 [4]	Gravel	0' 0 1/4"
2	Core Boundary	Layers Above Wrap	0' 0"
3	Membrane Layer	Roofing - EPDM Membrane	0' 0"
4	Substrate [2]	<By Category>	0' 0"
5	Structure [1]	<By Category>	0' 9"
6	Core Boundary	Layers Below Wrap	0' 0"

27.

Set Layer 4 Substrate [2] to **Plywood, Sheathing** with a thickness of **5/8″ [16]**.

	Name
	Plywood, Sheathing

plyw

Project Materials: All ▼

Search results for "plyw"

28.

Set Layer 5 Structure [1] to **Structure-Wood Joist/ Rafter Layer [Structure, Timber Joist/Rafter Layer]** with a thickness of **7 ½″ [200]**.

wood joi

Project Materials: All ▼

Search results for "wood joi"

	Name
	Structure, Wood Joist/Rafter Layer
	Structure, Wood Joist/Rafter Layer, Batt Insulation

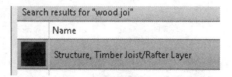

Search results for "wood joi"

	Name
	Structure, Timber Joist/Rafter Layer

29.

Press **OK**.

Layers

	Function	Material	Thickness	Wraps	Vari:
1	Finish 1 [4]	Gravel	0' 0 1/4"		
2	**Core Boundary**	**Layers Above Wrap**	**0' 0"**		
3	Membrane Lay	Roofing, EPDM Membr	0' 0"		
4	Substrate [2]	Plywood, Sheathing	0' 0 5/8"		
5	Structure [1]	Structure, Wood Joist/R	0' 7 1/2"		
6	**Core Boundary**	**Layers Below Wrap**	**0' 0"**		

Layers

	Function	Material	Thickness
1	Finish 1 [4]	Gravel	6.0
2	**Core Boundary**	**Layers Above Wrap**	**0.0**
3	Membrane Layer	Roofing, EPDM Me	0.0
4	Substrate [2]	Plywood, Sheathin	16.0
5	Structure [1]	Structure, Timber J	200
6	**Core Boundary**	**Layers Below Wrap**	**0.0**

30.

Graphics	⌃
Coarse Scale Fill Pattern	⊡
Coarse Scale Fill Color	■ Black

Press the Browse button in the Coarse Scale Fill Pattern to assign a fill pattern.

31.

Press **New** to define a new fill pattern.

32.

    Enable **Custom**.

    Press **Browse**.

33. Select the *gravel* hatch pattern in the exercise files.
    Press **Open**.

34. Change the Name to **Gravel**.

    Set the Import scale to **0.03**.

    Note how the preview changes.

    Press **OK**.

35.

    Highlight Gravel.

    Press **OK**.

36.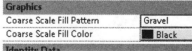

    Assign the **Gravel** pattern to the Coarse Scale Fill Pattern.

    Press **OK**.

    Press **OK** to close the Type Properties dialog box.

37.

    Select the **Green Check** under the Model panel to **Finish Roof**.

38. Zoom in to see the hatch
   pattern.

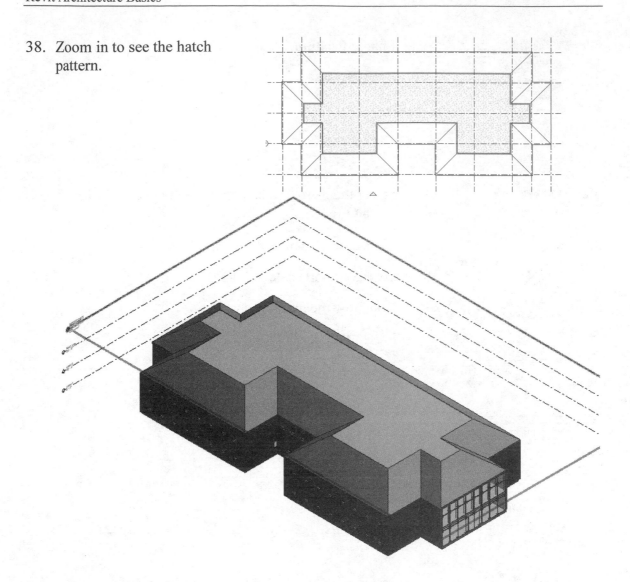

39. Save as *ex7-2.rvt*.

Exercise 7-3
Modifying a Roof Form

Drawing Name: ex7-2.rvt
Estimated Time: 10 minutes

This exercise reinforces the following skills:

- ❑ Shape Editing
- ❑ Window
- ❑ Load from Library
- ❑ Array

1. Open or continue working in *ex7-2.rvt*.

2. 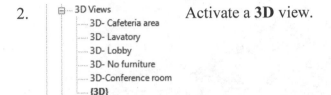 Activate a **3D** view.

3. 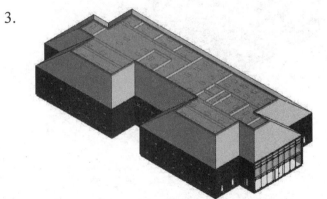 Select the tar and gravel roof.

   It will highlight in blue and appear to be transparent.

4. 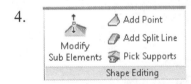 Select the **Add Point** tool on the **Shape Editing** panel on the ribbon.

5.  Place points as shown.

6.   Select **Modify Sub Elements**.

7. 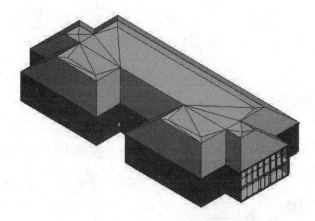 Select each point and set them to **-0′ 6″ [-15]**.

    These will be drain holes.

    Right click and select Cancel when you are done.

8. The roof surface will show the drainage areas.

9. Save as *ex7-3.rvt*.

Exercise 7-4
Adding Roof Drains

Drawing Name: ex7-3.rvt
Estimated Time: 10 minutes

This exercise reinforces the following skills:

- ❑ Component
- ❑ Load Family

1. Open or continue working in *ex7-3.rvt.*

2. 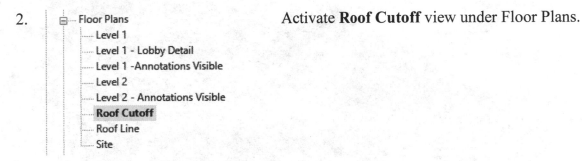   Activate **Roof Cutoff** view under Floor Plans.

3. `1/8" = 1'-0"` ☒ ⊞   Change the display settings to **Medium, Wireframe**.

4. ◰   Select **Place a Component** from the Build panel on the Architecture ribbon.
    Component

5. ◰   Select **Load Family** under the Mode panel.
    Load
    Family

6. 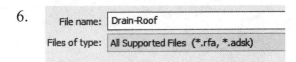   Locate the *Drain-Roof* from the Class Files downloaded from the publisher's website.

    Press **Open**.

7. There is no tag loaded for Plumbing Fixtures. Do you want to load one now?   If a dialog appears asking if you want to load a tag for Plumbing Fixtures, press **No**.

    [Yes] [No]

8. Place roof drains coincident to the points placed earlier.

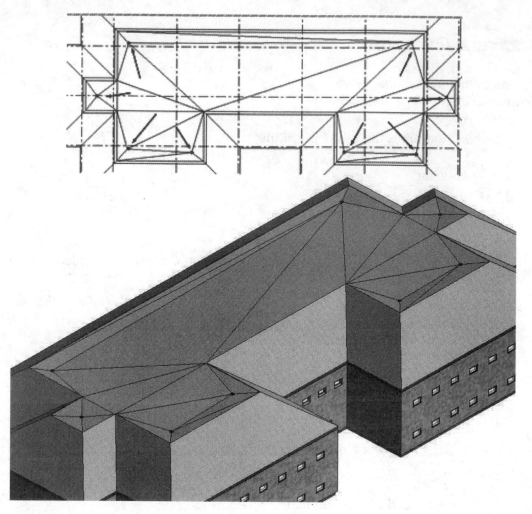

9. Save the file as *ex7-4.rvt*.

Exercise 7-5
Adding a Gable by Modifying a Roof Sketch

Drawing Name: ex7-4.rvt
Estimated Time: 10 minutes

This exercise reinforces the following skills:

    ❏   Modify a Roof Sketch

    A video for this exercise is available on my MossDesigns station on Youtube.com.

1.   Open or continue working in *ex7-4.rvt*.

    ⋯⋯ Roof Cutoff Activate the **Roof Line** view under Floor Plans.
    — **Roof Line**
    ⋯⋯ Site

2.   Select the Wood/Timber roof.

    Select **Edit Footprint** from the ribbon.

    This returns you to sketch mode.

3.   Select the **Split** tool on the Modify panel.

4.   Split the roof on the south side at the Grids B & C.

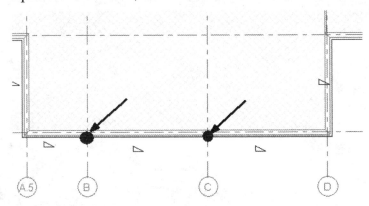

5.   Select the new line segment created by the split.

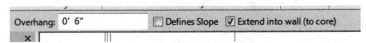

Uncheck **Defines Slope** on the option bar.

6.   Slope Arrow   Select the **Slope Arrow** tool on the ribbon.

7.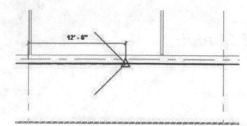

Select the left end point of the line and the midpoint of the line to place the slope.

8. 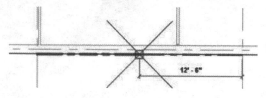 Select the **Slope Arrow** tool on the ribbon.

9.

Select the right end point of the line and the midpoint of the line to place the slope.

10. The sketch should appear as shown.

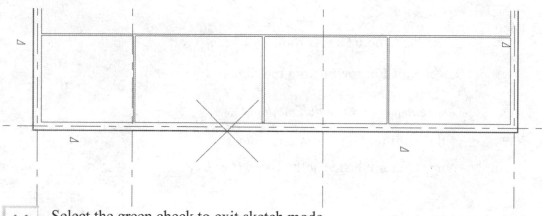

11. Select the green check to exit sketch mode.

12. Switch to a 3D view and orbit to inspect the new gable.

Note that the wall automatically adjusted.

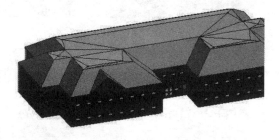

13. Save as *ex7-5.rvt*.

Additional Projects

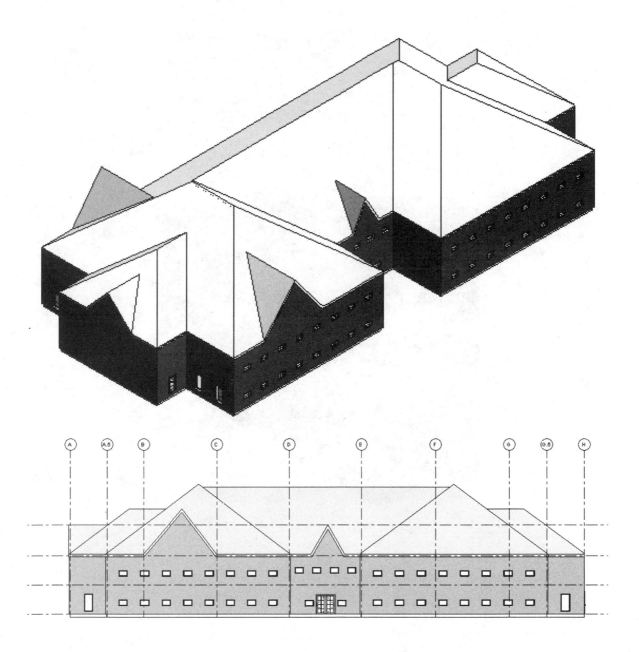

1) Create a roof with several dormers.

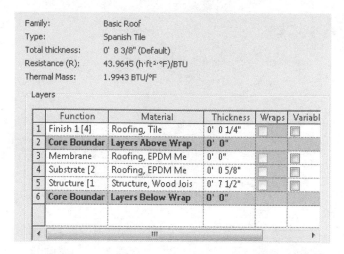

Family: Basic Roof
Type: Spanish Tile
Total thickness: 0' 8 3/8" (Default)
Resistance (R):   43.9645 (h·ft²·°F)/BTU
Thermal Mass: 1.9943 BTU/°F

Layers

	Function	Material	Thickness	Wraps	Variabl
1	Finish 1 [4]	Roofing, Tile	0' 0 1/4"	☐	☐
2	**Core Boundar**	**Layers Above Wrap**	**0' 0"**		
3	Membrane	Roofing, EPDM Me	0' 0"		☐
4	Substrate [2	Roofing, EPDM Me	0' 0 5/8"		☐
5	Structure [1	Structure, Wood Jois	0' 7 1/2"	☐	☐
6	**Core Boundar**	**Layers Below Wrap**	**0' 0"**		

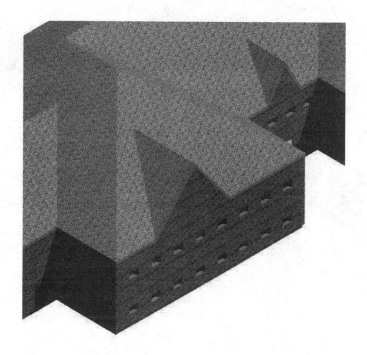

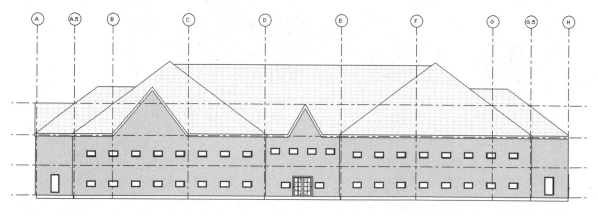

2) Create a roof family using Spanish tile material.

Lesson 7 Quiz

True or False

1. A roof footprint is a 2-D sketch.
2. When creating a roof by extrusion, the sketch must be closed.
3. When placing a roof by footprint, you must specify the work plane to place the sketch.
4. When you create a roof by extrusion, you can extend the extrusion toward or away from the view.
5. The Join/Unjoin Roof tool can be used to join two roofs together.

Multiple Choice

6. Roofs can be created using the following options:
   Select three:

   A. Roof by footprint
   B. Roof by extrusion
   C. Roof by face
   D. Roof by sketch

7. The Join/Unjoin roof tool is located:

   A. On the Modify panel of the Modify ribbon.
   B. On the Build panel of the Home ribbon.
   C. On the Geometry panel of the Modify ribbon.
   D. On the right shortcut menu when the roof is selected.

8. When creating a roof sketch you can use any of the following EXCEPT:

   A. Pick wall
   B. Lines
   C. Arcs
   D. Circles

9. In order to attach a wall to a roof that is above the wall, you should use this option:

   A. Base Attach
   B. Top Attach
   C. Edit Profile
   D. Join/Unjoin

10. The Properties pane for a roof displays the roof:

     A. Area
     B. Weight
     C. Elevation
     D. Function

ANSWERS:
    1) T; 2) F; 3) F; 4) T; 5) T; 6) A, B, & C; 7) C; 8) D; 9) B; 10) A

Lesson 8
Elevations, Details & Plans

An "elevation" is a drawing that shows the front or side of something. A floor plan, by contrast, shows a space from above – as if you are looking down on the room from the ceiling. Thus, you see the tops of everything, but you cannot view the front, side or back of an object. An elevation gives you the chance to see everything from the other viewpoints.

Elevations are essential in kitchen design, as well as other detailed renovations. Without elevation drawings, you cannot see the details of your new cabinetry, the size of each drawer or the location of each cabinet. A floor plan simply cannot communicate all of this information adequately.

While an elevation is not required for every renovation or redecorating project, they are very useful when designing items like a fireplace, bathroom vanities, bars, or any location with built-in cabinetry, such as an office or entertainment space. The information shown on an elevation drawing will give you a chance to make small changes to the design before anything is built or ordered – you don't want to be surprised during the installation!

While every detail isn't typically shown on an elevation (such as the exact cabinet door style you plan to use), the major elements will be there, including cabinet locations, the direction each cabinet door opens (hint: look at the "arrows" on the doors – the arrow points to the hinge, so you know which way the door opens!), appliance locations, height of cabinets and more. On elevations intended for use during the preliminary design stage of your project, you will find much less detail. Drawings intended for use as a guide for construction will include numerous notes and dimensions on the page.

Because Revit is a BIM software, when you update elements in the plan view, the elevation views will automatically update and vice-versa. This is called "bi-directional associativity."

Exercise 8-1
Creating Elevation Documents

Drawing Name: ex7-5.rvt
Estimated Time: 10 minutes

This exercise reinforces the following skills:

❑ Sheet
❑ Add View
❑ Changing View Properties

1. Open *ex7-5.rvt*.

2. Activate the **View** ribbon.

3. Select **Sheet** from the Sheet Composition panel to add a new sheet.

4. Select **D 22 x 34 Horizontal [A3 metric]** title block. Press **OK**.

   D 22 x 34 Horizontal
   E1 30 x 42 Horizontal : E1 30x42 Horizontal

In the Properties pane:

Approved By	M. Instructor
Designed By	J. Student
Checked By	M. Instructor
Drawn By	J. Student
Sheet Number	A201
Sheet Name	Exterior Elevations

Change Approved By to your instructor's name.
Change Designed By to your name.
Change Checked By to your instructor's name.
Change the Sheet Number to **A201**.
Change the Sheet Name to **Exterior Elevations**.
Change Drawn by to your name.

5. Elevations (Building Elevation)
   - East
   - North
   - South
   - South Lobby
   - West

Locate the North and South Elevations in the Project Browser.

6.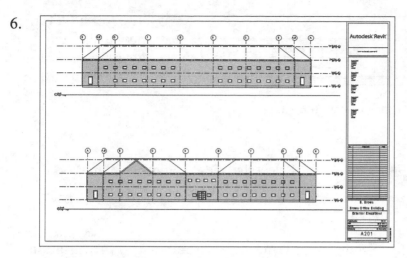

Drag and drop the North and South Elevations from the browser onto the sheet. The top elevation will be the North elevation. The bottom elevation will be the South Elevation.

You may notice that some entities are visible in the view, which you do not want to see, like the stairs inside.

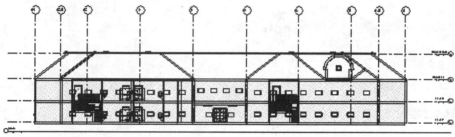

Let's look at the **North** Elevation first.

7.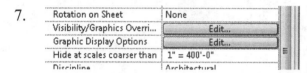

Select the **North** Elevation View on the sheet.

In the Properties pane:
Select the **Edit** button next to Graphic Display Options.

8.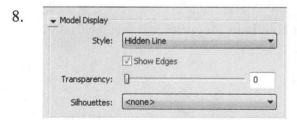

Under Model Display:

Set Style to **Hidden Line**.

Press **OK**.

9.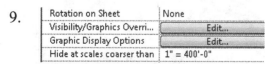

Select **Edit** next to Visibility/Graphics Overrides.

10.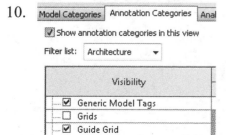

Under Annotation Categories, disable Grids and Sections.

Press **OK**.

11.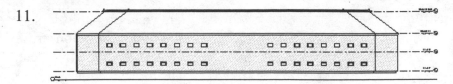

The North Elevation now looks more appropriate.

12. Select the **South** Elevation View on the sheet.

In the Properties pane:

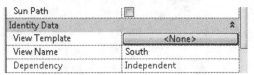

Select the View Template column.

13.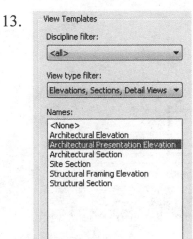

Set the View Template to **Architectural Presentation Elevation**.
Press **Apply**.

14.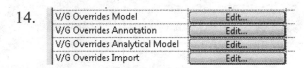

Select **Edit** next to V/G Overrides Annotation.

15.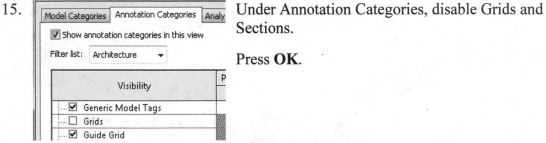

Under Annotation Categories, disable Grids and Sections.

Press **OK**.

16.

V/G Overrides Import	Edit...
V/G Overrides Filters	Edit...
Model Display	Edit...
Shadows	Edit...
Sketchy Lines	Edit...
Lighting	Edit...
Photographic Exposure	Edit...
Background	Edit...
Far Clipping	No clip

Select **Edit** next to Shadows.

17.

Enable **Cast Shadows**.
Enable **Show Ambient Shadows**.
Press OK.

Close the dialog.

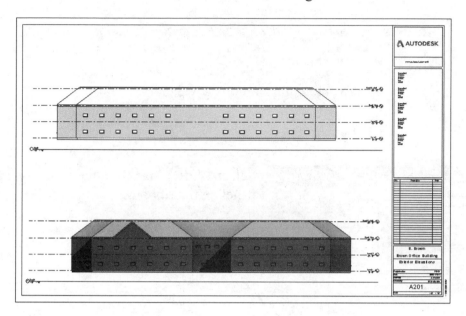

18. Save as *ex8-1.rvt*.

Exercise 8-2
Using Line Work

Drawing Name: ex8-1.rvt
Estimated Time: 15 minutes

This exercise reinforces the following skills:

- Activate View
- Linework tool
- Line Styles
- Line Weight

1. Open *ex8-1.rvt*.

2. Activate the **Exterior Elevations** sheet.

3. 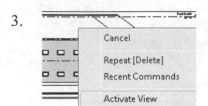 Select the top elevation view (**North**).
Right click and select **Activate View**.

Activate View works similarly to Model Space/Paper Space mode in AutoCAD.

4. Modify Select the **Modify** ribbon.

5. Select the **Linework** tool from the View panel.

6. 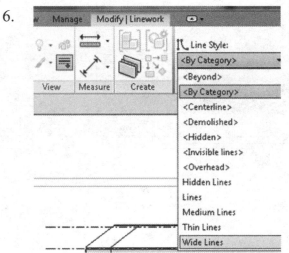 Select **Wide Lines** from the Line Style panel on the ribbon.

7. 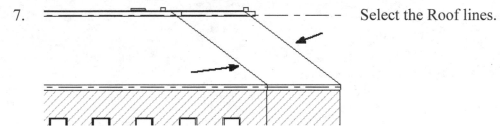 Select the Roof lines.

8. 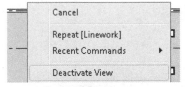 Right click in the graphics area.
Select **Deactivate View**.

9. Activate the **Manage** ribbon.

10. 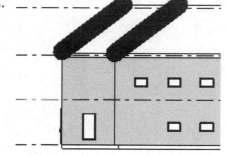 Go to **Settings → Additional Settings → Line Styles**.

Move the dialog out of the way so you can see what happens when you apply the new line weight.

11.

| Thin Lines | 1 |
| Wide Lines | 16 |

Set the Line Weight for Wide Lines to **16**.
Press **Apply**.

12. 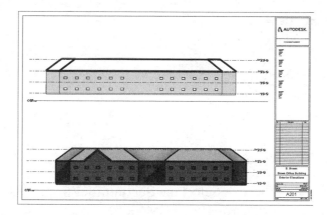 Our rooflines have dramatically changed.

13.

Medium Lines	3
Thin Lines	1
Wide Lines	12

Set the Wide Lines Line Weight to **12**.
Press **Apply**.
Press **OK**.

14. Note that both views updated with the new line style applied.

15. Save the file as *ex8-2.rvt*.

You can use invisible lines to hide any lines you don't want visible in your view. Linework is specific to each view. If you duplicate the view, any linework edits you applied will be lost.

Wall Section Views

The wall section view is used to show the various wall construction materials and act as a guide for the construction of the wall.

Exercise 8-3
Creating a Section View

Drawing Name: ex8-2.rvt
Estimated Time: 15 minutes

This exercise reinforces the following skills:

- ❑ Activate View
- ❑ Add Section Symbol

1.   ⊡   Open *ex8-2.rvt*.

2.   Activate **Level 1**.

    ⊞⋯ Structural Plans
    ⊟⋯ Floor Plans
        ⋯⋯ **Level 1**
        ⋯⋯ Level 1 - Lobby Detail
        ⋯⋯ Level 1 -Annotations Visible
        ⋯⋯ Level 2
        ⋯⋯ Level 2 - Annotations Visible
        ⋯⋯ Roof Cutoff
        ⋯⋯ Roof Line
        ⋯⋯ Site

3.   View   Activate the **View** ribbon.

4.   Select the **Section** tool from the Create panel.

    View Manage Modify
    3D   Section Callout
    View

5. 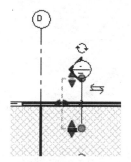   Place a section on the north exterior wall.

Adjust the clip range to show just the wall.

The first left click places the bubble.
The second click places the tail.
Use the flip orientation arrows to change the direction of the section view.

6. 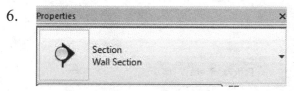   On the Properties panel:

Select **Wall Section** from the Type Selector drop-down list.

7.   Sections (Wall Section   The new section appears in
        Section 1   your Project Browser.

8.   Sections (Wall Section)
        North Exterior Wall

Rename the section view **North Exterior Wall**.

You can rename the view in the Properties panel or in the browser.

Activate the section view.

9.   1/8" = 1'-0"   Set the Detail Level to **Fine**.
Set the Visual Style to **Consistent Colors**.

10.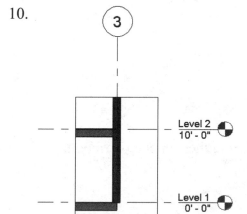

Adjust the crop region to display the wall as shown.
Select the crop region and use the bubble grips to re-size.

11. 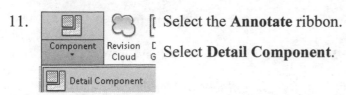 Select the **Annotate** ribbon.

Select **Detail Component**.

12.  Select **Load Family**.

13.

Browse to the *Detail Items\Div 01-General* folder.

Select the *Break Line [M_Break Line]*.

14.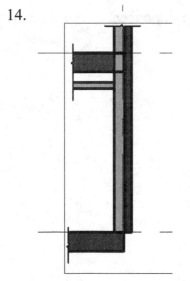

Add break lines to indicate that the top of the wall continues up, and the floors and ceiling continue toward the right.

Press the SPACE BAR to rotate the break line symbol.

15. Select the **Filled Region** tool from the Annotation ribbon.

16. 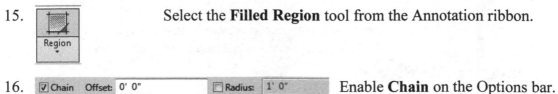 Enable **Chain** on the Options bar.

17. 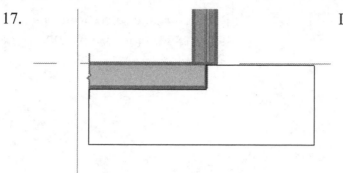 Draw the outline shown.

18. Press the Green Check to finish the filled region.

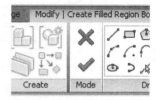

19. 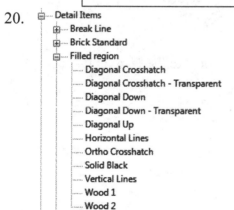 A filled region with a diagonal crosshatch pattern is placed.

20. Detail Items
    - Break Line
    - Brick Standard
    - Filled region
        - Diagonal Crosshatch
        - Diagonal Crosshatch - Transparent
        - Diagonal Down
        - Diagonal Down - Transparent
        - Diagonal Up
        - Horizontal Lines
        - Ortho Crosshatch
        - Solid Black
        - Vertical Lines
        - Wood 1
        - Wood 2

    Locate the **Filled region** under *Families/Detail Items* in the Project Browser.

21. 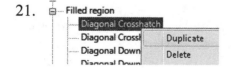 Highlight the **Diagonal Crosshatch**. Right click and select **Duplicate**.

22.

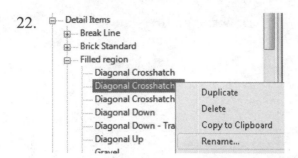

Right click and select **Rename** to rename the copied fill region.

23.    Rename **Gravel**.

24.

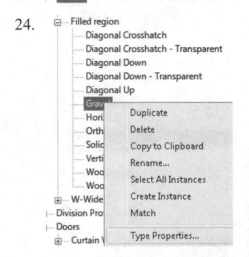

Highlight **Gravel**.

Right click and select **Type Properties**.

25.

Parameter	Value	
Graphics		
Foreground Fill Pattern	Diagonal crosshatch [Drafting]	
Foreground Pattern Color	■ Black	
Background Fill Pattern		
Background Pattern Color	■ Black	
Line Weight	1	
Masking	☑	

Select the **Foreground Fill Pattern**.

26.

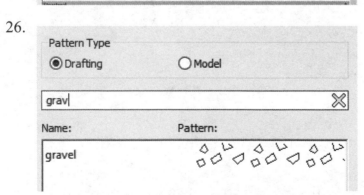

Locate the **Gravel** hatch pattern.

Press **OK**.

This fill pattern was defined in Exercise 7-2.

27.

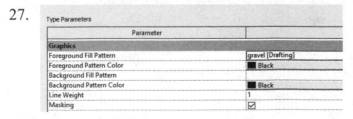

Type Parameters

Parameter	
Graphics	
Foreground Fill Pattern	gravel [Drafting]
Foreground Pattern Color	■ Black
Background Fill Pattern	
Background Pattern Color	■ Black
Line Weight	1
Masking	☑

Verify that the Gravel fill pattern has been assigned.

Press **OK**.

28.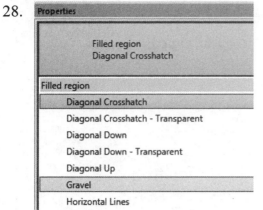

Select the Filled Region in the view.
Select by clicking on the outline.

Then select **Gravel** from the Type Selector.

29.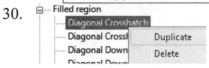

If you select the filled region you can use the grips to adjust the size.

30.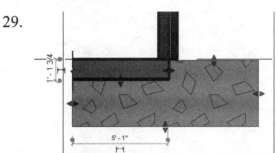

Highlight the **Diagonal Crosshatch** filled region in the Project Browser.

Right click and select **Duplicate**.

31.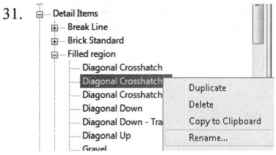

Right click and select **Rename** to rename the copied fill region.

32.

Rename **Earth**.

33.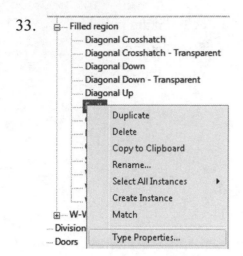

Highlight **Earth**.

Right click and select **Type Properties**.

34.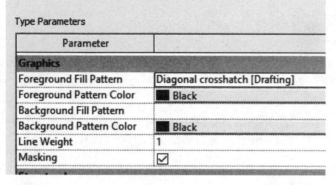

Select the **Foreground Fill Pattern**.

35.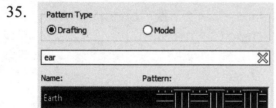

Select **the Earth** hatch pattern.
Press **OK**.

Exit the dialog.

36.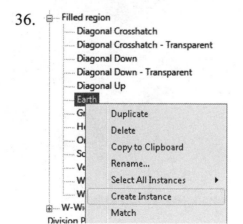

Highlight the Earth filled region in the Project Browser.

Right click and select **Create Instance**.

37.

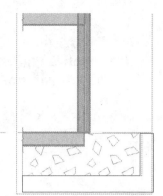

Draw an outline for the earth filled region.

Select the Green Check to complete.

38.

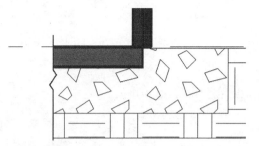

Adjust the break line to extend past the earth filled region.

39.

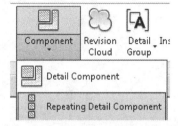

Select **Repeating Detail Component**.

40.

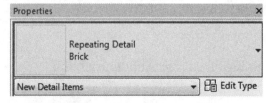

In the Properties Pane:
Verify that the Repeating Detail is **Brick**.
Select **Edit Type**.

41.

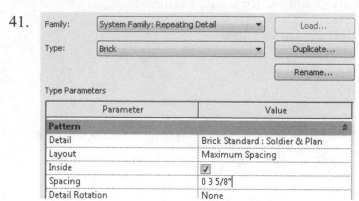

Set the Detail to **Brick Standard: Soldier & Plan**.
Set the Layout to **Maximum Spacing**.
Set the Spacing to **3 5/8″ [92]**.

Press **OK**.

42.

Select the interior bottom side for the start point of the brick detail.

Select the interior top side for the end point.

43. ⊞ Hide the Crop Region.

44. 🏷 Material Tag   Select the **Material Tag** tool from the Annotate ribbon.

45.

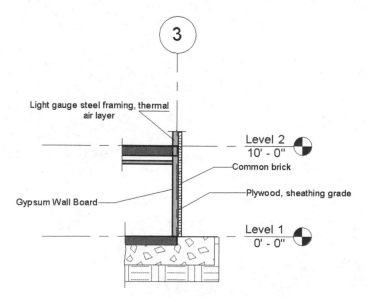

Place the material callouts.

The material displays the description of the material.

You should see a preview of the tag before you click to place.

46. Activate the **View** ribbon.

47. Select **New Sheet** from the Sheet Composition panel.

Sheet

48. Select titleblocks:

D 22 x 34 Horizontal
E1 30 x 42 Horizontal : E1 30x42 Horizontal
None

Select **D 22 x 34 Horizontal [A1 Metric]** titleblock.
Press **OK**.

49. In the Properties pane:

Approved By	M. Instructor
Designed By	J. Student
Checked By	M. Instructor
Drawn By	J. Student
Sheet Number	AD501
Sheet Name	Details

Change the Sheet Number to **AD501**.

Change the Sheet Name to **Details**.

Modify the Drawn By field with your name.

Modify the Approved By field with your instructor's name.

50. Drag and drop the North Exterior Wall section view onto the sheet.

51.

Viewports (1)	▼	Edit Type
Graphics		
View Scale	1/4" = 1'-0"	
Scale Value 1:	48	

Select the view.
In the Properties pane:
Set the View Scale to **1/4″ = 1′-0″**.

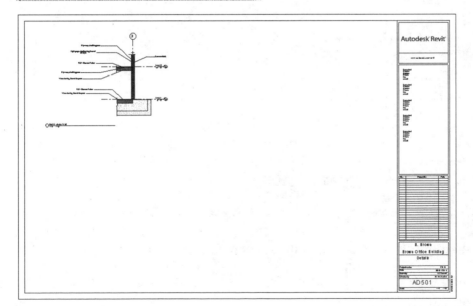

52. 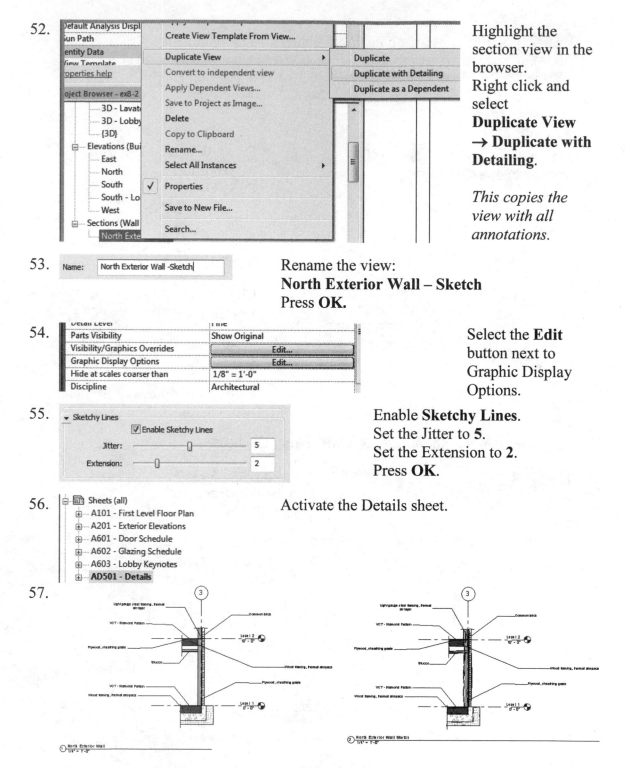 Highlight the section view in the browser.
Right click and select
Duplicate View → Duplicate with Detailing.

This copies the view with all annotations.

53. Name: North Exterior Wall -Sketch

Rename the view:
North Exterior Wall – Sketch
Press **OK.**

54.

Parts Visibility	Show Original
Visibility/Graphics Overrides	Edit...
Graphic Display Options	Edit...
Hide at scales coarser than	1/8" = 1'-0"
Discipline	Architectural

Select the **Edit** button next to Graphic Display Options.

55.

Sketchy Lines
☑ Enable Sketchy Lines
Jitter: 5
Extension: 2

Enable **Sketchy Lines.**
Set the Jitter to **5.**
Set the Extension to **2.**
Press **OK.**

56.
Sheets (all)
   A101 - First Level Floor Plan
   A201 - Exterior Elevations
   A601 - Door Schedule
   A602 - Glazing Schedule
   A603 - Lobby Keynotes
   AD501 - Details

Activate the Details sheet.

57.

Drag and drop the sketch version of the section next to the first section.

Compare the views.

58. Save the file as *ex8-3.rvt.*

Any blue symbol will behave like a hyperlink on a web page and can be double clicked to switch to the referenced view. (Note: The blue will print black on your plots if you want it to.)

Exercise 8-4
Modifying Keynote Styles

Drawing Name: ex8-3.rvt
Estimated Time: 15 minutes

This exercise reinforces the following skills:

- ❏ Keynotes
- ❏ Materials
- ❏ Edit Family

1. 📂 Open or continue working in *ex8-3.rvt*.

2. Activate the **Exterior Elevations** sheet.

⊟ 🗐 Sheets (all)
   ⊞ A101 - First Level Floor Plan
   ⊞ **A201 - Exterior Elevations**
   ⊞ A601 - Door Schedule
   ⊞ A602 - Glazing Schedule
   ⊞ A603 - Lobby Keynotes
   ⊞ AD501 - Details

3. Activate the Southern Elevation view.
Select the view.
Right click and select **Activate View**.

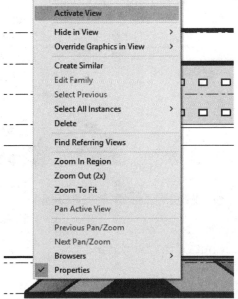

If you do not add the notes in Model Space, then the notes will not automatically scale and move with the view.

4. 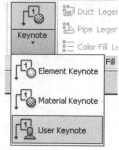 Activate the **Annotate** ribbon.

5. Select the **User Keynote** tool from the Tag panel.

Note: Revit only uses fonts available in the Windows fonts folder. If you are an AutoCAD user with legacy shx fonts, you need to locate ttf fonts that are equivalent to the fonts you want to use and load them into the Windows font folder. To convert shx fonts over to ttf, check with www.tcfonts.com.

6. 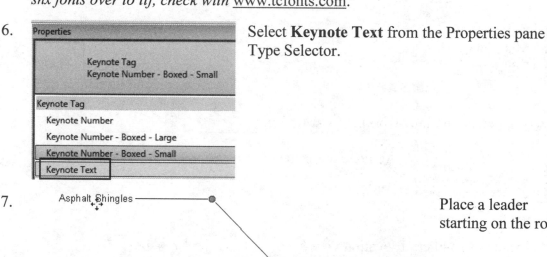 Select **Keynote Text** from the Properties pane Type Selector.

7. Place a leader starting on the roof.

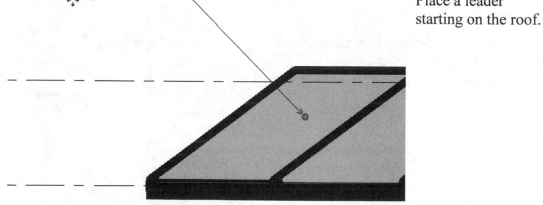

8. Locate the **Asphalt Shingles** material.

Press **OK**.

Cancel out of the command.

Architects are particular about the size of text, and often about the alignment. Notes should not interfere with elevations, dimensions, etc. You can customize the appearance of keynotes.

9. 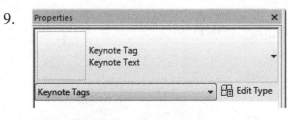 Select the keynote that was placed.

Select **Edit Type** from the Properties panel.

10. 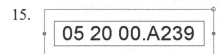 Select **Duplicate**.

11. Enter **Keynote Text - Custom**.

Press **OK**.

12. 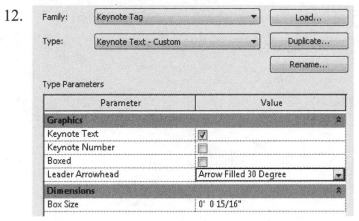 Change the Leader Arrowhead to **Arrow Filled 30 Degree**.

Press **OK**.

13. 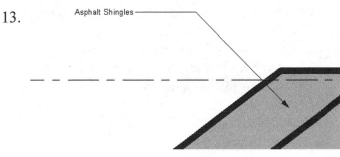 Apply the new type to the keynote.

Notice that the arrowhead changes.

14. 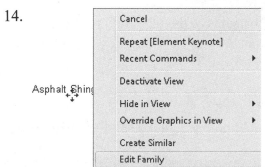 Select the keynote.
Right click and select **Edit Family**.

15. Select the text.

05 20 00.A239

16. Select **Edit Type**.

17. Select **Duplicate**.

18. Change the name to **3/32″ City Blueprint**.

Press **OK**.

Name: | 3/32" City Blueprint |

19.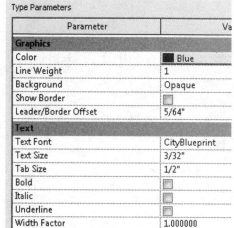

Change the Color to **Blue**.

Change the Text Font to **City Blueprint**.

Press **OK**.

20. Change the label text to use the new label type using the Type Selector.

21. *There are actually two labels overlapping each other. One label is the Keynote number and one is the Keynote text. Move the label above its current position to reveal the label below.*

05 20 00.A239

0

22. Select the keynote text label.

Using the Type Selector, change the label text to use the **3/32" City Blueprint**.

23. 05 20 00.A239

O

The two labels should both use the **3/32" City Blueprint type.**

Move the keynote number label back to its original position overlapping the keynote text label.

24. Save the file as *Keynote Tag CityBlueprint*.

Select **Load into Project and Close** under the Family Editor panel.

25. If this dialog appears:

Place a check next to the project and press **OK**.

26. Return to the **Exterior Elevations** sheet.

Activate the South Elevation view.

27. Select the keynote placed for the roof.

28. Using the Type Selector:

Locate the **Keynote Tag City Blueprint**.

Select the **Keynote Text** type for the Keynote Tag City Blueprint family.

29. 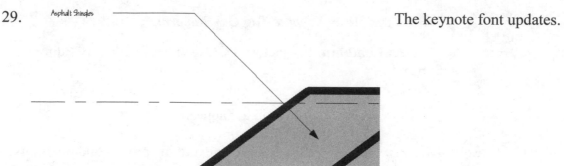   The keynote font updates.

30. 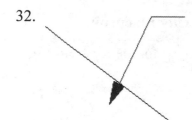 Activate the Manage ribbon.

Object Styles   Select **Object Styles** under Settings.

31.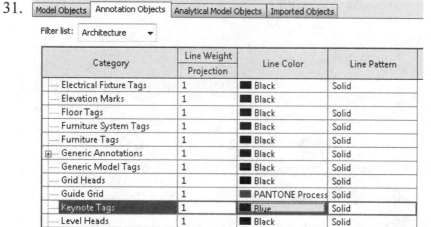

Select the Annotation Objects tab.
Set the Line Color for Keynote Tags to **Blue**.

Press **OK**.

| Model Objects | Annotation Objects | Analytical Model Objects | Imported Objects |

Filter list: Architecture ▼

Category	Line Weight Projection	Line Color	Line Pattern
Electrical Fixture Tags	1	■ Black	Solid
Elevation Marks	1	■ Black	
Floor Tags	1	■ Black	Solid
Furniture System Tags	1	■ Black	Solid
Furniture Tags	1	■ Black	Solid
⊞ Generic Annotations	1	■ Black	Solid
Generic Model Tags	1	■ Black	Solid
Grid Heads	1	■ Black	Solid
Guide Grid	1	■ PANTONE Process	Solid
Keynote Tags	1	■ Blue	Solid
Level Heads	1	■ Black	Solid

32.   Asphalt Shingles   The leader changes to blue.

33. Add additional keynotes.

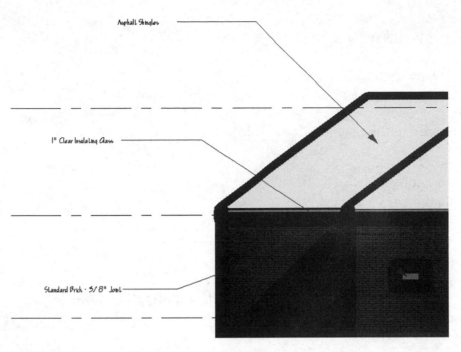

Asphalt Shingles

1" Clear Insulating Glass

Standard Brick - 3/8" Joint

34. Deactivate the view.

35. Save the file *ex8-4.rvt*.

Once you assign an elevation view to a sheet, the elevation marker will update with the sheet number.

Exercise 8-5
Adding Window Tags

Drawing Name: ex8-4.rvt
Estimated Time: 5 minutes

This exercise reinforces the following skills:

- Tag All Not Tagged
- Window Schedules
- Schedule/Quantities
- Schedule Properties

1. 📂   Open *ex8-4.rvt*.

2. 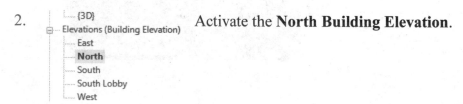   Activate the **North Building Elevation**.

3.   Activate the Annotate ribbon.

    Tag
    All   Select the **Tag All** tool from the Tag panel.

4. 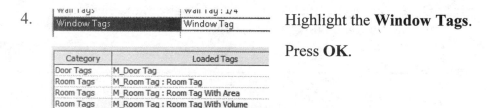   Highlight the **Window Tags**.

    Press **OK**.

5. 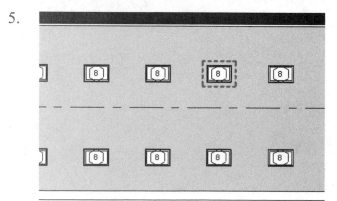   Tags will appear on all the windows.

    The window tag uses the window type as an identifier, not the window mark. So, all windows of the same type will display the same tag number.

6. Save the file as *ex8-5.rvt*.

Exercise 8-6
Changing Window Tags from Type to Instance

Drawing Name: ex8-5.rvt
Estimated Time: 20 minutes

This exercise reinforces the following skills:

    ❏ Family Types
    ❏ Parameters
    ❏ Tags

In Exercise 8-5, we added a window tag that was linked to window type. So, all the tags displayed the same number for all windows of the same type. Some users want their tags to display by instance (individual placement). That way they can specify in their schedule the location of each window as well as the type. In this exercise, you learn how to modify Revit's window tag to link to an instance instead of a type.

1.   🗁 Open *ex8-5.rvt.*

2.   ⊟ Elevations (Building Elevation)   Activate the **North Building Elevation**.
          East
          North
          South
          South - Lobby
          West

3.   Select one of the window tags.

    Make sure you don't select a window.
    You can use the TAB key to cycle through the selections.

The Window Tag will display in the Properties pane if it is selected.

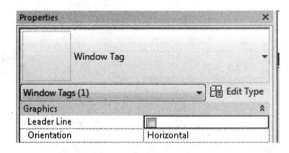

4. Right click and select **Edit Family**.

5. Select the label/text located in the center of the tag.
Select **Edit Label** on the ribbon bar.

6. Highlight the Type Mark listed as the Label Parameters.

This is linked to the type of window placed.

Select the **Remove** button.

7. 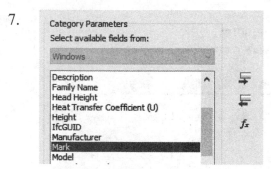 Locate **Mark** in the Category Parameters list.
This parameter is linked to the instance or
individual window.

Select the **Add** button.

8. You should see the Mark parameter listed.

Press **OK**.

9.
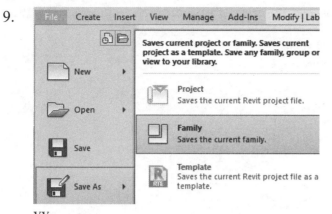

On the Applications Menu:

Go to **File → Save As → Family**.

vv

10.

| File name: | Window Tag - Instance.rfa |
| Files of type: | Family Files (*.rfa) |

Save the file with a new file name – **Window Tag_Instance** – in your class work folder.

11.

Load into Project and Close

Select **Load into Project and Close** to make the new tag available in the building model.

12.

Load into Projects

Check the open Projects/Families you want to load the edited Family into

☑ ex6-5.rvt

If this dialog appears:

Place a check next to the project file.

Press **OK**.

13.
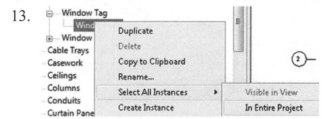

In the Project Browser:
Locate the **Window Tag** under the Annotation Symbols category.

Right click and select **Select All Instances → In Entire Project**.

14.

Properties

Window Tag

Window Tag
  Window Tag
Window Tag - Instance
  Window Tag - Instance

The window tags will be selected in all views.

In the Properties Pane:

Use the Type Selector to select the **Window Tag - Instance**.

15. Activate the North Exterior elevations.

Zoom in to see that the windows are now renumbered using the Mark parameter.

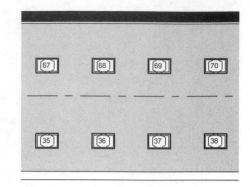

16. Save as *8-6.rvt*.

Exercise 8-7
Creating a Plan Region View

Drawing Name: plan region view.rvt (This can be located in the Class Files downloaded for the text.)

Estimated Time: 20 minutes

Thanks to John Chan, one of my Revit students at SFSU, for this project!

This exercise reinforces the following skills:

- ❑ Plan Region View
- ❑ Split Level Views
- ❑ Linework

Some floor plans are split levels. In those cases, users need to create a Plan Region View in order to create a good floor plan.

1. 📂 Open *plan region view.rvt*.

   If you switch to a 3D view, you see that this is a split-level floor plan.

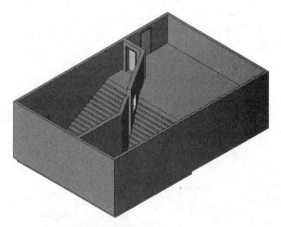

2. Floor Plans
   - **1/F Level 1**
   - 1/F Level 2
   - 2/F
   - Site

   Activate the **1/F Level 1** view under floor plans.

3.　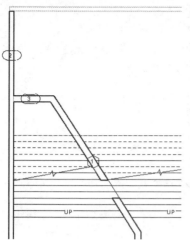　You see the door and window tags, but no doors or windows.

4.　| View |　Activate the **View** ribbon.

5.　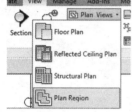　Select **Plan Views** → **Plan Region** tool on the Create panel.

6.　Select the **Rectangle** tool from the ribbon.

7.　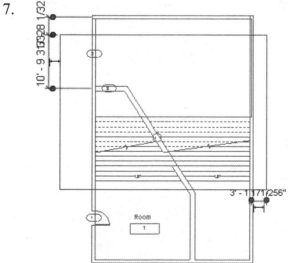　Draw a rectangle around the region where you want the doors and windows to be visible.

Be sure to extend the rectangle across the entire plan or the stairs may not appear properly.

8.　Select the **Edit** button next to View Range on the Properties pane.

9. 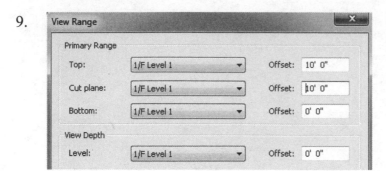 Set the Offset for the Top
plane to **10′ 0″**.
Set the Cut Plane Offset to
10′ 0″.

Press **OK**.

Note: Your dimension values may be different if you are using your own model.

10.  Select the **Green Check** under the Mode panel to **Finish Plan Region**.

11. 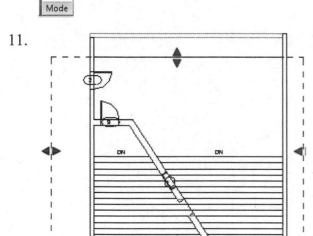 The doors and window are now visible.

12. To turn off the visibility of the Plan Region rectangle in the view, type **VG**.

Select the Annotations tab.
Disable **Plan Region**.
Press **OK**.

13. 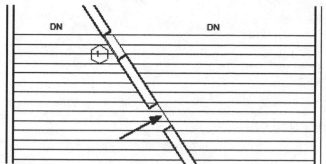 There is a line where the two stairs connect.

We can use the Linework tool to make this line invisible.

14. **Modify** Activate the Modify ribbon.

15. Select the **Linework** tool under the View panel.

16. Select **Invisible Lines** from the Line Style drop-down list.

Line Style:
<Invisible lines>

17. Select the lines you wish to make invisible.

You may need to pick the line more than once as there are overlapping lines.

18. The lines are now invisible.

19. Close without saving.

Exercise 8-8
Creating a Drafting View

Drawing Name: ex8-6.rvt
Estimated Time: 40 minutes

This exercise reinforces the following skills:

- ❑ Detail Components
- ❑ Filled Regions
- ❑ Notes

1.   📂   Open *ex8-6.rvt*.

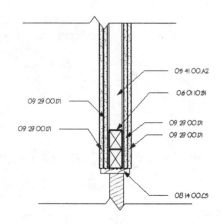

You will create this view of a Head Detail for an interior door using detail components, filled regions, and key notes.

Drafting views are 2D views which are not generated from the Revit model. They are re-usable views which can be used in any project to detail construction views. Once the view is imported or created, you link it to a call out on a model view.

2.   Activate the **View** ribbon.

       Select **Drafting View**.

3.   Type **Head Detail – Single Flush Door** for the name.

Press **OK**.

A blank view is opened.

New Drafting View	
Name:	Head Detail - Single Flush Door
Scale:	1 1/2" = 1'-0"
Scale value 1:	8

4.   Activate the **Insert** Ribbon.

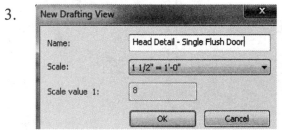

       Select **Load Family**.

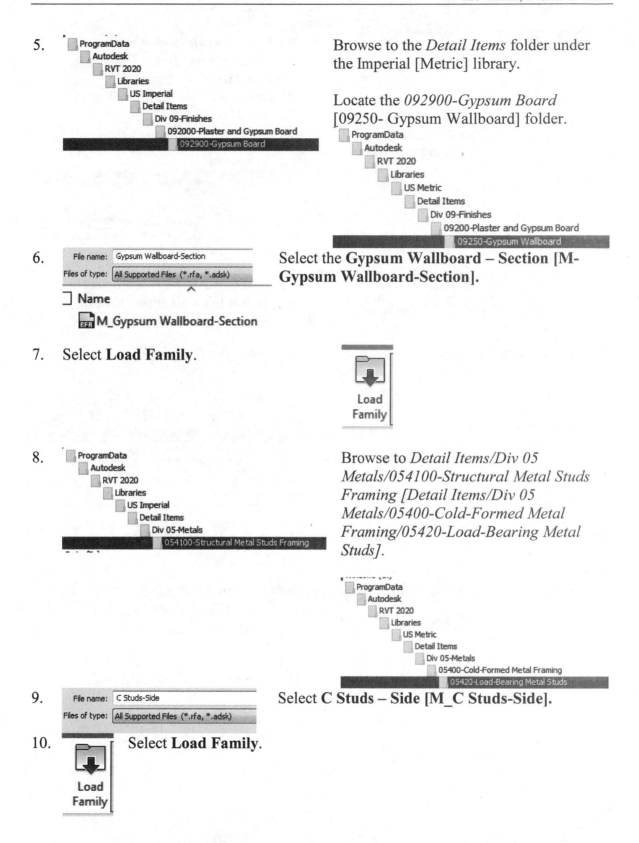

5. Browse to the *Detail Items* folder under the Imperial [Metric] library.

Locate the *092900-Gypsum Board* [09250- Gypsum Wallboard] folder.

6. Select the **Gypsum Wallboard – Section [M-Gypsum Wallboard-Section]**.

7. Select **Load Family**.

8. Browse to *Detail Items/Div 05 Metals/054100-Structural Metal Studs Framing [Detail Items/Div 05 Metals/05400-Cold-Formed Metal Framing/05420-Load-Bearing Metal Studs]*.

9. Select **C Studs – Side [M_C Studs-Side]**.

10. Select **Load Family**.

11.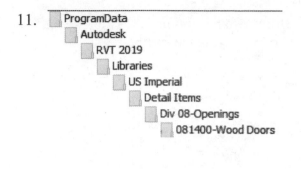

Browse to *Detail Items/Div 08-Openings/081400-Wood Doors [Detail Items/Div 08-Doors and Windows/08200-Wood and Plastic Doors/08210-Wood Doors]*.

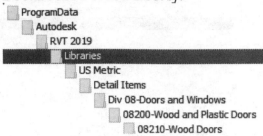

12.

| File name: | Wood Door Frame-Section |
| Files of type: | All Supported Files (*.rfa, *.adsk) |

Select **Wood Door Frame – Section [M_Wood Door Frame – Section].**

When you place the detail components, be sure to use the Type Selector to select the correct size of component to be used.

The structure of the interior wall used in our project:

Layers		EXTERIOR SIDE	
	Function	Material	Thickness
1	Finish 2 [5]	Gypsum Wall Board	0' 0 5/8"
2	Finish 2 [5]	Gypsum Wall Board	0' 0 5/8"
3	**Core Boundary**	**Layers Above Wrap**	**0' 0"**
4	Structure [1]	Metal - Stud Layer	0' 2 1/2"
5	**Core Boundary**	**Layers Below Wrap**	**0' 0"**
6	Finish 2 [5]	Gypsum Wall Board	0' 0 5/8"
7	Finish 2 [5]	Gypsum Wall Board	0' 0 5/8"

13.
Component

Select the **Detail Component** tool from the Annotate ribbon.

14.

Properties ✕

Gypsum Wallboard-Section 5/8"

Select **Gypsum Wallboard – Section 5/8″ [M-Gypsum Wallboard-Section 16 mm]** on the Properties palette.

15. Place two instances of gypsum wall board by drawing two vertical lines 5' 0" tall.

16. Component

Select the **Detail Component** tool from the Annotate ribbon.

17.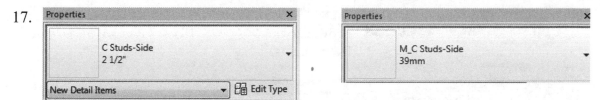

Select **C Studs – Side 2 1/2″ [M_C Studs- Side 39 mm]** on the Properties palette.

18. Draw the stud layer on the left side of the gypsum sections.

Start on the bottom and end at the top.

19. Add two more layers of 5/8″ gypsum wall board on the left side of the stud.

The detail view should show two layers of 5/8″ gypsum board on either side of the metal stud.

You can also mirror the gypsum wall board to the other side using the midpoint of the stud layer, if you prefer.

20. Select the **Detail Component** tool from the Annotate ribbon.

Component

21. Select the **Load Family** tool from the ribbon.

Load
Family
Mode

22. 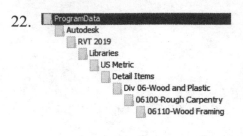 Browse to the *061100 – Wood Framing folder* under Detail Items.

23. 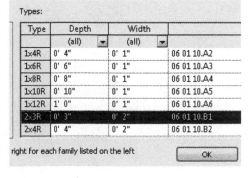 Select **Rough Cut Lumber – Section [M-Rough Cut Lumber - Section]**.

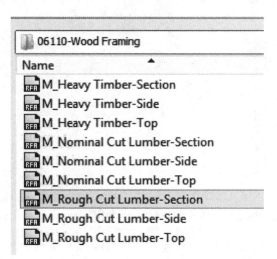

24. Locate the **2x3R [50 x 75mm]** size.

Press **OK**.

Type	Depth (all)	Width (all)	
1x4R	0' 4"	0' 1"	06 01 10.A2
1x6R	0' 6"	0' 1"	06 01 10.A3
1x8R	0' 8"	0' 1"	06 01 10.A4
1x10R	0' 10"	0' 1"	06 01 10.A5
1x12R	1' 0"	0' 1"	06 01 10.A6
2x3R	0' 3"	0' 2"	06 01 10.B1
2x4R	0' 4"	0' 2"	06 01 10.B2

right for each family listed on the left

Type	Depth (all)	Width (all)
25 x 150mm	0' 5 29/32"	0' 0 63/64"
25 x 200mm	0' 7 7/8"	0' 0 63/64"
25 x 250mm	0' 9 27/32"	0' 0 63/64"
25 x 300mm	0' 11 13/16"	0' 0 63/64"
50 x 75mm	0' 2 61/64"	0' 1 31/32"
50 x 100mm	0' 3 15/16"	0' 1 31/32"
50 x 150mm	0' 5 29/32"	0' 1 31/32"

25. 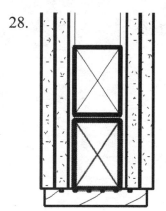 Place the headers between the gypsum board assemblies at the bottom.

26. 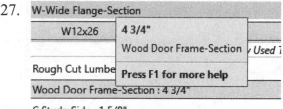 Select the **Detail Component** tool from the Annotate ribbon.

Component

27. Select **Wood Door Frame Section 4 3/4″ [M-Wood Door Frame - Section 110 mm]** from the Properties palette.

28. Place the door frame below the other detail components.

29. Select the **Filled Region** tool.

Region

30. 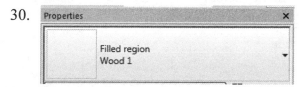 Select **Filled region: Wood 1** from the Properties palette.

31. 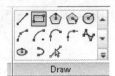 Select **rectangle** from the Draw panel on the ribbon.

32. Draw the rectangle below the door frame.

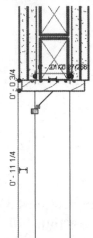

33. Select **Green Check** on the ribbon to exit filled region mode.

34. Place the door using a filled region with a wood fill pattern.

35. Add the keynotes.

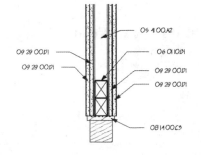

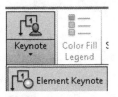

The element keynote tool will be able to identify all the detail components placed.

36.

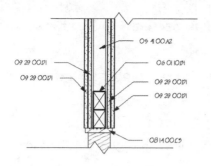

Add the break lines.

Component

Select the Detail Component tool.

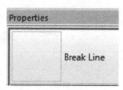

Select **Break Line** from the Properties palette.

37. Save as *ex8-8.rvt*.

Exercise 8-9
Adding a Callout

Drawing Name: ex8-8.rvt
Estimated Time: 10 minutes

This exercise reinforces the following skills:

□ Detail Components
□ Notes
□ Callouts

This exercise assumes that the user has created the drafting view in Ex8-8. Once you have created or imported a drafting view you can link it to the 3D Revit model. Keep in mind that any changes to the model are not reflected in the drafting view, so if you create a drafting view of a wall construction and then change the wall materials in the model, the drafting view does not update.

1. 📂 Open *ex8-8.rvt*.

2. Activate **Level 1**.

3. ⭕ Activate the **View** ribbon.

Section
Select the **Section** tool.

4.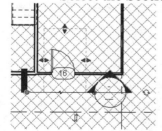
Place a section so it is creating a view of one of the single flush interior doors.

5. ⋮──── West
    ⋮⊟── Sections (Building Section)   Activate **(Building Section) Section 1** in the browser.
    ⋮   └── Section 1

6. ⊟── Sections (Building Section)   Rename the section view **Single Flush Door**
    ⋮   └── Single Flush Door   **Elevation**.

7.   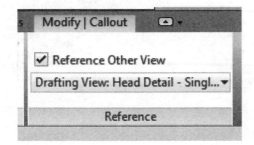   Adjust the crop region for the view.

8. On the View ribbon, select the **Callout** tool on the Create panel.

    Callout

9. On the Options bar:

    Enable **Reference other view**:
    Select the **Drafting View: Head Detail -
    Single Flush Door**.

10. 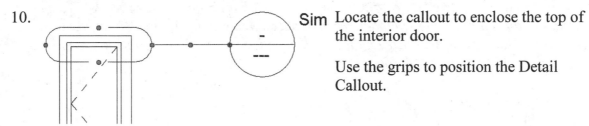   Sim   Locate the callout to enclose the top of
                                   the interior door.

                                   Use the grips to position the Detail
                                   Callout.

    *The detail number and sheet number will be filled out when you place the view on a
    sheet automatically.*

11. Double click on the callout bubble to open the detail view.

12. Save as *ex8-9.rvt*.

Exercise 8-10
Adding a Detail to a Sheet

Drawing Name: ex8-9.rvt
Estimated Time: 15 minutes

This exercise reinforces the following skills:

- Views
- Sheets
- Schedules
- Keynote Legends

This exercise assumes that the user has created the drafting view in Ex8-8.

1.   Open *ex8-9.rvt*.

2.   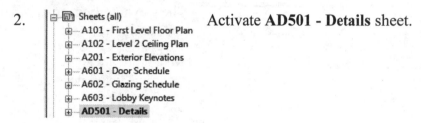   Activate **AD501 - Details** sheet.

3.   Add the Head Detail onto the sheet.

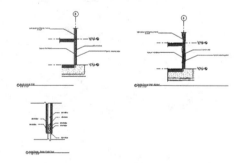

4.   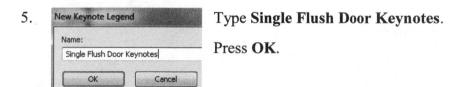   Activate the View ribbon.

Select **Legend → Keynote Legend**.

5.   Type **Single Flush Door Keynotes**.

Press **OK**.

6.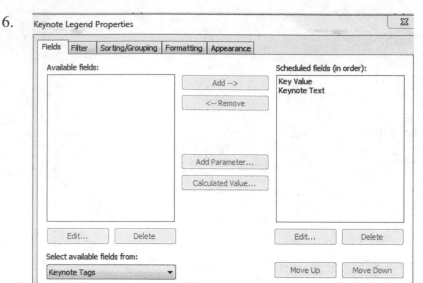

Press **OK**.

7.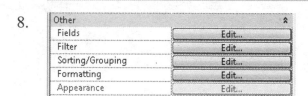

The Keynote Legend appears in the view.

Note there are some materials listed we don't need to call out for the door. All keynotes which have been added in the project are listed.

8.

Other	
Fields	Edit...
Filter	Edit...
Sorting/Grouping	Edit...
Formatting	Edit...
Appearance	Edit...

Select the **Edit** button next to Filter.

9.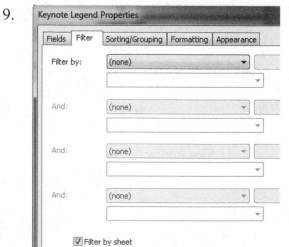

Enable **Filter by sheet** at the bottom of the dialog.

Press **OK**.

10. 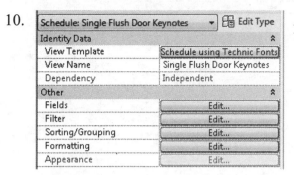 Apply the **Schedule using Technic Fonts** View Template.

11.
- ⊟ 🗐 Sheets (all)
  - ⊞ A101 - First Level Floor Plan
  - ⊞ A201 - Exterior Elevations
  - ⊞ A601 - Door Schedule
  - ⊞ A602 - Glazing Schedule
  - ⊞ A603 - LOBBY KEYNOTES
  - ⊞ **AD501 - Details**
- ⊟ 🗐 Families

Activate **AD501 - Details** sheet.

12.
- ⊟ 🗐 Legends
  - └ FINISH SCHEDULE KEYS
  - └ Single Flush Door Keynotes

Locate the Keynote Legend for the Single Flush Door.

13.

SINGLE FLUSH DOOR KEYNOTES	
KEY VALUE	KEYNOTE TEXT
05 41 00.A2	2-1/2" METAL STUD
06 01 10.B1	2x3R
08 14 00.C3	WOOD DOOR FRAME
09 29 00.D1	5/8" GYPSUM WALLBOARD

Drag and drop it onto the Details sheet.

Notice that the legend automatically changes to reflect the keynotes on the view on the sheet.

Legend views can be placed on multiple sheets.

14. ③ Head Detail - Single Flush Door
1 1/2" = 1'-0"

Zoom into the Head Detail.

The bubble is filled in with the number 3. This is the detail number.

15.
- 🗐 Schedules/Quantities
  - **Door Details**
  - Door Schedule
  - Glazing Schedule
  - Lobby Keynotes
  - Room Schedule

Activate the **Door Details** schedule.

16. Modify the detail number for the Head Detail for the interior single flush doors.

A	B	C	D		E	F	
			SIZE				
Door No	Door Type	W	H		THK	Head Detail	
1	36" x 84"	3' - 0"	7' - 0"		0' - 2"	HD3	JD4
2	34" x 84"	2' - 10"	7' - 0"		0' - 2"	HD3	JD4
3	34" x 84"	2' - 10"	7' - 0"		0' - 2"	HD3	JD4
4	34" x 84"	2' - 10"	7' - 0"		0' - 2"	HD3	JD4
5	72" x 78"	6' - 0"	6' - 6"		0' - 2"	HD3	JD3
6	72" x 78"	6' - 0"	6' - 6"		0' - 2"	HD3	JD3
7	36" x 84"	3' - 0"	7' - 0"		0' - 2"	HD3	JD4

17. Save as *ex8-10.rvt*.

Exercise 8-11
Importing a Detail View

Drawing Name: ex8-10.rvt
Estimated Time: 20 minutes

This exercise reinforces the following skills:

- ❑ Import CAD
- ❑ Sheets
- ❑ Detail Views

1. 📂   Open *ex8-10.rvt*.
   Activate Level 1.

2. 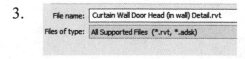   Activate the Insert ribbon.

   Select **Insert from File → Insert Views from File**.

   If your active view is a schedule or legend, you will not be able to insert a view from a file.

3. File name: Curtain Wall Door Head (in wall) Detail.rvt
   Files of type: All Supported Files (*.rvt, *.adsk)

   Locate the **Curtain Wall Door Head (in wall) Detail** file available in the downloaded Class Files. Press **Open**.

4. A preview of the detail will appear.

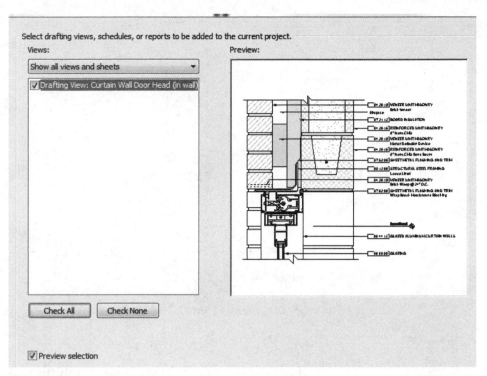

Press **OK**.

5.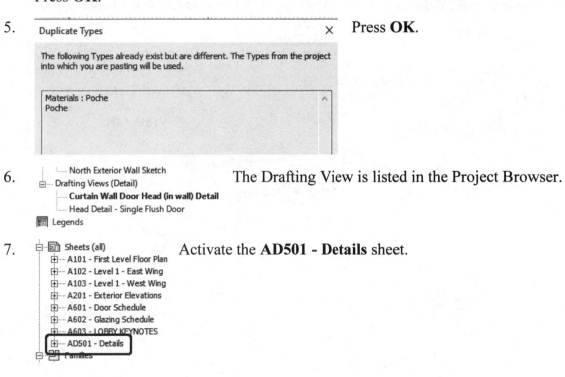

Press **OK**.

6. ⌐ North Exterior Wall Sketch
⊟ Drafting Views (Detail)
⌐ **Curtain Wall Door Head (in wall) Detail**
⌐ Head Detail - Single Flush Door
⊞ Legends

The Drafting View is listed in the Project Browser.

7. ⊟ Sheets (all)
⊞ A101 - First Level Floor Plan
⊞ A102 - Level 1 - East Wing
⊞ A103 - Level 1 - West Wing
⊞ A201 - Exterior Elevations
⊞ A601 - Door Schedule
⊞ A602 - Glazing Schedule
⊞ A603 - LOBBY KEYNOTES
⊞ AD501 - Details
⊟ Families

Activate the **AD501 - Details** sheet.

8.

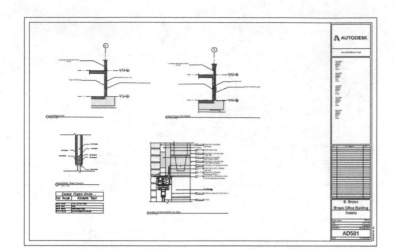

Drag and drop the curtain wall door head view onto the sheet.

9.

Activate the Insert ribbon.

Select **Insert from File→Insert Views from File**.

10.

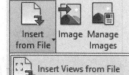

File name: Curtain Wall Door Jamb (in wall) Detail.rvt

Files of type: All Supported Files (*.rvt, *.adsk)

Locate the **Curtain Wall Door Jamb (in wall) Detail** file available in the downloaded Class Files.
Press **Open**.

11.

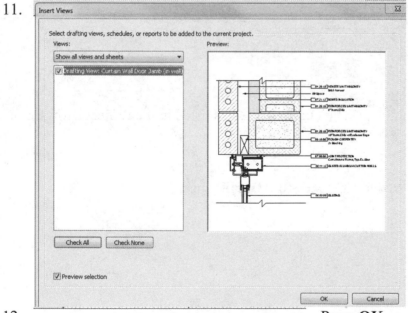

A preview of the detail will appear.

Press **OK**.

12.

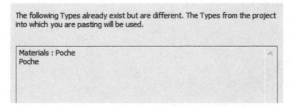

Duplicate Types ✕

The following Types already exist but are different. The Types from the project into which you are pasting will be used.

Materials : Poche
Poche

Press **OK**.

13.
    ┊──── North Exterior Wall Sketch
    ┊─ Drafting Views (Detail)
    ┊   ──── Curtain Wall Door Head (in wall) Detail
    ┊   **Curtain Wall Door Jamb (in wall) Detail**
    ┊   ──── Head Detail - Single Flush Door

The Drafting View is listed in the Project Browser.

14.
    ┊─ Sheets (all)
    ┊   ⊞─ A101 - First Level Floor Plan
    ┊   ⊞─ A102 - Level 2 Ceiling Plan
    ┊   ⊞─ A201 - Exterior Elevations
    ┊   ⊞─ A601 - Door Schedule
    ┊   ⊞─ A602 - Glazing Schedule
    ┊   ⊞─ A603 - Lobby Keynotes
    ┊   ⊞─ **AD501 - Details**

Activate the **AD501 - Details** sheet.

15.

Drag and drop the **Curtain Wall Jamb Detail** view onto the sheet.

16.

Note that the Curtain Wall Head Detail Number is 4.
Note that the Curtain Wall Jamb Detail is 5.

17. Save as *ex8-11.rvt*.

Reassigning a Callout to a new Drafting View

Drawing Name: ex8-11.rvt
Estimated Time: 15 minutes

This exercise reinforces the following skills:

- ❑ Callouts
- ❑ Drafting Views
- ❑ Detail Items
- ❑ Views

Starting in Revit 2015, users have the ability to reassign the link to a callout. In order to reassign a link, the user needed to have enabled Use Referring View.

1.  Open *ex8-11.rvt*.

2.

    Activate **Head Detail – Single Flush Door** detail view.

3.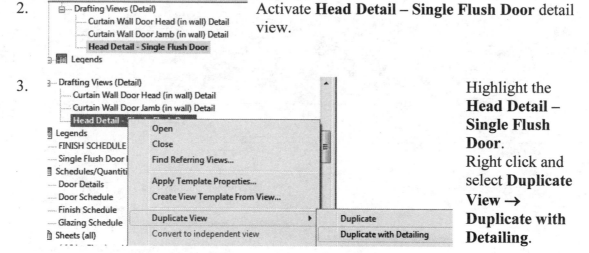

    Highlight the **Head Detail – Single Flush Door**.
    Right click and select **Duplicate View →**
    Duplicate with Detailing.

4.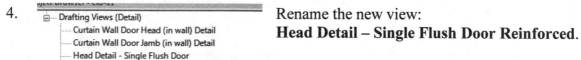

    Rename the new view:
    Head Detail – Single Flush Door Reinforced.

5. Select **Load Family** from the Insert ribbon.

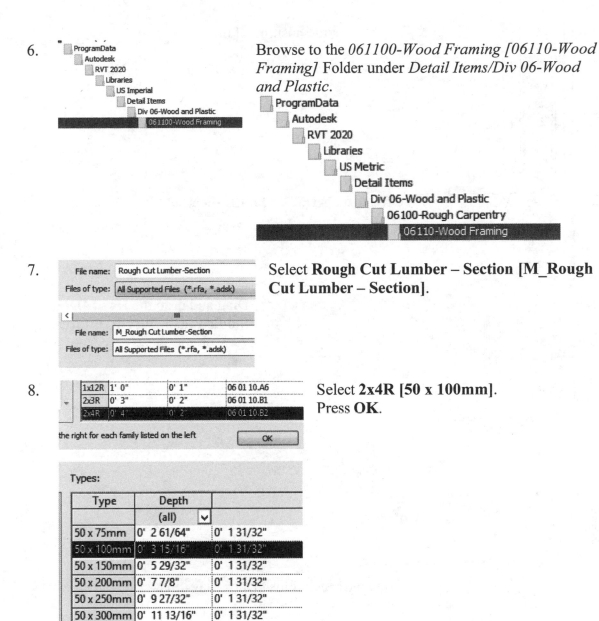

6. Browse to the *061100-Wood Framing [06110-Wood Framing]* Folder under *Detail Items/Div 06-Wood and Plastic*.

7. Select **Rough Cut Lumber – Section [M_Rough Cut Lumber – Section]**.

8. Select **2x4R [50 x 100mm]**.
   Press **OK**.

9. Hold down the CTL key and select the two rough cut sections.

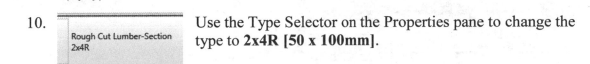

10. Use the Type Selector on the Properties pane to change the type to **2x4R [50 x 100mm]**.

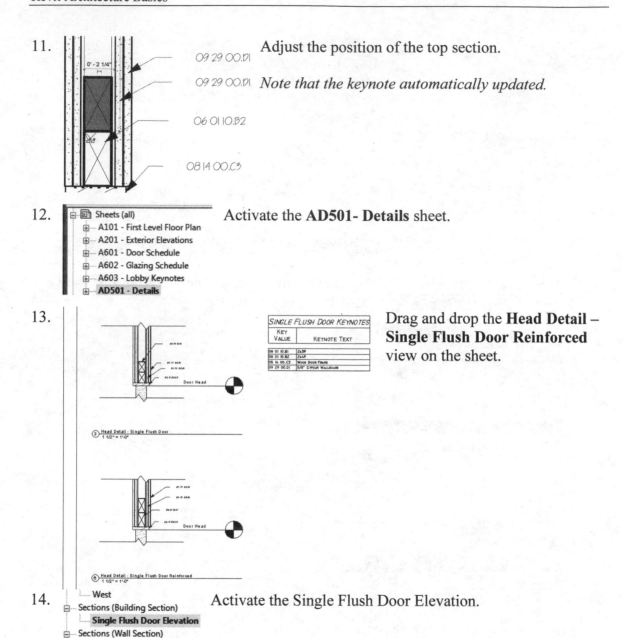

11. Adjust the position of the top section.

Note that the keynote automatically updated.

12. Activate the **AD501- Details** sheet.

13. Drag and drop the **Head Detail – Single Flush Door Reinforced** view on the sheet.

14. Activate the Single Flush Door Elevation.

15. Select the callout in the view.
On the Options bar:
Select the Drafting view for the **Head Detail – Single Flush Door Reinforced**.

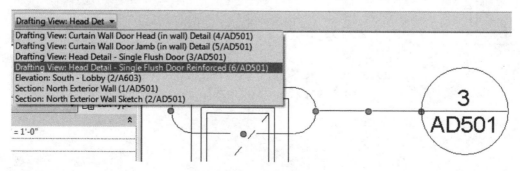

Note that the detail and sheet number are displayed for each view.

16.

The callout should update with the new view number.

17. Save as *ex8-12.rvt*.

Exercise 8-13
Using a Matchline

Drawing Name: ex8-12.rvt
Estimated Time: 30 minutes

This exercise reinforces the following skills:

- ❑ Matchline
- ❑ Sheets
- ❑ Views
- ❑ View Reference Annotation

1. Open *ex8-12.rvt*.

2. 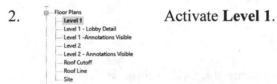   Activate **Level 1**.

3. Select the Grid tool from the Architecture ribbon.

4. On the Options ribbon, set the Offset to **12' 6"** **[3810mm]**.

5.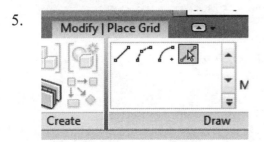

Select Pick Line mode on the Draw panel on the ribbon.

6. ─────────────── Place a grid line between Grids D & E.

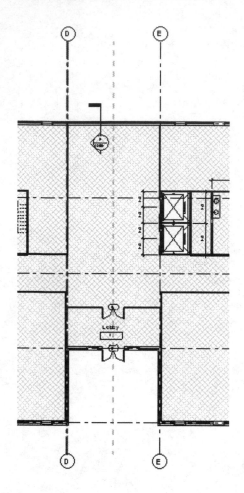

7. Rename the Grid bubble **D.5.**

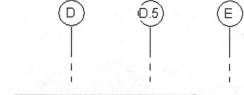

8.

Duplicate View ▶	Duplicate
Convert to independent view	Duplicate with Detailing
Apply Dependent Views...	Duplicate as a Dependent

Right click on **Level 1** in the browser.
Select **Duplicate View →**
Duplicate with Detailing.

9. ⊟ Floor Plans
   ── Level 1
   ── **Level 1 - East Wing**

Rename the view **Level 1-East Wing**.

10. 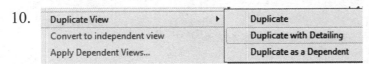 Right click on **Level 1** in the browser.
Select **Duplicate View →**
Duplicate with Detailing.

11. Rename the view **Level 1-West Wing**.

─ Floor Plans
├─── Level 1
├─── Level 1 - East Wing
├─── Level 1 - Lobby Detail
├─── **Level 1 - West Wing**

12. Activate the view **Level 1-West Wing** floor plan.

Extents	
Crop View	☑
Crop Region Visible	☑
Annotation Crop	☐

In the Properties pane:
Enable **Crop View**.
Enable **Crop Region Visible**.

13.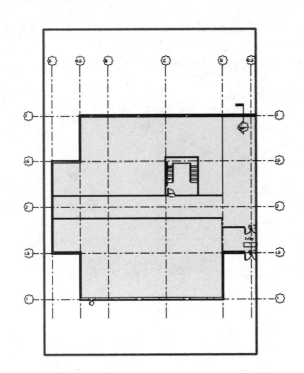

Use the grips on the crop region
to only show the west side of
the floor plan.

14.
─ Floor Plans
├─── Level 1
├─── **Level 1 - East Wing**
├─── Level 1 - Lobby Detail
├─── Level 1 - West Wing

Activate **Level 1 - East Wing**.

15.
Extents	
Crop View	☑
Crop Region Visible	☑
Annotation Crop	☐

In the Properties pane:
Enable **Crop View**.
Enable **Crop Region Visible**.

16.

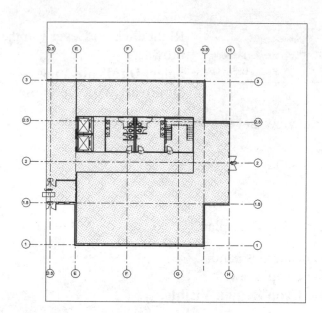

Use the grips on the crop region to only show the east side of the floor plan.

17. | View | Activate the **View** ribbon.

18. 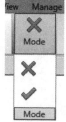 Under the Sheet Composition panel:

Select the **Matchline** tool.

19.

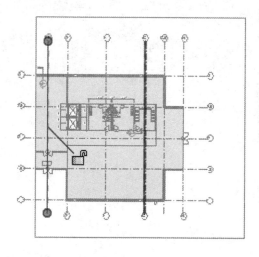

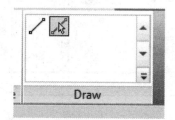

Activate Pick Line mode.

Select Grid D.5.

20. Select the **Green Check** under Mode to finish the matchline.

21. 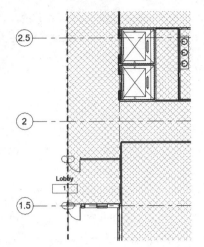 Adjust the crop region so that you see the match line.

22. | Extents | |
    | --- | --- |
    | Crop View | ☑ |
    | Crop Region Visible | |
    | Annotation Crop | |

    Disable **Crop Region Visible** in the Properties pane.

23. ┌ Level 1 - Lobby Detail
    │ **Level 1 - West Wing**
    └ Level 1- East Wing

    Activate the **Level 1 - West Wing** floor plan.

24. You should see a match line in this view as well.

25. | Extents | |
    | --- | --- |
    | Crop View | ☑ |
    | Crop Region Visible | |
    | Annotation Crop | |

    Disable **Crop Region Visible** in the Properties pane.

26. Add a new **Sheet** using the Sheet Composition panel on the View ribbon.

    Press **OK** to accept the default title block.

27. Place the **Level 1-East Wing** floor plan on the sheet.

    Adjust the scale so it fills the sheet.

    Adjust the grid lines, if needed.

    Turn off the visibility of the elevation markers.

28.

Sheet: Level 1 - East Wing	
Current Revision	
Approved By	M. Instructor
Designed By	J. Student
Checked By	M. Instructor
Drawn By	J. Student
Sheet Number	A102
Sheet Name	Level 1 - East Wing
Sheet Issue Date	01/25/16

Change the Sheet Number to **A102**.
Name the sheet **Level 1 - East Wing**.

Enter in your name and your instructor's name.

29. Add a new **Sheet** using the Sheet Composition panel on the View ribbon.

Press **OK** to accept the default title block.

30.

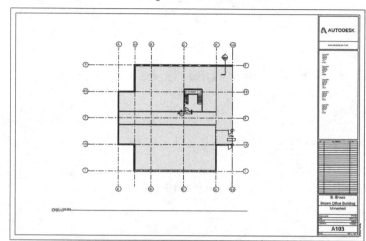

Place the **Level 1- West Wing** floor plan on the sheet.

Adjust the scale so it fills the sheet.

Turn off the visibility of the elevation markers.

31.

Approved By	M. Instructor
Designed By	J. Student
Checked By	M. Instructor
Drawn By	J. Student
Sheet Number	A103
Sheet Name	Level 1 - West Wing

Change the Sheet Number to **A103**.

Name the sheet **Level 1 -West Wing**.

Enter in your name and your instructor's name.

32.

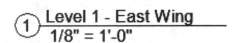

Zoom into the view title bar on the sheet.

Note the view is labeled 1 and is on Sheet A102.

33.

```
Floor Plans
    Level 1
    Level 1 - East Wing
    Level 1 - Lobby Detail
    Level 1 - West Wing
```

Activate the **Level 1** floor plan.

34.

(D.5)

You can see the matchline that was added to the view if you zoom into the D.5 grid line.

35.

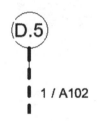

View Reference

Activate the View ribbon.

Select the **View Reference** tool on the Sheet Composition panel.

This adds an annotation to a view.

36.

(D.5)

1 / A102

Place the note on the right side of the grid line to correspond with the view associated with the right wing.

Cancel out of the command.

37.

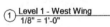
Level 1 - West Wing
1/8" = 1'-0"

Note that the **A103 – Level 1 – West Wing** sheet view is also labeled 1.

38.

```
Sheet    Title Block
View     Revisions
         Guide Grid
    Sheet Composition
```

Activate the View ribbon.

Select the **View Reference** tool on the Sheet Composition panel.

39.

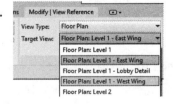

On the ribbon, you can select which view you want to reference.

Select the **Level 1 – West Wing** view.

40.

Place the note to the left of the grid line.

1 / A103   1 / A102

41. Save the file as *ex8-13.rvt*.

Exercise 8-14
Modifying a Crop Region

Drawing Name: ex8-13.rvt
Estimated Time: 15 minutes

This exercise reinforces the following skills:

- Crop Region
- Duplicate View
- Views

1.   📂   Open *ex8-13.rvt*.

2.

Floor Plans
　Level 1
　Level 1 - East Wing
　Level 1 - Lobby Detail
　Level 1 - West Wing
　Level 1 -Annotations Visible
　Level 2
　Level 2 - Annotations Visible
　Roof Cutoff
　Roof Line
　Site

Activate **Level 1**.

3.   Right click on **Level 1** in the browser.

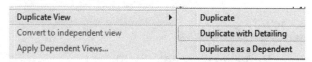

Select **Duplicate View →
Duplicate with Detailing**.

4.

Floor Plans
　Level 1
　Level 1 - East Wing
　Level 1 - Interior Plan
　Level 1 - Lobby Detail
　Level 1 - West Wing

Rename the view **Level 1 – Interior Plan**.

5.

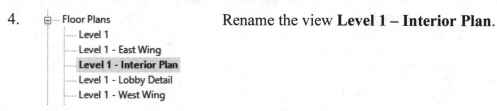

Turn on the visibility of the crop region.

6. 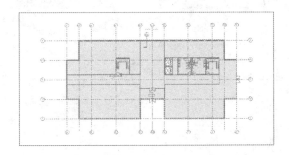 Select the crop region so it highlights.

7. Select **Edit Crop** on the ribbon.

8. The crop region is now a sketch.

9.

Add lines and trim to enclose the area indicated –
The right stairs, the lavatories, the elevators, and the lobby area.

10. Select the **Green check** to finish.

11. Error - cannot be ignored

Crop area sketch can either be empty or include one closed, not self-intersecting loop.

Show   More Info   Expand >>

If you see this error, the sketch is not closed or needs to be trimmed.

Zoom out to see if there are any stray lines.

Press **Continue** and fix the sketch.

12.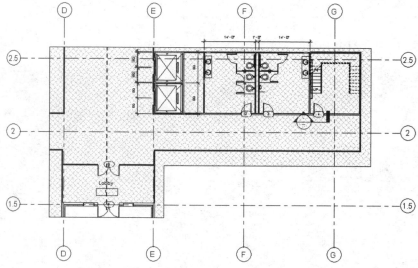

The new view is displayed.

Note that crop region sketches can only use lines, not arcs.

13. 1/8" = 1'-0"   Turn off visibility of the crop region.

14. Save as *ex8-14.rvt*.

Exercise 8-15
Updating a Schedule Using Shared Parameters

Drawing Name: ex8-14.rvt
Estimated Time: 5 minutes

This exercise reinforces the following skills:

- ❑ Sheet List Schedule
- ❑ Sheets

1.   Open *ex8-14.rvt.*

2. 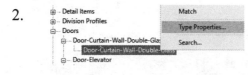 Locate the Curtain Wall-Double-Glass door in the Project Browser.

   Right click and select **Type Properties**.

3. Enter in the correct numbers for the Jamb and Head Details.

   Refer to the Details sheet for the detail view number.

   Press **OK**.

Type Parameters	
Parameter	
Construction	
Function	Exterior
Construction Type	
Head Detail	HD4
Jamb Detail	JD5
Threshold Detail	TD1

4. Schedules/Quantities
   Door Details
   Door Schedule
   Glazing Schedule
   Lobby Keynotes
   Room Schedule

   Activate the **Door Details** schedule.

5. Notice that the schedule has updated.

36" x 84"	3' - 0"	7' - 0"	0' - 2"	HD3	JD4	TD4
36" x 84"	3' - 0"	7' - 0"	0' - 2"	HD3	JD4	TD4
36" x 84"	3' - 0"	7' - 0"	0' - 2"	HD3	JD4	TD4
36" x 84"	3' - 0"	7' - 0"	0' - 2"	HD3	JD4	TD4
12000 x 2290 OPENING	2' - 11 1/2"	7' - 6 1/4"	0' - 1 1/4"	HD3	JD3	TD3
12000 x 2290 OPENING	2' - 11 1/2"	7' - 6 1/4"	0' - 1 1/4"	HD3	JD3	TD3
DOOR-CURTAIN-WALL-DOUBLE-GLASS	6' - 9"	8' - 0"		HD4	JD5	TDI

   Check the detail numbers for the curtain wall.

6. Save as *ex8-15.rvt.*

Exercise 8-16
Create a Sheet List

Drawing Name: ex8-15.rvt
Estimated Time: 20 minutes

This exercise reinforces the following skills:

- ❑   Sheet List Schedule
- ❑   Sheets

1.   📂   Open *ex8-15.rvt*.

2.   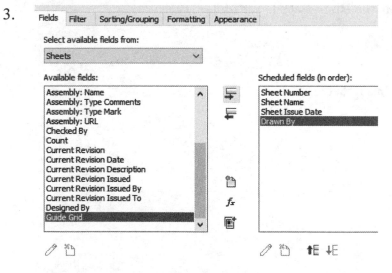   Activate the **View** ribbon.

Select **Schedule→Sheet List**.

3.   Add the following fields in this order:
- •   Sheet Number
- •   Sheet Name
- •   Sheet Issue Date
- •   Drawn By

4.   Select the Sorting/Grouping tab.
Sort by **Sheet Number**.

5.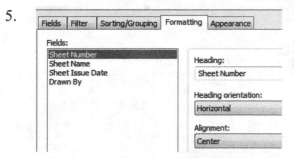

Select the Formatting tab.

Highlight the Sheet Number field.
Change the Alignment to **Center**.

6.

Highlight the Sheet Issue Date.
Change the Heading to **Issue Date**.
Change the Alignment to **Center**.

7.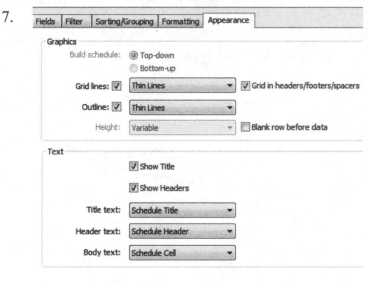

Select the Appearance tab.
Enable Grid lines.
Enable Outline
Enable Grid in headers/
footer/spacers.
Disable blank row before data.
Enable Show Title.
Enable Show Headers.
Change the Title text to
Schedule Title.
Change the Header Text to
Schedule Header.
Change the Body text to
Schedule Cell.
Press **OK**.

8.

A	B	C	D
SHEET NUMBER	SHEET NAME	SHEET ISSUE DATE	DRAWN BY
A101	FIRST LEVEL FLOOR PLAN	05/29/18	J. STUDENT
A102	LEVEL 1 - EAST WING	06/01/18	J. STUDENT
A103	LEVEL 1 - WEST WING	06/01/18	J. STUDENT
A201	EXTERIOR ELEVATIONS	05/31/18	J. STUDENT
A601	DOOR SCHEDULE	05/31/18	J. STUDENT
A602	GLAZING SCHEDULE	05/31/18	J. STUDENT
A603	LOBBY KEYNOTES	05/31/18	J. STUDENT
AD501	DETAILS	05/31/18	J. STUDENT

\<SHEET LIST\>

The Sheet List appears.

This is a good way to check if you added your name as the Author to all sheets.

Can you find a sheet without your name?

9. Activate the **View** ribbon.

 Select N**ew Sheet**.
Select the D-size title block.
Press **OK**.

10. Drag and drop the sheet list schedule on to the sheet.

SHEET LIST			
SHEET NUMBER	SHEET NAME	SHEET ISSUE DATE	DRAWN BY
A101	FIRST LEVEL FLOOR PLAN	05/29/18	J. STUDENT
A102	LEVEL 1 - EAST WING	06/01/18	J. STUDENT
A103	LEVEL 1 - WEST WING	06/01/18	J. STUDENT
A104	UNNAMED	06/01/18	J. STUDENT
A201	EXTERIOR ELEVATIONS	05/31/18	J. STUDENT
A601	DOOR SCHEDULE	05/31/18	J. STUDENT
A602	GLAZING SCHEDULE	05/31/18	J. STUDENT
A603	LOBBY KEYNOTES	05/31/18	J. STUDENT
AD501	DETAILS	05/31/18	J. STUDENT

11.

Approved By	M. Instructor
Designed By	J. Student
Checked By	M. Instructor
Drawn By	J. Student
Sheet Number	0.0
Sheet Name	Cover Sheet
Sheet Issue Date	06/01/18
Appears In Sheet List	☐

Fill in the properties for the sheet.
Sheet Number should be **0.0**.
Sheet Name should be **Cover Sheet**.
Uncheck **Appears in Sheet List**.

12. Save as *ex8-16.rvt*.

Exercise 8-17
Create a PDF Document Set

Drawing Name: ex8-16.rvt
Estimated Time: 10 minutes

Revit comes with a PDF writer called **PDF Complete**.
This exercise reinforces the following skills:

❑   Printing

1.   📂   Open *ex8-16.rvt*.

2. 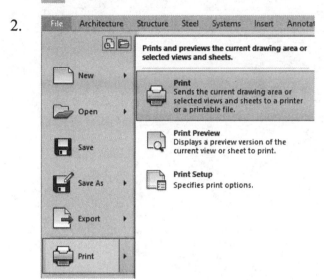   In the Application Menu:

Select **Print→Print**.

3. 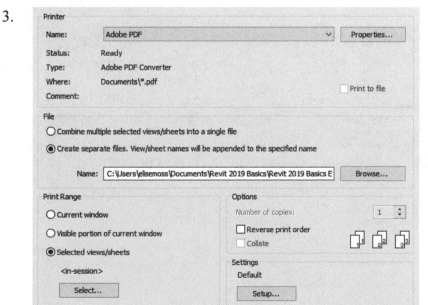   Select **Adobe PDF** as
the printer.

Enable **combine
multiple selected
views/ sheets into a
single file**.

Enable **Selected
views/sheets**.

Press the **Select** button to select which sheets to print.

4.   On the bottom of the dialog, enable **Sheets** only.

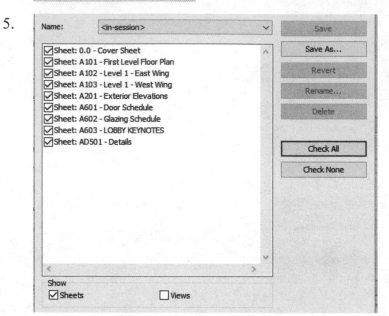

5.   Select **Check All** to select all the sheets in your project.

Press **OK**.

6.   Do you want to save these settings for use in a future Revit session?

Press **No**.

7.   Select the **Setup** next to Settings.

8.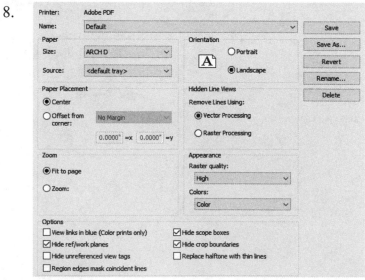

Set the Size to **ARCH D**.
Set the Orientation to **Landscape**.
Set Zoom to **Fit to Page**.
Enable **Center** for Paper Placement.

Press **OK**.

9.   Press **OK** to print.

10.  Browse to the folder where you want to store the pdf file.
Name the file **Bill Brown project**.
Press **Save**.

You may want to save the file with your last name so your instructor can identify your work.

11. 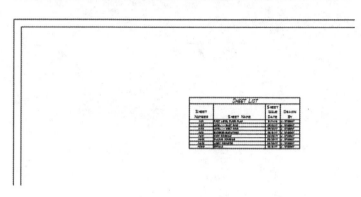 Locate the file and preview it before you turn it in or send it to someone.

Notes:

Additional Projects

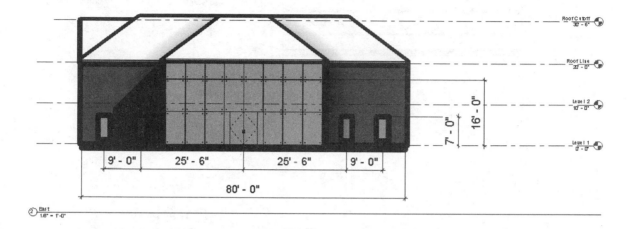

1) Create East & West Elevations view on a sheet.
   Add dimensions to the North Elevation view.

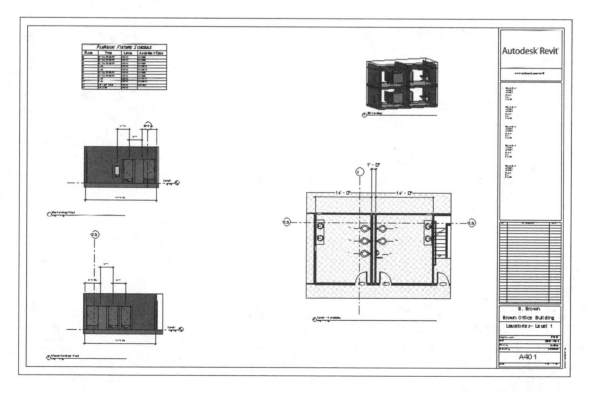

2) Create a sheet for the level 1 Lavatories. The sheet should have a plumbing
   fixture schedule, interior views for the men's and women's lavatories, a cropped
   view of the floor plan and the 3D view as shown.

3) Create Head, Threshold, and Jamb Details for the remaining doors and windows.

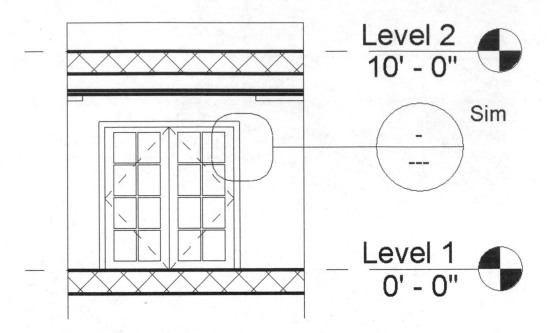

4) Create a sheet with elevations for each door and window.

5) Create a sheet with detail views for the doors and windows.

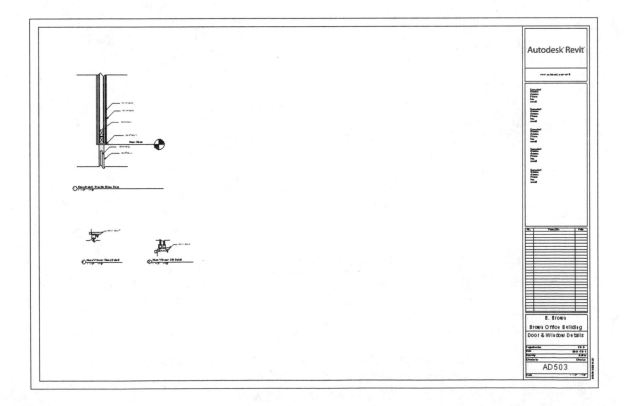

Lesson 8 Quiz

True or False

1. You have to deactivate a view before you can activate a different view.
2. When adding dimensions to an elevation, you add horizontal dimensions, not vertical dimensions.
3. The Linework tool is used to change the line style of a line in a view.
4. If you change the line style of a line in one view, it will automatically update in all views.
5. Double clicking on a blue symbol will automatically activate the view linked to that symbol.

Multiple Choice

6. To add a view to a sheet, you can:

   A. Drag and drop the view name from the browser onto the sheet
   B. Go to View→New→Add View
   C. Select 'Add View' from the View Design Bar
   D. All of the above

7. To add a sheet, you can:

   A. Select the New Sheet tool from the View ribbon
   B. Right click on a sheet and select New Sheet
   C. Highlight Sheets in the Browser, right click and select 'New Sheet'
   D. Go to File→New Sheet

8. To control the depth of an elevation view (visible objects behind objects):

   A. Change to Hidden Line mode
   B. Adjust the location of the elevation clip plane
   C. Adjust the Section Box
   D. Change the View Underlay

9. The keyboard shortcut key for Linework is:

   A. L
   B. LW
   C. LI
   D. LK

10. The text leader option NOT available is:

    A. Leader with no shoulder
    B. Leader with shoulder
    C. Curved Leader
    D. Spline Leader

11. The number 1 in the section symbol shown indicates:

    A. The sheet number
    B. The sheet scale
    C. The elevation number
    D. The detail number on the sheet

12. The values in the section symbol are automatically linked to:

    A. The browser
    B. The sheet where the section view is placed
    C. The text entered by the user
    D. The floor plan

13. To control the line weight of line styles:

    A. Go to Additional Settings→Line Styles
    B. Go to Line Styles→Properties
    C. Pick the Line, right click and select 'Properties'
    D. Go to Tools→Line Weights

14. To re-associate a callout to a new/different detail view:

    A. Delete the callout and create a new callout linked to the new detail view.
    B. Select the callout and use the drop-down list of available drafting views on the Option bar to select a new/different view.
    C. Duplicate the old detail view and then rename it. Name the desired detail view with the name used by the callout.
    D. Activate the desired detail view. In the properties pane, select the Callout to be used for linking.

ANSWERS:

1) T; 2) F; 3) T; 4) T; 5) T; 6) A & B; 7) A & C; 8) B; 9) B; 10) D; 11) D; 12) B; 13) A; 14) B

Lesson 9

Rendering

Rendering is an acquired skill. It takes practice to set up scenes to get the results you want. It is helpful to become familiar with photography as much of the same lighting and shading theory is applicable. Getting the right scene is a matter of trial and error. It is a good idea to make small adjustments when setting up a scene as you often learn more about the effects of different settings.

Exercise 9-1
Create a Toposurface

Drawing Name: 8-16.rvt
Estimated Time: 15 minutes

This exercise reinforces the following skills:

❑ Site
❑ Toposurface

Before we can create some nice rendered views, we need to add some background scenery to our model.

1. 📂   Open *ex8-16.rvt.*

2. 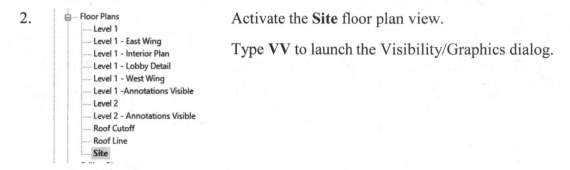   Activate the **Site** floor plan view.

    Type **VV** to launch the Visibility/Graphics dialog.

On the Annotations tab:
Turn off the visibility of grids, elevations, matchline, and sections.

3. 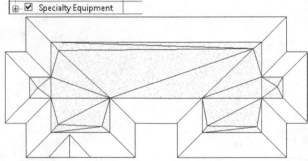 Turn off the visibility of the Project Base Point and the Survey Point (located under Site) on the Model Categories tab.

Press **OK**.

4. The site plan should just display the roof and building.

5. Massing & Site   Activate the **Massing & Site** ribbon.

6. Select the **Toposurface** tool from the Model Site panel.

   Toposurface

7. Place Point   Use the **Place Point** tool from the Tools panel to create an outline of a lawn surface.

8. Pick the points indicated to create a lawn expanse.
   You can grab the points and drag to move into the correct position.

   You are creating a toposurface around the existing building.

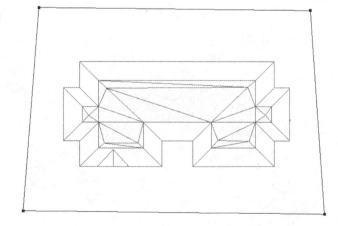

9. Select the **green check** on the Surface panel to **Finish Surface**.

10. 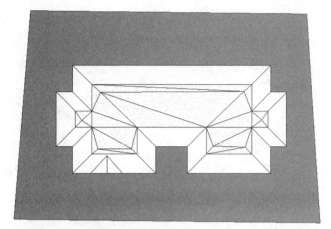 Select the toposurface.

11. 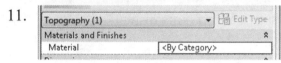 Left click in the **Materials** column on the Properties pane.

12. Locate the **grass** material.

Select and press **OK**.

13. Verify that Grass is now listed as the material for the toposurface.

14. 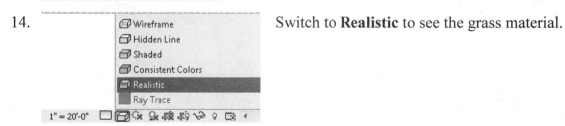 Switch to **Realistic** to see the grass material.

Ray trace is used to test render your view. Do not use this option unless you want to wait a significant amount of time while your system renders.

15. Save as *ex9-1.rvt*.

Exercise 9-2
Create a Split Region

Drawing Name: ex9-1.rvt
Estimated Time: 20 minutes

This exercise reinforces the following skills:

  ❑ Site
  ❑ Toposurface
  ❑ Split a Toposurface

1.   Open *ex9-1.rvt.*

2.   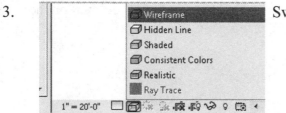 Activate the **Site** floor plan.

3.    Switch to a **Wireframe** display.

4.   Activate the **Massing & Site** ribbon.

     Split Surface
     Select the **Split Surface** tool on the Modify Site panel.

     Select the toposurface.

5.   Select the Pick Line tool from the Draw panel.

6.   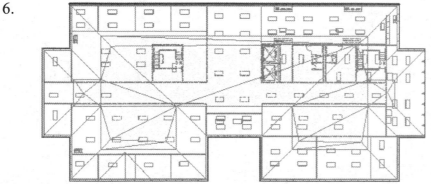 Select the outline of the building.

7.   *If you zoom into the area where the curtain wall ends at the top or bottom of the curtain wall, you will see a slight gap in the outline.*

Use the TRIM tool to close the outline at the top and bottom of the curtain wall.

8.   Left click in the Material field on the Properties pane.

9.   Set the Material to **Concrete - Cast in-Place gray**.

Press **OK**.

10.   Select the **green check** on the Mode panel.

12.   If you get this error message, it means that the sketch is not closed or there are overlapping lines.

Close the dialog and inspect the sketch to correct any errors.

13.   Switch to **Realistic** to see the new material.

14.   If you orbit the 3D model, you will see the building foundation below the lawn.

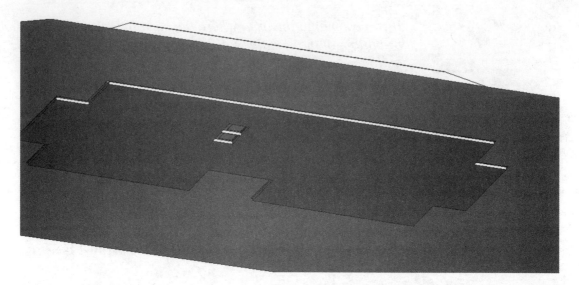

The openings are the shaft openings for the elevators.

15. Save as *ex9-2.rvt*.

Exercise 9-3
Create a Building Pad

Drawing Name: ex9-2.rvt
Estimated Time: 30 minutes

This exercise reinforces the following skills:

- ❑ Site
- ❑ Toposurface
- ❑ Building Pad
- ❑ Materials
- ❑ Importing a Fill Pattern

1.   Open *ex9-2.rvt*.

2.   Roof Cutoff
     Roof Line
     Site   Activate the **Site** floor plan.

3.   Wireframe
     Hidden Line
     Shaded
     Consistent Colors
     Realistic
     Ray Trace
     1" = 20'-0"   Switch to a **Wireframe** display.

4.   Select the **Building Pad** tool on the Model Site panel.
     Building
     Pad

5. 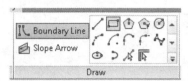 Select the **Rectangle** tool from Draw panel on the ribbon.

6. Use the **Rectangle** tool on the Draw panel to create a sidewalk up to the building.

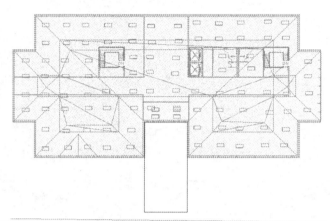

7. Select **Edit Type** on the Properties palette.

8. Select **Duplicate**.

9. Enter **Walkway** in the Name field. Press **OK**.

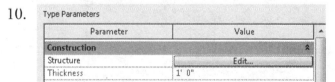

10. Select **Edit** under Structure.

11. Left click in the Material column to bring up the Material Library.

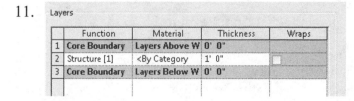

12. 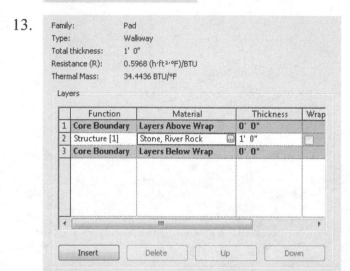 Locate the **Stone, River Rock** material.

Select and press **OK**.

13. Verify that the **Stone, River Rock** material is assigned.

Press **OK**.

14. Left click in the Coarse Scale Fill Pattern field.

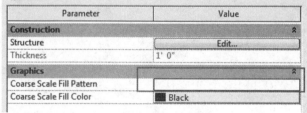

15. 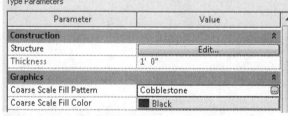 Assign the **Cobblestone** fill pattern.

Press **OK**.

16. Set the Height Offset from Level to **2″** on the Properties palette.

This allows the walkway to be slightly set above the grass and will make it more visible.

17. Select the **green check** on the Surface panel to **Finish Building Pad**.

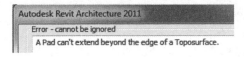

If you see this error, you need to adjust the rectangle sketch so it is entirely on the toposurface.

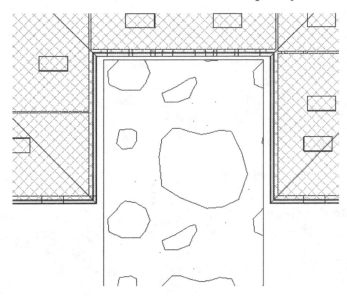

The rectangle is placed slightly away from the building, so that it is entirely on the toposurface, not the foundation.

18. Switch to a **3D** View.

19. Set the display to **Realistic**.

> Wireframe
> Hidden Line
> Shaded
> Consistent Colors
> Realistic
> Ray Trace

1" = 20'-0"

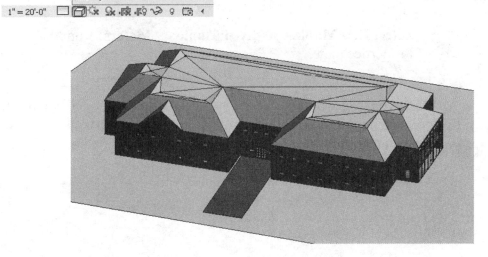

20. Save as *ex9-3.rvt*.

Exercise 9-4
Add Site Components

Drawing Name: ex9-3.rvt
Estimated Time: 15 minutes

This exercise reinforces the following skills:

- Site
- Planting
- Entourage

1.   Open *ex9-3.rvt.*

2.   Activate the **Site** floor plan.

3.   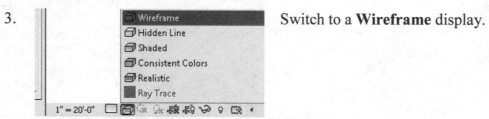 Switch to a **Wireframe** display.

4.   Activate the **Massing & Site** ribbon.

5.   Select the **Site Component** tool on the Model Site panel.

6.   RPC Tree - Deciduous Select **Red Maple - 30′** [**Red Maple - 9 Meters**] from the
    Red Maple - 30′   Properties pane.

When you select the Site Component tool instead of the Component tool, Revit automatically filters out all loaded components except for Site Components.

---

7. Place some trees in front of the building.

   To add plants not available in the Type Selector drop-down, load additional families from the Planting folder.

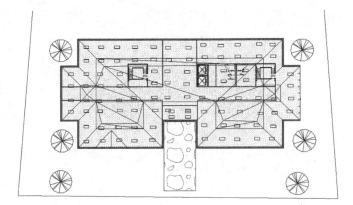

8. 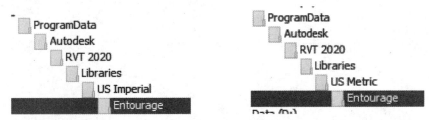 Select the **Site Component** tool on the Model Site panel.

9. Select **Load Family** from the Mode panel.

10. Browse to the *Entourage* folder.

ProgramData	ProgramData
Autodesk	Autodesk
RVT 2020	RVT 2020
Libraries	Libraries
US Imperial	US Metric
Entourage	Entourage

11. File name: RPC Female.rfa   Locate the **RPC Female [M_RPC Female.rfa]** file.

    File name: M_RPC Female.rfa   Press **Open**.

12. RPC Female Cathy   Set the Female to **Cathy** on the Properties pane.

13. 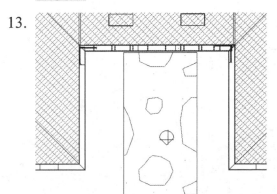   Place the person on the walkway.

14. If you zoom in on the person, you only see a symbol – you don't see the real person until you perform a Render. You may need to switch to Hidden Line view to see the symbol for the person.

The point indicates the direction the person is facing.

15. Rotate your person so she is facing the building.

16. Save the file as *ex9-4.rvt*.

 Make sure Level 1 or Site is active or your trees could be placed on Level 2 (and be elevated in the air). If you mistakenly placed your trees on the wrong level, you can pick the trees, right click, select Properties, and change the level.

Exercise 9-5
Defining Camera Views

Drawing Name: ex9-4.rvt
Estimated Time: 15 minutes

This exercise reinforces the following skills:

- Camera
- Rename View
- View Properties

1. Open *ex9-4.rvt*.

2. Activate **Site** floor plan.

3. Activate the **View** ribbon.

4. 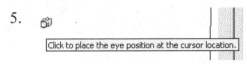 Select the **3D View→Camera** tool from the Create panel on the View ribbon.

5. 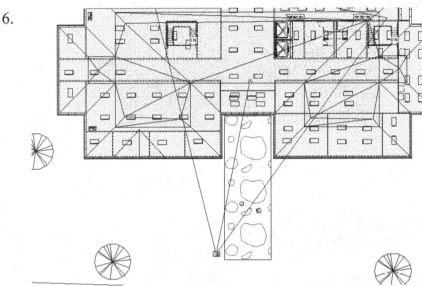 If you move your mouse in the graphics window, you will see a tool tip to pick the location for the camera.

6. Aim the camera towards the front entrance to the building.

7. A window opens with the camera view of your model.

The person and the tree appear as stick figures because the view is not rendered yet.

8.  Change the Model Graphics Style to **Realistic**.

9. Our view changes to a realistic display.

10. | Perspective 🖼🗔⚙️○🐘🔍🐘🔍◇♀☐🗔 | Turn ON Sun and Shadows.

11. The sun will not display because Relative to View is selected in Sun Settings. What would you like to do?

➔ Use the specified project location, date, and time instead.
   Sun Settings change to <In-session, Still> which uses the specified project location, date, and time.

➔ Continue with the current settings.
   Sun Settings remain in Lighting mode with Relative to View selected. Sun will not display.

Cancel

Select **Use the specified project location, date and time instead** to enable the sun.

12.
🗔 Save Orientation and Lock View
🗔 Restore Orientation and Lock View
🗔 Unlock View

Perspective 🖼🗔⚙️○🐘🔍🐘🔍◇♀☐🗔 ◄

If you are happy with the view orientation, select the 3D Lock tool on the bottom of the Display Bar and select **Save Orientation and Lock View**.

13.
Camera	
Rendering Settings	Edit...
Locked Orientation	☐
Projection Mode	Perspective
Eye Elevation	5' 6"
Target Elevation	5' 6"
Camera Position	Explicit

You should see the Locked Orientation enabled in the Properties pane.

14.
⊟ 3D Views
  3D View 1
  3D- Cafeteria area
  3D- Lavatory
  3D- Lobby
  3D- No furniture

If you look in the browser, you see that a view has been added to the 3D Views list.

15. Highlight the **3D View 1**.

Right click and select **Rename**.

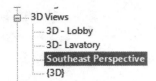

Rename to **Southeast Perspective**.

The view we have is a perspective view – not an isometric view.
Isometrics are true scale drawings. The Camera View is a perspective view with vanishing points. Isometric views have no vanishing points. Vanishing points are the points at which two parallel lines appear to meet in perspective.

16. Save the file as *ex9-5.rvt*.

Additional material libraries can be added to Revit. You may store your material libraries on a server to allow multiple users access to the same libraries. You must add a path under Settings→Options to point to the location of your material libraries. There are many online sources for photorealistic materials; www.accustudio.com is a good place to start.

Take a minute to add more plants or adjust your people before you proceed.

The same view with lighting, additional plants and other items added. Additional families are provided in the exercise files to be used to stage your building.

Many school computers will not have the memory necessary to perform ray tracing. Be prepared to shut down Revit if it hangs up or freezes. Save your work before performing any rendering in case your computer hangs or freezes.

Exercise 9-6
Ray Trace

Drawing Name: ex9-5.rvt
Estimated Time: Processing time depends on your workstation

This exercise reinforces the following skills:

☐   Ray Trace

1.   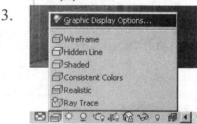   Open *ex9-5.rvt*.

2.   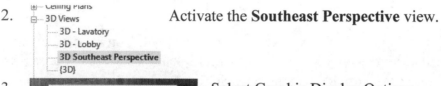   Activate the **Southeast Perspective** view.

3.   Select Graphic Display Options.

4.   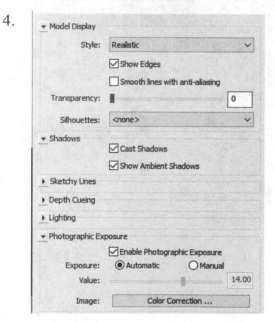   Enable Show Ambient Shadows.
     Adjust the Lighting.
     Enable Photographic Exposure.

5.  Select **Color Correction** button next to Image.

Adjust the sliders and observe all the image changes on the screen.

Press **OK** to close the dialogs.

6.  Select **Ray Trace**.

Ray Trace creates a rendering.

7.  *The idea of ray trace is that you can adjust your view (zoom, pan, orbit) while rendering your model. In order for Ray Trace to work properly, you need a high-performance system.*

Do not select this option if you do not want to wait for the processing.

8. Switch back to **Realistic** mode.

9. Save as *ex9-6.rvt*.

Exercise 9-7
Rendering Settings

Drawing Name: ex9-6.rvt
Estimated Time: 20 minutes

This exercise reinforces the following skills:

- ❑ Rendering
- ❑ Settings
- ❑ Save to Project

You can also assign a specific time, date, location, lighting, and environment to your model.

1.   Open *ex9-6.rvt*.

2.   Activate the **Southeast Perspective** view.

3.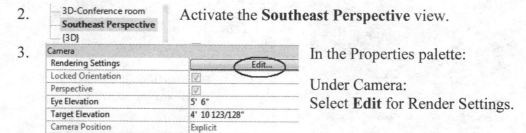

In the Properties palette:

Under Camera:
Select **Edit** for Render Settings.

4.

Set the Quality Setting to **Medium**.

Press **OK**.

5.   Activate the **Manage** ribbon.

Select **Location** on the Project Location panel.

6.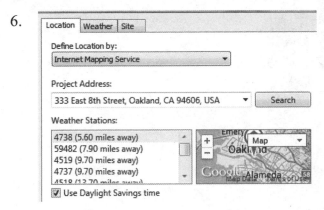

Set the Project Address to **333 E. 8ᵗʰ Street, Oakland, CA.**

The map will update to the location.

Press **OK**.

To use an address, you need to have an internet connection. If you don't have an internet connection, you can select a location from the Default City list.

Note that the shadows adjust to the new location.

7. Perspective | Select the **Rendering** tool located on the bottom of the screen.

8. Lighting — Scheme: Exterior: Sun only — Sun Setting: <In-session, Still> | Select the **...** browse button next to **Sun Setting**.

9. Solar Study — Still / Single Day / Multi-Day / Lighting | Set the Solar Study to **Still**.

10. Settings — Location: 333 E 8th St, Oakland, CA 94 — Date: 6/ 1/2018 — Time: 2:04 PM — Use shared settings | Set the Date and Time.

Press **OK**.

11. Render | Select the **Render** button.

12.    Your window will be rendered.

13. Save to Project...   Select **Save to Project**.

14. 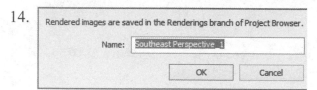   Press **OK** to accept the default name.

Rendered images are saved in the Renderings branch of Project Browser.

Name: Southeast Perspective_1

OK   Cancel

15. Renderings
    Southeast Perspective_1

Under Renderings, we now have a view called **Southeast Perspective_1**.

You can then drag and drop the image onto a sheet.

16. Display
    Show the model

Select the **Show the Model** button.

Close the Rendering dialog.

17. Our window changes to Realistic – not Rendered mode.

18. Save the file as *ex9-7.rvt*.

➤ The point in the person symbol indicates the front of the person. To position the person, rotate the symbol using the Rotate tool.
➤ Add lights to create shadow and effects in your rendering. By placing lights in strategic locations, you can create a better photorealistic image.

Exercise 9-8
Render Region

Drawing Name: ex9-7.rvt
Estimated Time: 15 minutes

This exercise reinforces the following skills:

- ❑ Components
- ❑ Model Text
- ❑ Render Region
- ❑ Load Family

1.   Open *ex9-7.rvt*.

2.   ⊟— Floor Plans
        — Level 1
        — Level 1 - Lobby Detail
        — Level 1 - Interior Plan
        — Level 1-East Wing
        — Level 1-West Wing
        — Level 2
        — Roof Cutoff
        — Roof Line
        — **Site**

        Activate the **Site** plan.

3.  Select **Component→Place a Component** on the Architecture ribbon.

4.   Select **Load Family**.

5. ☐ ProgramData
    ☐ Autodesk
       ☐ RVT 2020
         ☐ Libraries
           ☐ US Imperial
             ☐ Site
               Accessories

    Browse to the *Site/Accessories* folder under the US Imperial library.

6. File name: Building Sign
Files of type: All Supported Files (*.rfa, *.adsk)

    Select the **Building Sign** family.
Press **Open**.

7. Place the Building Sign in front of the building. Cancel out of the command.

8. Activate the **Southeast Perspective** 3D view.

   Site

   Ceiling Plans

   3D Views

      3D- Lavatory

      3D- Lobby

      Southeast Perspective

      {3D}

9. Select the sign in the view and adjust its location.

10. With the sign selected:
Select **Edit Type** in the Properties pane.

11. Change the Sign Text to **Autodesk**.
Press **OK**.

12. The sign updates.
The text on the sign was created using Model Text.

13. Select the **Render** tool in the task bar.

14. Place a check next to Region.

   Render   ☑ Region

15. 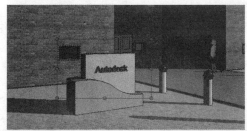 A window will appear in the view. Select the window and use the grips to position the window around the building sign.
Press **Render**.

16.  Just the sign is rendered.

You can use Render Region to check a portion of the view to see how it renders.

Close the Render dialog.

17. Save as ex9-8.rvt.

Exercise 9-9
Space Planning

Drawing Name: ex9-8.rvt
Estimated Time: 20 minutes

This exercise reinforces the following skills:

- ❑ Components
- ❑ Duplicate View
- ❑ View Settings
- ❑ Load Family

1. 📂 Open *ex9-8.rvt*.

2. ⊟ Floor Plans
   └── Level 1
   └── Level 1 - East Wing
   └── Level 1 - Interior Plan
   └── Level 1 - Lobby Detail
   └── **Level 1 - Lobby Space Planning**
   └── Level 1 - West Wing
   └── Level 1 -Annotations Visible

   Select the **Level 1- Lobby Detail** Floor Plan.
   Right click and select **Duplicate View → Duplicate**.

3. Rename the new view **Level 1 - Lobby Space Planning**.

4. ☑ Electrical Fixture Tags
   ☐ Elevations
   ☑ Floor Tags
   ☑ Furniture System Tags
   ☑ Furniture Tags
   ☑ Generic Annotations
   ☑ Generic Model Tags
   ☐ Grids

   Type **VV** to launch the Visibility/Graphics dialog.

   Turn **off** the visibility of the elevation markers, matchlines, and grids.

   Press **OK** to close the dialog.

5. Component

   Select the **Component→Place a Component** tool on the Build Panel from the Architecture ribbon.

6. Load Family   Model In-place
   Mode

   Select **Load** Family from the Mode panel.

7. File name: "Television-Plasma" "Chair-Corbu" "Table-Coffee"
   Files of type: All Supported Files (*.rfa, *.adsk)

   Locate the *Television-Plasma, Chair-Corbu,* and *Table-Coffee* in the exercise files.

8. Hold down the **CTRL** key to select more than one file.

   Press **Open**.

9. Place the chair and table in the room.

   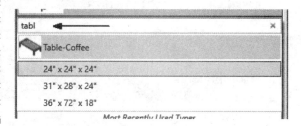

   You can type in the family you are looking for in the Type Selector so you don't have to spend time scrolling down searching for it.

   If you press the SPACE bar before you click to place you can rotate the object.

10. Place the *Television-Plasma* on the East wall using the **Component** tool.

11.

In the Properties pane:
Set the Offset to **3′ 6″**.

This will locate the television 3′ 6″ above Level 1.

12.

Select the **Component→Place a Component** tool on the Build Panel from the Architecture ribbon.

13.

Select **Load** Family from the Mode panel.

14.

Browse to the *Entourage* folder.

15.

Select the **RPC Male** family.
Press **Open**.

16.

Place Dwayne in the lobby.

Rotate him so he is facing the television frame/east wall.

17.

Activate the **Level 1** Ceiling Plan.

18.

Select the **Component→Place a Component** tool on the Build Panel from the Architecture ribbon.

19.

Select **Load** Family from the Mode panel.

20. 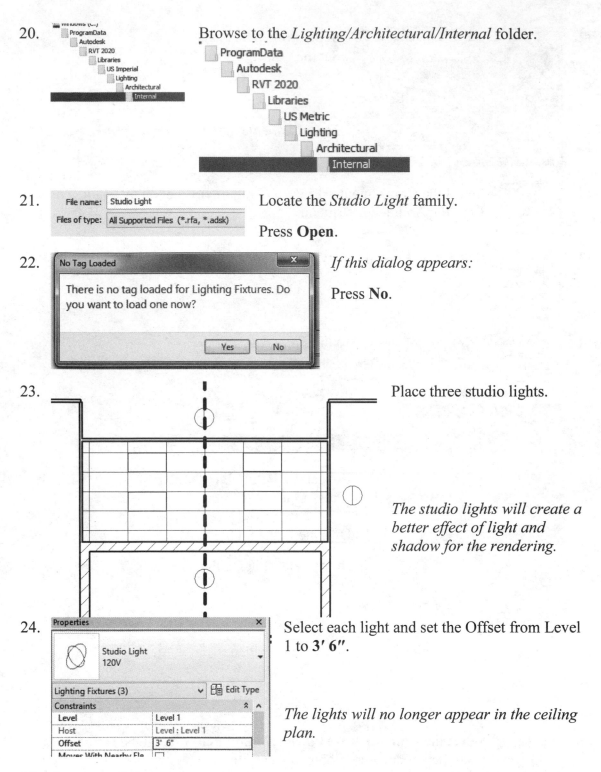 Browse to the *Lighting/Architectural/Internal* folder.

ProgramData
  Autodesk
    RVT 2020
      Libraries
        US Metric
          Lighting
            Architectural
              Internal

21. File name: Studio Light
Files of type: All Supported Files (*.rfa, *.adsk)

Locate the *Studio Light* family.

Press **Open**.

22. **No Tag Loaded**

There is no tag loaded for Lighting Fixtures. Do you want to load one now?

Yes   No

If this dialog appears:

Press **No**.

23. Place three studio lights.

The studio lights will create a better effect of light and shadow for the rendering.

24. **Properties**

Studio Light
120V

Lighting Fixtures (3)   Edit Type

Constraints	
Level	Level 1
Host	Level : Level 1
Offset	3' 6"
Moves With Nearby Ele	☐

Select each light and set the Offset from Level 1 to **3' 6"**.

The lights will no longer appear in the ceiling plan.

25. Save as *ex9-9.rvt*.

Exercise 9-10
Building Sections

Drawing Name: ex9-9.rvt
Estimated Time: 10 minutes

This exercise reinforces the following skills:

- ❑   Section Views
- ❑   Elevations
- ❑   Crop Regions

1.   📂   Open *ex9-9.rvt*.

2.      Activate the **Level 1 - Lobby Space Planning** floor plan.

3.   `View  Manag`   Activate the **View** ribbon.

     Select **Create→Section**.

4.   Properties   Verify **Building Section** is enabled on the Properties palette.

5.   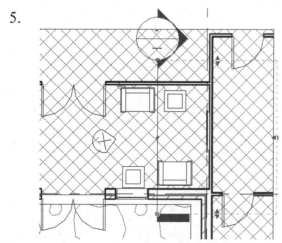   Place a section so it is looking at the east wall with the television.

6. 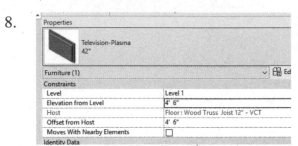 Rename the Section **Lobby East Elevation**.

7. 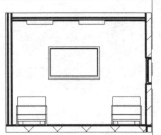 Activate the **Lobby East Elevation** section.

8. 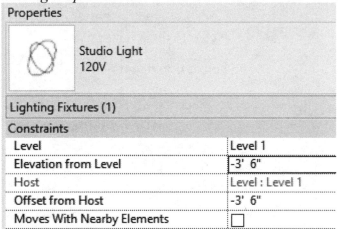 If the television is sitting on the floor, select the television.

On the Properties panel, set the Elevation from Level to 4' 6".

9. Adjust the crop region so only the lobby area is visible.

Note the position of the studio light. Adjust the elevation of the light so it is positioned on the frame. You can move it simply by selecting the light and moving it up or down.

Properties	
Studio Light 120V	
Lighting Fixtures (1)	
Constraints	
Level	Level 1
Elevation from Level	-3' 6"
Host	Level : Level 1
Offset from Host	-3' 6"
Moves With Nearby Elements	☐

10. Save as *ex9-10.rvt*.

Exercise 9-11
Decals

Drawing Name: ex9-10.rvt
Estimated Time: 20 minutes

This exercise reinforces the following skills:

- ❏ Decal Types
- ❏ Decals
- ❏ Set Workplane

1.   Open *ex9-10.rvt*.

2.   Activate the **Lobby East Elevation** Section.

3.   Activate the **Insert** ribbon.

4.   Select **Decal→Decal Types** from the Link panel.

5.   Select **Create New Decal** from the bottom left of the dialog box.

6.   Type a name for your decal.

   The decal will be assigned an image. I have included several images that can be used for this exercise in the downloaded Class Files.

   You can select the image of your choice or use your own image.

   Press **OK**.

7.   Select the Browse button to select the image.

8.  Browse to where your exercise files are located.

9. Select *soccer.jpg* or the image file of your choice.
Press **Open**.

10. 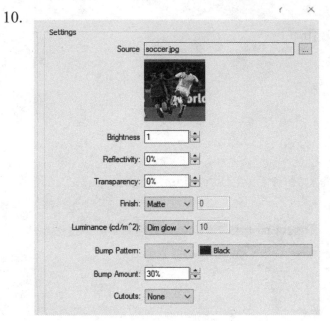 You will see a preview of the image.
Set the Finish to **Matte**.
Set the Luminance **to Dim glow**.

Press **OK**.

11. 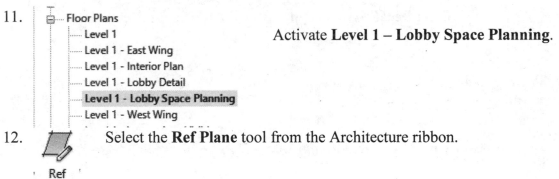 Activate **Level 1 – Lobby Space Planning**.

12. Select the **Ref Plane** tool from the Architecture ribbon.

13. 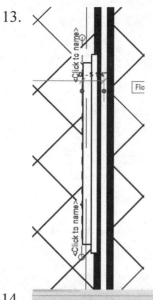 Place a reference plane in front of the television 5.25" from the wall.

14. 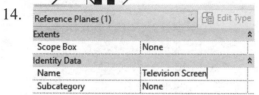 Change the name of the reference plane to **Television Screen** in the Property Pane.

15.  On the Architecture ribbon, select **Set Work Plane.**

16. 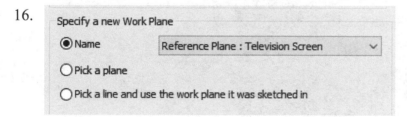 Select the **Television Screen** reference plane from the Name drop-down list.

Press **OK**.

17. Select **Lobby East Elevation** from the dialog.

Press **Open View**.

18. Activate the Insert ribbon.

Select **Decal→Place Decal** on the Link panel.

19. Place the decal on the television frame.

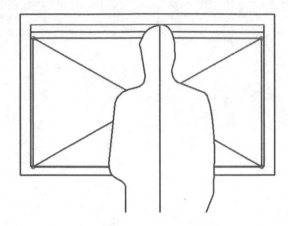

20. On the Options bar:

Disable **Lock Proportions**.
Set the Width to **3′ 7″**.
Set the Height to **2′ 1″**.

21.

Position the decal so it is inside the frame.

22. Save as *ex9-11.rvt*.

Exercise 9-12
Creating a 3D Camera View (Reprised)

Drawing Name: ex9-11.rvt
Estimated Time: 20 minutes

This exercise reinforces the following skills:

- 3D Camera
- Rename View
- Navigation Wheel
- Render

1.   📂   Open *ex9-11.rvt*.

2.      Activate the **Level 1 - Lobby Space Planning** floor plan.

3. | View | Activate the View ribbon.

4.   Select the **3D View→Camera** tool from the Create panel.

5.   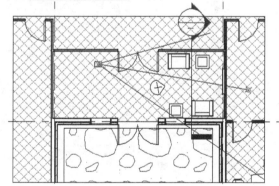   Place the camera in the upper left of the waiting room.

Rotate the camera and target the sitting area of the waiting room.

6.

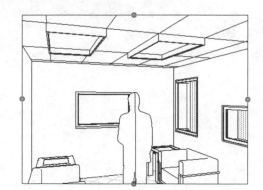

A window will appear with our interior scene.

7.

Highlight the 3D view in the Project Browser.

Right click and select **Rename**.

8.

Rename the view **3D- Lobby Interior**.

9.

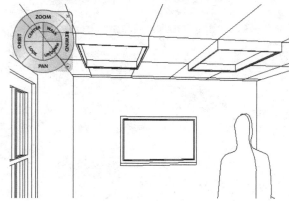

Select **Mini Tour Building Wheel** from View toolbar.

Use the Navigation Wheel to adjust your scene. Try the LOOK and WALK buttons.

10. Adjust the camera view.

You can also use this tool to 'walk through your model'.

11. Click on the Sun on the Display toolbar.
Turn Sun Path ON.

Sun Settings...

Sun Path Off

Sun Path On

Perspective

12. The sun will not display because Relative to View is selected in Sun Settings. What would you like to do?

→ Use the specified project location, date, and time instead.
Sun Settings change to <In-session, Still> which uses the specified project location, date, and time.

→ Continue with the current settings.
Sun Settings remain in Lighting mode with Relative to View selected. Sun will not display.

Select the **Use the specified project location date and time instead** option.

13. Perspective   Turn **Shadows On.**

14. Activate the **Rendering** dialog.

15. 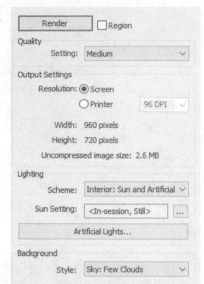 Set the Quality Setting to **Medium**.

Set the Lighting Scheme to **Interior: Sun and Artificial Light**.

Press **Render**.

16. The interior scene is rendered.

A bright light coming from the south window and the door is from the studio light. Repositioning the light might make the rendering better.

17. Adjust Exposure...   Select the **Adjust Exposure** button on the Rendering Dialog.

18.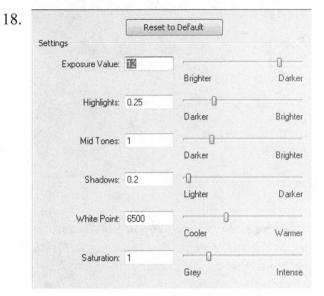

If necessary:

Adjust the controls and select **Apply**.

Readjust and select **Apply**.

Press **OK**.

19.   Press **Save to Project**.

20. 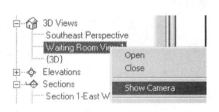 Press **OK**.

21. Close the Rendering Dialog.

22. Save the file as *ex9-12.rvt*.

Cameras do not stay visible in plan views. If you need to re-orient the camera location, you can do so by activating your plan view, then right clicking on the name of the 3D view you want to adjust and selecting the 'Show Camera' option.

Exercise 9-13
Rendering Using Autodesk 360

Drawing Name: ex9-12.rvt
Estimated Time: 5 minutes

This exercise reinforces the following skills:

- ❑ Autodesk 360
- ❑ Rendering

If you have an Autodesk 360 account, you can render for free (if you are a student) or a limited number of renderings for free if you are a professional. By rendering using the Cloud, you can continue to work in Revit while the Rendering is being processed. Creating an Autodesk 360 account is free. An internet connection is required to use Autodesk 360 for rendering.

1.   Open *ex9-12.rvt.*

2. 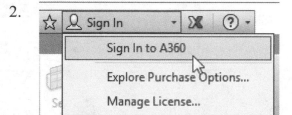   Click the **Sign In** button next to Help.

If you are already signed in, you can skip these steps.

3.    Sign in using your name and password.

4.   If sign-in is successful, you will see your log-in name in the dialog.

5.  Activate the **3D Lobby Interior** view.

3D Views
 3D Lobby Interior
 3D- Cafeteria area
 3D- Lavatory
 3D- Lobby
 3D- No furniture
 3D-Conference room
 Southeast Perspective
 {3D}

6. Activate the **View** ribbon.

Select **Render in Cloud**.

er Render
in Cloud

7. Press **Continue.**

Welcome to Rendering in the Cloud for Revit 2020

To begin rendering in the cloud:

Step 1
On the next screen, select 3D views and start rendering.

Step 2
The service will notify you when your images are ready.

Step 3
Select "View > Render Gallery" to view, download, or do more with your completed renderings online.

☐ Don't show this message next time   Continue

8. AUTODESK. RENDERING

Select 3D views to render in the Cloud

3D View	3D -Lobby Interior
Output Type	Still Image
Render Quality	Final
Image Size	Medium (1 Mega Pixel)
Exposure	Advanced
File Format	JPEG (High Quality)

☐ Alpha (Transparency Background)

1 credits required   Render

Estimated wait time ⏱ <10 minutes ⓘ

Set the Render Quality to **Final**.
Set the Image Size to **Medium**.
Set the Exposure to **Advanced**.
Set the File Format to **JPEG (High Quality)**.

Press **Render**.

Notice that you can enable email notification for when your rendering is completed.

9. smoss@peralt... ▾ ✗ ? ▾
    Sign Out
    Account Details
    View Completed Renderings

When the rendering is done, select **View Completed Renderings** in the drop-down under your sign-in name.

10.

The rendering will be displayed.
To view a larger image, simply click on the thumbnail.

11. Select **Download** from the menu to download the file.

    Download image as PNG.

    Note the editing tools available inside Autodesk 360 for your renderings.

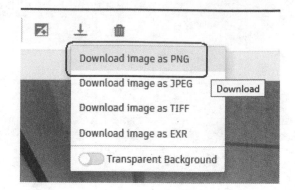

12. Save as *ex9-13.rvt*.

Exercise 9-14
Placing a Rendering on a Sheet

Drawing Name: ex9-13.rvt
Estimated Time: 5 minutes

This exercise reinforces the following skills:

- View Properties
- View Titles

1. Open *ex9-13.rvt*.

2. 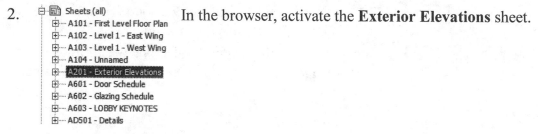   In the browser, activate the **Exterior Elevations** sheet.

3. 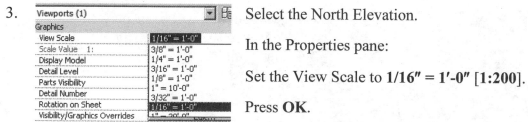   Select the North Elevation.

    In the Properties pane:

    Set the View Scale to **1/16″ = 1′-0″** [1:200].

    Press **OK**.

The scale on the South Elevation is set in the view template.

4.   Activate the **South** building elevation.

   - Elevations (Building Elevation)
       - East
       - North
       - **South**
       - South - Lobby
       - West

5.   Select the **View Template** in the Properties panel.

Sun Path	☐
Identity Data	⌃
View Template	Architectural Presentation Elevation
View Name	South
Dependency	Independent
Title on Sheet	
Sheet Number	A201

6.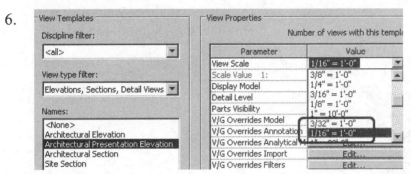

   Change the View Scale to **1/16″ = 1′-0″**.

   Press **Apply** and **OK**.

7.   Activate the **Exterior Elevations** sheet.

   - Sheets (all)
       - A101 - First Level Floor Plan
       - A102 - Level 1 - East Wing
       - A103 - Level 1 - West Wing
       - **A201 - Exterior Elevations**
           - Elevation: North
           - Elevation: South

8.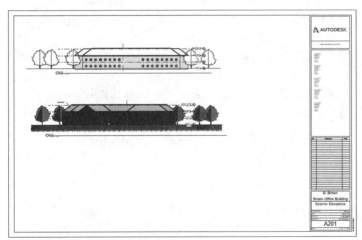

   Zoom to Fit.

   Shift the elevation views to the left to make room for an additional view.

   Note you see a toposurface in the South elevation view because one was added for the exterior rendering.

9.   In the browser, locate the **Southeast Perspective** view under *Renderings*.

   - Renderings
       - 3D - Lobby Interior_1
       - Southeast Perspective_1
   - Drafting Views (Detail)

   Drag and drop it onto the sheet.

10.

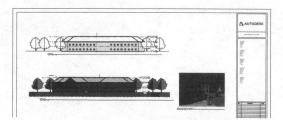

You see how easy it is to add rendered views to your layouts.

11.

To adjust the length of the view title, click on the view. This will activate the grips and allow you to adjust the length and location of the title.

12.

Select the rendering image that was just placed on the sheet.

13.

In the Properties pane:

Select **Edit Type**.

14.

Note that the current type of Viewport shows the title.

Select **Duplicate**.

Parameter	Value
Graphics	
Title	View Title
Show Title	Yes
Show Extension Line	☑
Line Weight	1
Color	■ Black
Line Pattern	Solid

Family: System Family: Viewport   Load...

Type: Viewport 1   Duplicate...   Rename...

Type Parameters

15.

Name the new Viewport type: **Viewport-No Title No Line**.

Press **OK**.

16. 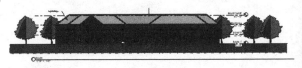 SetView Title to **None**.
Uncheck **Show Extension Line**.

Set Show Title to **No**.

Press **OK**.

17. The rendering viewport has no title or line.

The elevation viewport has a title and line.

18. Activate the **Lobby Keynotes** sheet.

```
Sheet List
Sheets (all)
  0.0 - Cover Sheet
  A101 - First Level Floor Plan
  A102 - Level 1 - East Wing
  A103 - Level 1 - West Wing
  A201 - Exterior Elevations
  A601 - Door Schedule
  A602 - Glazing Schedule
  A603 - LOBBY KEYNOTES
  AD501 - Details
```

19. Activate the **Insert** ribbon.

Image   Select the **Image** tool.

20. Locate the rendering that was downloaded from the Autodesk 360 account.

| File name: | ex9-12.rvt_2019-Jun-02_03-01-13PM-000_3D_-_Lobby_Interior_png.png |
| Files of type: | All Image Files (*.bmp, *.jpg, *.jpeg, *.pdf, *.png, *.tif) |

Press **Open**.

There is also an image file located in the exercise files.

21. Place the image on the sheet and resize.

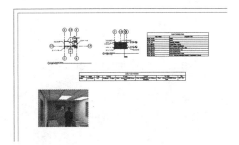

Note that it does not have a viewport associated with it because it is not a view. It is an image file.

22. Save the file as *ex9-14.rvt*.

You can free up memory resources for rendering by closing any windows you are not using. Windows with 3D or renderings take up a considerable amount of memory.

Exercise 9-15
Placing a Path for a Walkthrough

Drawing Name: i_Urban_House.rvt
Estimated Time: 10 minutes

This exercise reinforces the following skills:

❏ Creating a Walkthrough view

1. File name: i_Urban_House.rvt   Locate the *i_Urban_House.rvt* file. This file is included in the Class Files download. Select **Open**.

2. Floor Plans / **FIRST FLOOR** / GROUND FLOOR   Activate the **First Floor** floor plan view.

3. View   Activate the **View** ribbon.

4. Select the **Walkthrough** tool under the 3D View drop-down.

5. 

Verify that **Perspective** is enabled in the Status Bar.

This indicates that a perspective view will be created.
The Offset indicates the height of the camera offset from the floor level.

6. 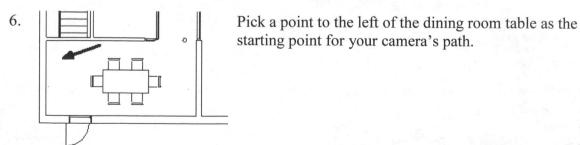   Pick a point to the left of the dining room table as the starting point for your camera's path.

7. 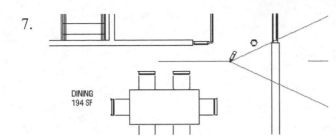   Place a second point at the start of the hallway.

8. 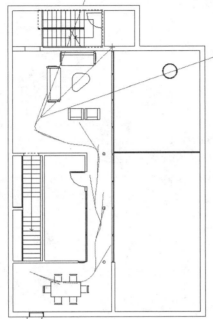   Place points at the locations indicated by the arrows to guide your walkthrough.

9.    Select **Walkthrough→Finish Walkthrough** from the ribbon.

10. In the Project Browser, you will now see a Walkthrough listed under Walkthroughs.

11. Save as *ex9-15.rvt*.

Exercise 9-16
Playing the Walkthrough

Drawing Name: ex9-15.rvt
Estimated Time: 5 minutes

This exercise reinforces the following skills:

□ Playing a Walkthrough

1. Open *ex9-15.rvt*.

2. Highlight the Walkthrough in the Browser.
Right click and select **Open**.

3. Highlight the Walkthrough in the Browser.
Right click and select **Show Camera**.

4. Select **Edit Walkthrough** on the ribbon.

5. Look at the Options bar.

Set the Frame number to **1.0**.

6. Activate the **Edit Walkthrough** ribbon.

7. Select **Open Walkthrough**.

8.  Press **Play** on the ribbon.

9. ⬜ 300 ⬜ Press the button that displays the total frame number. Your frame number value may be different.

10. Total Frames: [50] Change the Total Frames value to **50**.
    Press **Apply** and **OK**.

11. ▷ Press **Play**.

    Play

    Notice that fewer frames speeds up the animation.

12. Save as *ex9-16.rvt*.

Exercise 9-17
Editing the Walkthrough Path

Drawing Name: ex9-16.rvt
Estimated Time: 15 minutes

This exercise reinforces the following skills:

- ❑ Show Camera
- ❑ Editing a Walkthrough
- ❑ Modifying a Camera View

1. Open *ex9-16.rvt*.

2. ⊟ Floor Plans
    ┆└─ **FIRST FLOOR**   Activate the **First Floor** floor plan view.

3. 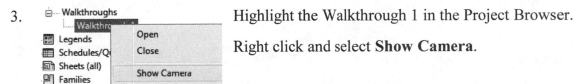 Highlight the Walkthrough 1 in the Project Browser.

    Right click and select **Show Camera**.

4.  Select **Edit Walkthrough** from the ribbon.

5. | Extents | |
    | --- | --- |
    | Crop Region Visible | ☑ |
    | Far Clip Active | ☐ |

    Disable **Far Clip Active**.

6. 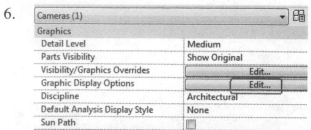 Select **Edit** next to Graphic Display Options.

7. 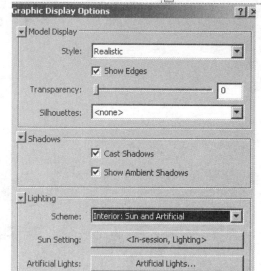 Set the Style to **Realistic**.
Enable **Show Edges**.

Enable **Cast Shadows**.
Enable **Show Ambient Shadows**.
Set the Lighting to **Interior: Sun and Artificial.**
Press **Apply** and **OK.**

8. Double left click on **Walkthrough 1** to open the view.

9. Select **Edit Walkthrough** from the ribbon.

10. 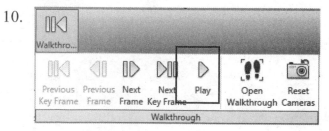 Press **Play**.

11. Save as *ex9-17.rvt*.

The appearance of shaded objects is controlled in the material definition.
To stop playing the walkthrough, press **ESC** at any frame point.

Exercise 9-18
Creating an Animation

Drawing Name: ex9-17.rvt
Estimated Time: 15 minutes

This exercise reinforces the following skills:

- ❑ Show Camera
- ❑ Editing a Walkthrough
- ❑ Modifying a Camera View

1. Open *ex9-17.rvt*.

2. 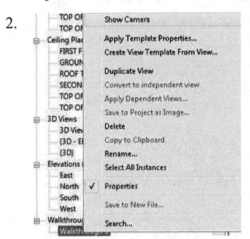 Highlight the Walkthrough 1 in the Project Browser.

   Right click and select **Show Camera**.

3. Select **Edit Walkthrough** on the ribbon.

4. `Frame   1.0 of 50`   Set the Frame number to **1.0**.

5. 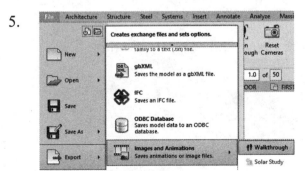 Go to **File→Export→ Images** and **Animations→ Walkthrough**.

6.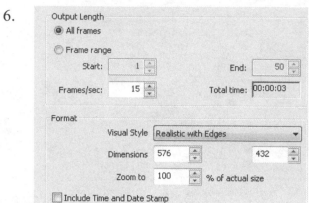

Set the Visual Style to **Realistic with Edges**.

Press **OK**.

You can create an animation using Rendering but it takes a long time to process.

7.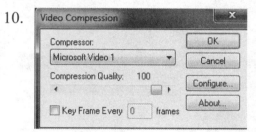

Locate where you want to store your avi file. You can use the Browse button (…) next to the Name field to locate your file.

8. Name your file *ex9-18 Walkthrough 1.avi*.

9. Press **Save**.

10.

Select **Microsoft Video 1**.

This will allow you to play the avi on RealPlayer or Microsoft Windows Media Player.
If you select a different compressor, you may not be able to play your avi file.

Press **OK**.

11.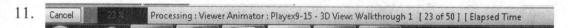

A progress bar will appear in the lower right of your screen to keep you apprised of how long it will take.

12. ex9-18 Walkthrough 1
00:00:20
9.01 MB

Locate the file using Explorer.
Double click on the file to play it.

Use Windows Media Player to play the file.

13. Save as *ex9-18.rvt*.

Additional Projects

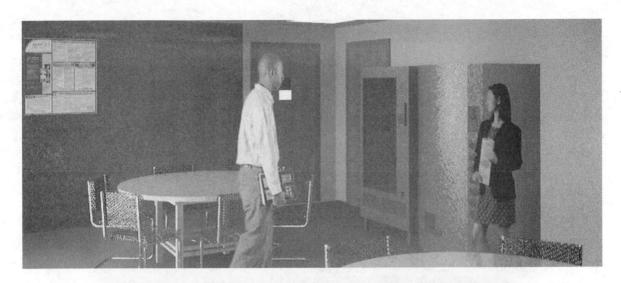

1) Create a rendering of the cafeteria area.

2) Create a rendering of one of the office cubicles.
   The figure is the Female Entourage family - Cynthia. The publisher's website
   (www.sdcpublications.com/downloads/978-1-63057-263-1) has the images and
   the families for the monitor, phone, keyboard, and CPU. Image files can be
   used in the rendering.

3) Add exterior lighting using the bollard lights in the Revit library to both sides of
   the walkway to the building.
   Set the day to December 11 and the time at 7:30 pm.
   Adjust the exposure after the rendering is done.

4) Create a walkthrough avi of one of your rendered scenes.

Lesson 9 Quiz

True or False

1. The Walkthrough tool on the View ribbon allows you to create an avi file.
2. The Environment button on Render Settings dialog allows you to set the time of day.
3. The Sun button on Render Settings dialog allows you to set the time of day.
4. When you create a Rendering, you will be able to see how your people will appear.
5. Rendering is a relatively quick process.
6. Renderings can be added to sheets.
7. You can adjust the number of frames in a walkthrough.
8. You cannot create walkthrough animations that go from one level to another.
9. Once a walkthrough path is placed, it cannot be edited.
10. You cannot modify the view style of a walkthrough from wireframe to shaded.

Multiple Choice

11. The folder, which contains RPC People, is named:

    A. Site
    B. People
    C. Entourage
    D. Rendering

12. The Camera tool is located on the _____ ribbon.

    A. Rendering
    B. Site
    C. Home
    D. View

13. To see the camera in a view:

    A. Launch the Visibility/Graphics dialog and enable Camera in the Model Categories.
    B. Highlight the view in the browser, right click and select Show Camera.
    C. Go to View→Show Camera.
    D. Mouse over the view, right click and select Show Camera.

14. In order to save your rendering, use:

    A. File→Export
    B. Save to Project
    C. Capture Rendering
    D. Export Image

15. When editing a walkthrough, the user can modify the following:

    A. Camera
    B. Path
    C. Add Key Frame
    D. Remove Key Frame
    E. All of the above

16. Decals are visible in the display mode:

    A. Hidden
    B. Wireframe
    C. Realistic
    D. Raytrace

17. In order to render using the Cloud, the user needs:

    A. Internet access
    B. An umbrella
    C. A Revision Cloud
    D. A valid ID

18. When placing a building pad, the following elements are required:

    A. You must be in a site plan view
    B. There must be a toposurface
    C. There must be a building pad family defined
    D. There must be a building pad material defined

ANSWERS:
1) F; 2) F; 3) T; 4) T; 5) F; 6) T; 7) T; 8) F; 9) F; 10) F; 11) C; 12) D; 13) B; 14) B; 15) E; 16) C; 17) A; 18) B

Lesson 10
Customizing Revit

Exercise 10-1
Creating an Annotation Symbol

File: north arrow.dwg (located in the Class Files download)
Estimated Time: 30 minutes

This exercise reinforces the following skills:

- ❏ Import AutoCAD Drawings
- ❏ Full Explode
- ❏ Query
- ❏ Annotation Symbol

Many architects have accumulated hundreds of symbols that they use in their drawings. This exercise shows you how to take your existing AutoCAD symbols and use them in Revit.

1. Locate the north arrow.dwg in the Class Files downloaded from the publisher's website.

2. 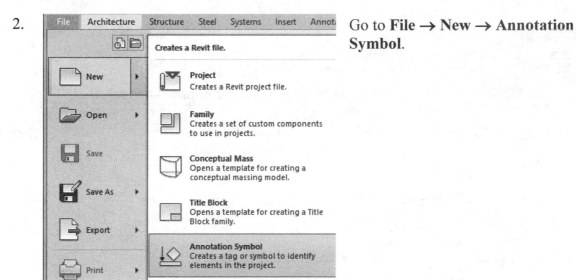 Go to **File → New → Annotation Symbol**.

3. Highlight the *Generic Annotation.rft* file under Annotations.

   Press **Open**.

4. Activate the Insert ribbon.

   Select **Import CAD**.

5. north arrow.dwg
    DWG Files (*.dwg)   Locate the *north arrow.dwg* file in the Class Files download.

6.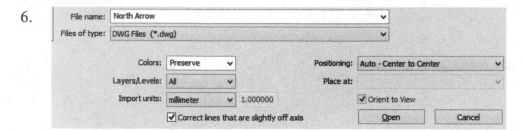

Under Colors: Enable **Preserve**.
Under Layers: Select **All**.
Under Import Units: Select **millimeter**.
Under Positioning: Select **Auto – Center to Center**.

This allows you to automatically scale your imported data.
Note that we can opt to manually place our imported data or automatically place using the center or origin.

Press **Open**.

7. The Import tool looks in Paper space and Model space. You will be prompted which data to import.
    Press **Yes**.

    > Revit
    >
    > Import detected no valid elements in the file's Paper space. Do you want to import from the Model space?
    >
    > Yes   No

8.   **Zoom In Region** to see the symbol you imported.

9. Pick the imported symbol.

    Query   Select **Query** from the ribbon.

10. 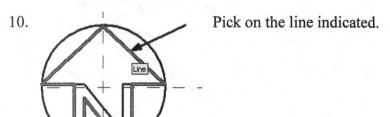   Pick on the line indicated.

    Line

11.

Parameter	Value
Type	Line
Block Name	N/A
Layer/Level	0
Style By	Layer/Level

Layer/Level
Delete   Hide in view   OK

A dialog appears that lists what the selected item is and the layer where it resides.

Press **OK**.

Right click and select **Cancel** to exit Query mode.

12. 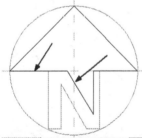 Pick the imported symbol.

Select **Full Explode**.

We can now edit our imported data.

13. Activate the Create ribbon.

Select the **Filled Region** tool on the Detail panel from the Architecture ribbon.

14. ☑ Chain   Offset: 0' 0"   ☐ Radius: 1' 0"   Enable **Chain** on the status bar.

15. 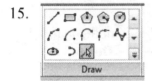 Use the **Pick** tool on the Draw panel to select the existing lines to create a filled arrowhead.

16. Use the TRIM tool to connect the two lines indicated and close the region outline.

17.

Properties   ✕

Filled region
Solid Black

Filled region   Edit Type

On the Properties palette:
Verify that the Region is set to **Solid Black**.

18.   Select the **Green Check** under the Mode panel to **Finish Region**.

19. 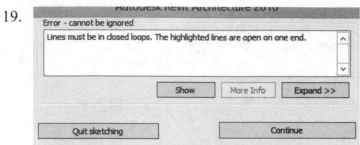   If you get an error message, check that you have a single closed loop.

20. 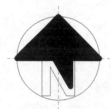   Using grips, shorten the horizontal and vertical reference planes if needed.

You will need to unpin the reference planes before you can adjust the link.

Be sure to pin the reference planes after you have finished any adjustments. Pinning the reference planes ensures they do not move.

21. 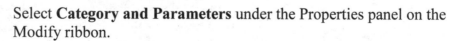 Select the note that is in the symbol. Right click and select **Delete**.

22. 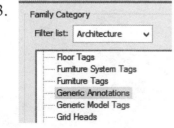 Select **Category and Parameters** under the Properties panel on the Modify ribbon.

23. Note that **Generic Annotations** is highlighted. Press **OK**.

24. 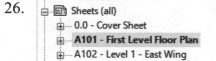 Save the file as a new family called **north arrow**.

25. Open *ex9-14.rvt*.

26.
    - Sheets (all)
      - 0.0 - Cover Sheet
      - **A101 - First Level Floor Plan**
      - A102 - Level 1 - East Wing

    Activate the **First Level Floor Plan** sheet.

27. [Annotate] Activate the **Annotate** ribbon.

28. Select the **Symbol** tool.

    Symbol

29. Select **Load Family**.

    Load
    Family
    Mode

30. File name: north arrow.rfa   Load the *north arrow.rfa* file that was saved to your student
    Files of type: Family Files (*.rfa)   folder.

31. Place the North Arrow symbol in your drawing.

    ① Level 1
    1/8" = 1'-0"

32. Save *ex9-14.rvt* as *ex10-1.rvt*.

Exercise 10-2
Creating a Custom Title Block

Estimated Time: 60 minutes

This exercise reinforces the following skills:

- Titleblock
- Import CAD
- Labels
- Text
- Family Properties

1. 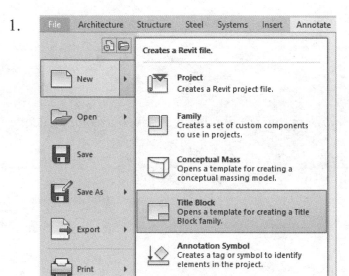 Go to **File → New → Titleblock**.

2. Select **New Size**.

   Press **Open**.

   File name: New Size

   Files of type: Family Template Files (*.rft)

3. 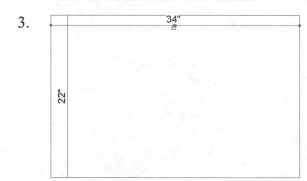 Pick the top horizontal line.
   Select the dimension and change it to **22″**.

   Pick the right vertical line.

   Select the dimension and change it to **34″**.

   I modified the dimension family to use a larger size font, but you don't have to do this.

The title block you define includes the sheet size. If you delete the title block, the sheet of paper is also deleted. This means your title block is linked to the paper size you define.

You need to define a title block for each paper size you use.

4.

| Zoom In Region |
| Zoom Out (2x) |
| Zoom To Fit |

Right click in the graphics window and select **Zoom to Fit**.

You can also double click on the mouse wheel.

5. Activate the Insert ribbon.

Select **Import→Import CAD**.

6.

File name: Architectural Title Block
Files of type: DWG Files (*.dwg)

Locate the *Architectural Title Block* in the exercise files directory.

7.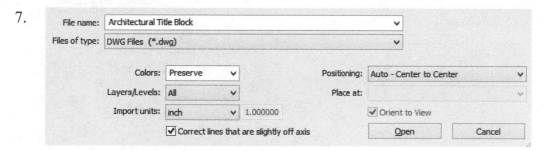

Set Colors to **Preserve**.
Set Layers to **All**.
Set Import Units to **Inch**.
Set Positioning to: **Auto - Center to Center**.

Press **Open**.

8.

Import detected no valid elements in the file's Paper space. Do you want to import from the Model space?

| Yes | No |

Press **Yes**.

9.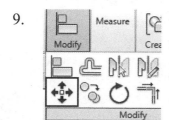

Select the title block.

Use the **Move** tool on the Modify panel to reposition the titleblock so it is aligned with the existing Revit sheet.

10.

Select the imported title block so it is highlighted.

Select **Explode → Full Explode** from the Import Instance panel on the ribbon.

11.

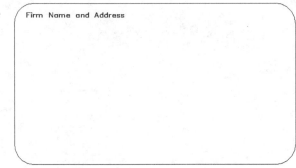

Firm Name and Address

We will place an image in the rectangle labeled Firm Name and Address.

12.

Activate the **Insert** ribbon.
Select the **Import→Image** tool.

13.

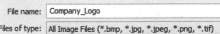

File name: Company_Logo

Files of type: All Image Files (*.bmp, *.jpg, *.jpeg, *.png, *.tif)

Open the *Company_Logo.jpg* file from the downloaded Class Files.

14.

Firm Name and Address

Company

Logo

Place the logo on the right side of the box.

Scale and move it into position.

To scale, just select one of the corners and drag it to the correct size.

Revit can import jpg, jpeg, or bmp files. Imported images can be resized, rotated, and flipped using activated grips.

15. **A**
Text

Select the **Text** tool from the **Create** Ribbon.

16.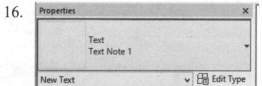

Select **Edit Type**.

17.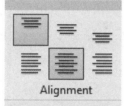

Change the Text Size to **1/8″**.

Press **OK**.

Revit can use any font that is available in your Windows font folder.

18.

Select **CENTER** on the Format panel.

19.

Type in the name of your school or company next to the logo.

 The data entered in the Value field for labels will be the default value used in the title block. To save edit time, enter the value that you will probably use.

20. **A**
Label

Select the **Label** tool from the Create ribbon.

Labels are similar to attributes. They are linked to file properties.

21. 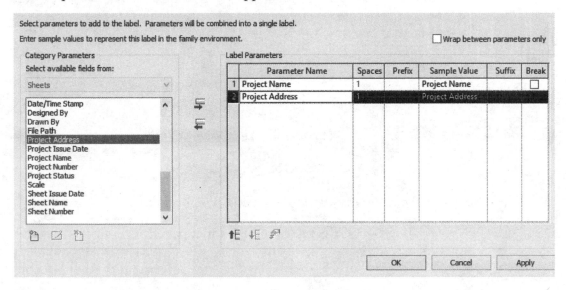 Left pick in the Project Name and Address box.

22. Press **OK**.
Left click to release the selection.
Pick to place when the dashed line appears.

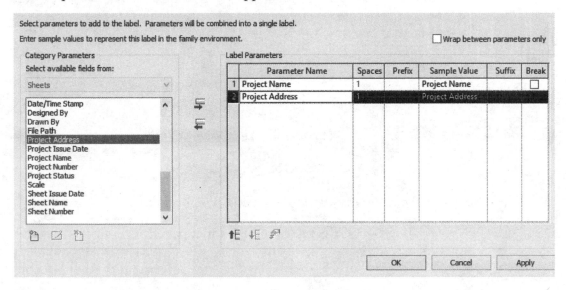

Select **Project Name**.
Use the **Add** button to move it into the Label Parameters list.
Add **Project Address**.
Press **OK**.

23.  Cancel out of the Label command.

Position the label.
Use the grips to expand the label.

24. 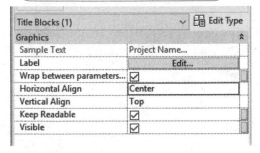 Select the label.

In the Properties panel:

Enable **Wrap between parameters**.

Verify that the Horizontal Align is set to Center.

25. 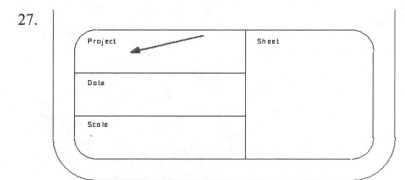   Use Modify→Move to adjust the position of the project name and address label.

26. Select the **Label** tool.

27. 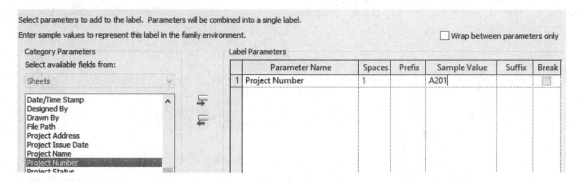   Left click in the Project box.

28. Locate **Project Number** in the Parameter list.
Move it to the right pane.

Select parameters to add to the label. Parameters will be combined into a single label.

Enter sample values to represent this label in the family environment.   ☐ Wrap between parameters only

Category Parameters

Select available fields from:

Sheets

Label Parameters						
	Parameter Name	Spaces	Prefix	Sample Value	Suffix	Break
1	Project Number	1		A201		☐

Date/Time Stamp
Designed By
Drawn By
File Path
Project Address
Project Issue Date
Project Name
Project Number
Project Status

Add a sample value, if you like.

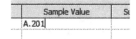

Press **OK**.

29.    Left click to complete placing the label.

Select and reposition as needed.

30. Select the **Label** tool from the Create ribbon.

31. Left click in the Date box.

32.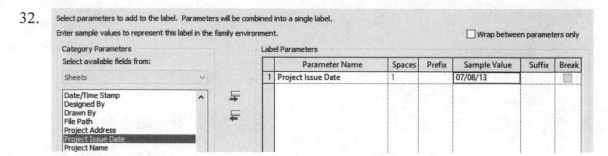

Highlight **Project Issue Date**.

Press the Add button.

In the Sample Value field, enter the default date to use.

Press **OK**.

33.

Project

A201

Date

07/13/18

Scale

Position the date in the date field.

34. Label

Select the **Label** tool.

35.

Project

A201

Date

07/08/13

Scale

Left click in the Scale box.

36.

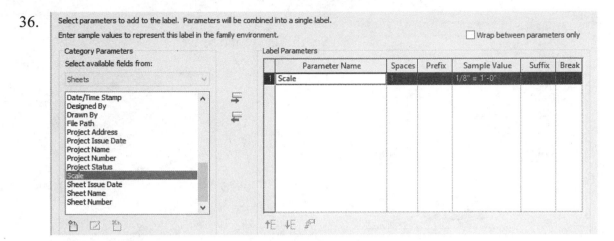

Highlight **Scale**.

Press the Add button.

Press **OK**.

37. Highlight the scale label that was just placed and select **Edit Type** on the Properties panel.

38. Set the Background to **Transparent**.

Press **OK**.

39. Note that the border outline and the block text are no longer hidden by the label background.

40. Select the scale label.

On the Properties panel, select **Edit Type**.

41. 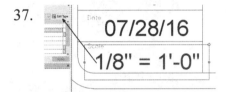 Select **Duplicate**.

42. Rename **Tag – Small**.

Press **OK**.

43.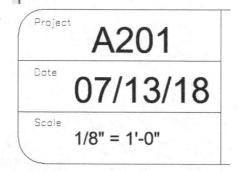

Set the Text Size to **1/8"**.

Press **OK**.

44.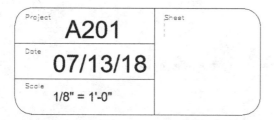

Adjust the lines, text and labels so the Sheet label looks clean.

45. Label Select the **Label** tool.

46.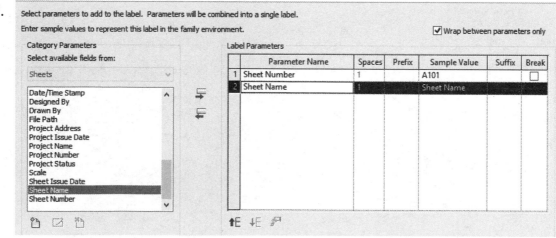

Left click in the Sheet box.

47.

Select parameters to add to the label. Parameters will be combined into a single label.

Enter sample values to represent this label in the family environment.   ☑ Wrap between parameters only

Category Parameters

Select available fields from:

Sheets

- Date/Time Stamp
- Designed By
- Drawn By
- File Path
- Project Address
- Project Issue Date
- Project Name
- Project Number
- Project Status
- Scale
- Sheet Issue Date
- Sheet Name
- Sheet Number

Label Parameters

	Parameter Name	Spaces	Prefix	Sample Value	Suffix	Break
1	Sheet Number	1		A101		☐
2	Sheet Name	1		Sheet Name		

Highlight **Sheet Number**.

Press the Add button.

Highlight **Sheet Name**.

Press the Add button.

Press **OK**.

48. 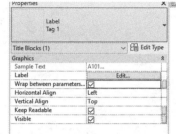 Enable **Wrap between parameters only** in the upper right corner of the dialog.

49. 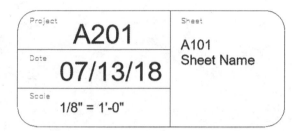 Position the Sheet label.

Use the Tag-Small for the new label.

50. Save the file as *Title block 22 x 34.rfa*.

 You can select Shared Parameters for use in your title block.

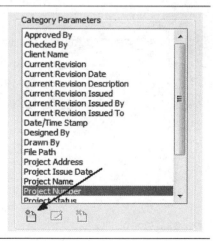

Exercise 10-3
Using a Custom Title Block

File: ex10-1.rvt
Estimated Time: 10 minutes

This exercise reinforces the following skills:

- Title block
- Import CAD
- Labels
- Text
- Family Properties

1. Open *ex10-1.rvt*, if it is not already open.

2.  Activate the Insert ribbon.

    Select **Load Family** in the Load from Library panel.

3.  Browse to the folder where you saved the custom title block.

    Select it and press **Open**.

4. 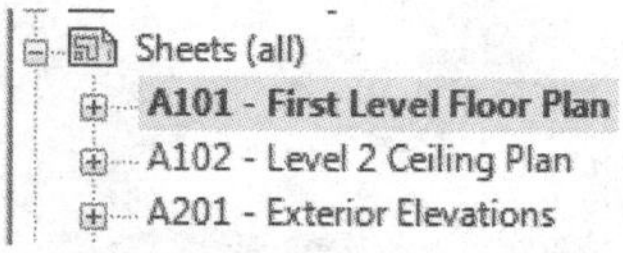 Activate the **First Level Floor Plan** sheet.

5. 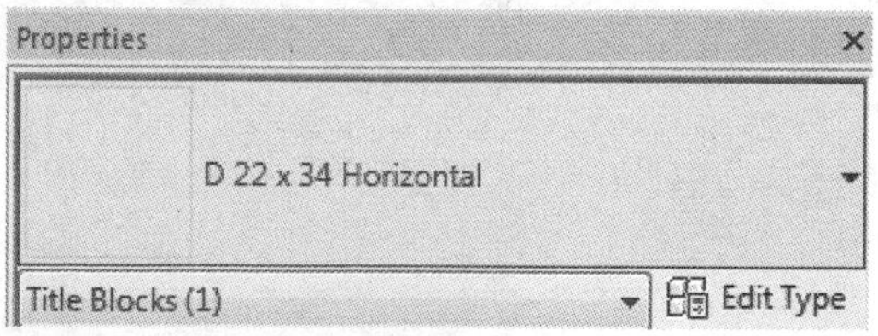 Select the title block so it is highlighted.

    You will see the name of the title block in the Properties pane.

6. 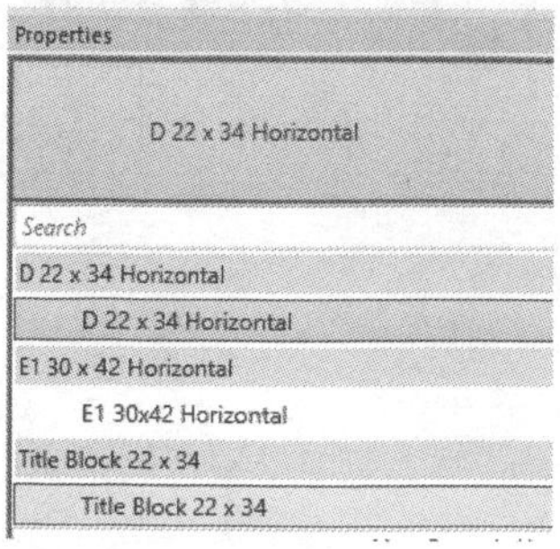 Select the **Title Block 22 x 34** using the Type Selector.

7.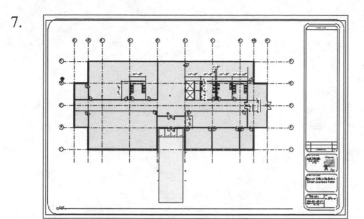
You may need to reposition the view on the title block.

8.

Zoom into the title block.

Note that the parameter values have all copied over.

9. Save the file as *ex10-3.rvt*.

It is a good idea to save all your custom templates, families, and annotations to a directory on the server where everyone in your team can access them.

Exercise 10-4
Creating a Line Style

Drawing Name: ex10-3.rvt
Estimated Time: 10 minutes

This exercise reinforces the following skills:

    ❑   Line Styles

1.   Open *ex10-3.rvt*.

2.   Activate the **Site** Floor plan.

3.   Activate the **Manage** ribbon.

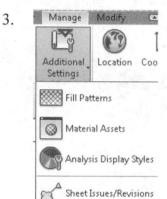

    Go to **Settings→Additional Settings→Line Styles**.

4.   Select **New** under Modify Subcategories.

5.   Enter **Sewer** in the Name field.

    Press **OK**.

6.

Lines	1	RGB 000-166-000	Solid
Medium Lines	3	Black	Solid
Sewer	1	Black	Solid
Thin Lines	1	Black	Solid
Wide Lines	12	Black	Solid

Sewer appears in the list.

Note Revit automatically alphabetizes any new line styles.

7.

Lines	1	RGB 000-166-000	Solid
Medium Lines	3	Black	Solid
Sewer	3	Blue	Dash dot dot
Thin Lines	1	Black	Solid

Set the Line Weight to **3**.
Set the Color to **Blue**.
Set the Line Pattern to **Dash Dot Dot**.

Press **OK**.

8.　Activate the Site Plan.

Floor Plans
　　Level 1 - Interior Plan
　　Level 1 - Lobby Detail
　　Level 1 - Lobby Space Planning
　　Level 1 - No Annotations
　　Level 1 East Wing
　　Level 1 Floorplan
　　Level 1 West Wing
　　Level 2
　　Level 2 - No Annotations
　　Roof Cutoff
　　Roof Line
　　Site

9.　**Annotate**　Activate the **Annotate** ribbon.

10.　Detail Line　Select the **Detail Line** tool under Detail.

11.　Line Style:

Thin Lines

<Beyond>
<Centerline>
<Demolished>
<Hidden>
<Overhead>
Hidden Lines
Lines
Medium Lines
Sewer
Thin Lines
Wide Lines

Select **sewer** from the drop-down list in the Properties palette.

You can also set the Line Style from the ribbon drop-down list.

12.
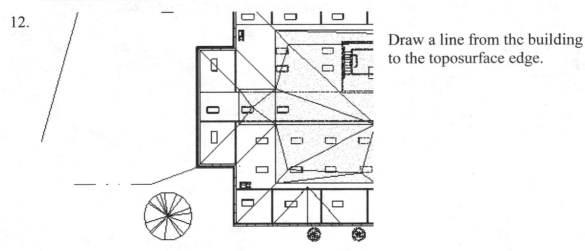

Draw a line from the building to the toposurface edge.

13.　Save the file as *ex10-4.rvt*.

Revit has three line tools: *Lines*, *Detail Lines*, and *Linework*. All of them use the same line styles, but for different applications. The *Lines* command draws model lines on a specified plane and can be seen in multiple views (i.e., score joints on an exterior elevation). The *Detail Lines* command draws lines that are view specific and will only appear on the view where they were placed (i.e., details). The *Linework* tool is used to change the appearance of model-generated lines in a view.

Exercise 10-5
Defining Keyboard Shortcuts

Estimated Time: 5 minutes

This exercise reinforces the following skills:

❑ Keyboard Shortcuts

1.

Activate the View ribbon.

Select **User Interface→Keyboard Shortcuts** from the Windows panel.

2. 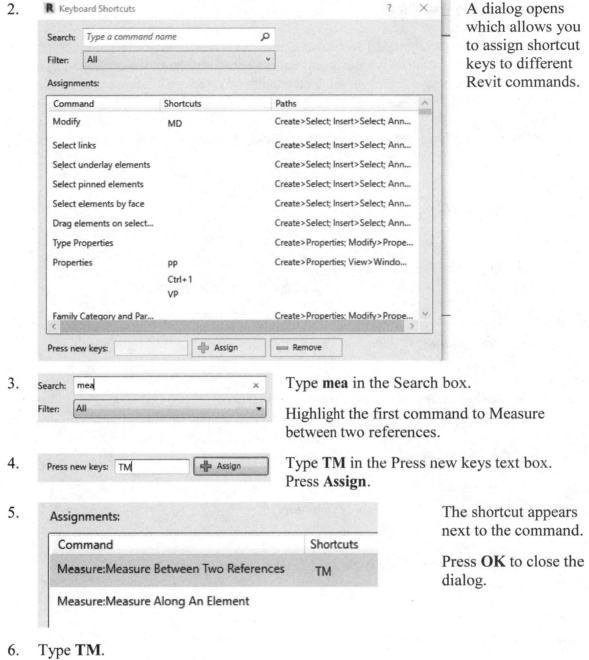 A dialog opens which allows you to assign shortcut keys to different Revit commands.

3. Type **mea** in the Search box.

Highlight the first command to Measure between two references.

4. Type **TM** in the Press new keys text box. Press **Assign**.

5. The shortcut appears next to the command.

Press **OK** to close the dialog.

6. Type **TM**.

Total Length appears on the Options bar, and your cursor is in Measure mode.

Creating Custom Families

One of the greatest tools inside of Revit is the ability to create custom families for common components, such as windows and doors. These are more powerful than blocks because they are completely parametric. You can create one set of geometry controlled by different sets of dimensions.

The steps to create a family are the same, regardless of whether you are creating a door, window, furniture, etc.

Step 1:
Select the appropriate family template to use.

Step 2:
Define sub-categories for the family.
Sub-categories determine how the object will appear in different views.

Step 3:
Lay out reference planes.

Step 4:
Dimension planes to control the parametric geometry.

Step 5:
Label dimensions to become type or instance parameters.

Step 6:
Create your types using the 'FamilyTypes' tool.

Step 7:
Activate different types and verify that the reference planes shift correctly with the dimensions assigned.

Step 8:
Label your reference planes.

Step 9:
Create your geometry and constrain to your reference planes using Lock and Align.

Step 10:
Activate the different family types to see if the geometry reacts correctly.

Step 11:
Save family and load into a project to see how it performs within the project environment.

Exercise 10-6
Defining Reference Plane Object Styles

Estimated Time: 15 minutes

This exercise reinforces the following skills:

- ❑ Reference Planes
- ❑ Object Styles

Users have the ability to create different styles of reference planes. This is especially useful when creating families as reference planes are used to control family geometry.

1.   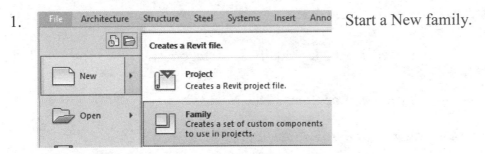   Start a New family.

2.   File name: Generic Model

     Files of type: Family Template Files (*.rft)   Select the Generic Model template.

Note that there are already two reference planes in the file. The intersection of the two planes represents the insertion point for the family. Both reference planes are pinned in place so they cannot be moved or deleted.

3.   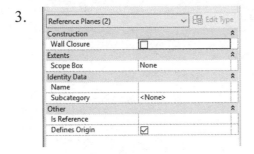   Hold down the CTRL key and select the two reference planes.

Note in the Properties pane **Defines Origin** is enabled.

This is how you can control the insertion point for the family. The intersection of any two reference planes may be used to define the origin.

4.

Activate the Manage ribbon.

Select **Object Styles** from the Settings panel.

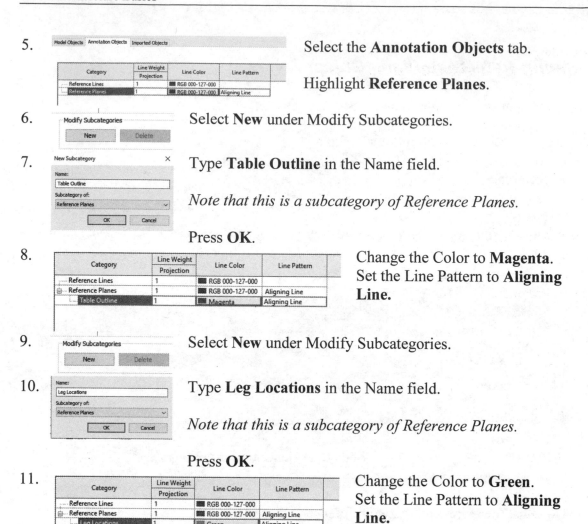

5.   Select the **Annotation Objects** tab.

         Highlight **Reference Planes**.

6.   Select **New** under Modify Subcategories.

7.   Type **Table Outline** in the Name field.

         Note that this is a subcategory of Reference Planes.

         Press **OK**.

8.   Change the Color to **Magenta**. Set the Line Pattern to **Aligning Line.**

9.   Select **New** under Modify Subcategories.

10.   Type **Leg Locations** in the Name field.

         Note that this is a subcategory of Reference Planes.

         Press **OK**.

11.   Change the Color to **Green**. Set the Line Pattern to **Aligning Line.**

12. Press **OK** to close the dialog.

13. Save as *Coffee Table.rfa.*

Exercise 10-7
Creating a Furniture Family

Estimated Time: 120 minutes
File: Coffee Table.rfa

This exercise reinforces the following skills:

- ❑ Standard Component Families
- ❑ Templates
- ❑ Reference Planes
- ❑ Align
- ❑ Dimensions
- ❑ Parameters
- ❑ Sketch Tools
- ❑ Solid extrusion
- ❑ Materials
- ❑ Types

1. Select the Reference Plane tool from the Datum panel on the Create ribbon.

    Reference Plane

2. On the ribbon, set the Subcategory to **Table Outline**.

3.  Draw four reference planes: two horizontal and two vertical on each side of the existing reference planes.

    These will act as the outside edges of the table top.

    Note they are a different color because they are using the reference plane style you defined earlier.

4.   Add a continuous aligned dimension horizontally and vertically so they can be set equal.

5.

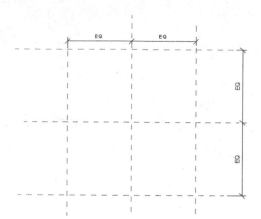

To place the dimension, select the reference planes by clicking in order without clicking to place until all three planes have been selected.

Then, left click on the EQ toggle to set equal.

6.

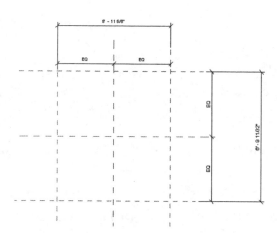

Place an overall horizontal dimension.

Place an overall vertical dimension.

7.

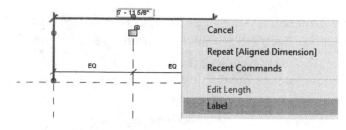

Select the overall horizontal dimension.

Right click and select **Label**.

Label: <None> ▼ ☐ Instance Parameter

You can also select Label from the Options bar.

8. <None> ▼

<None>
<Add parameter...>

Select **Add Parameter**...

9.

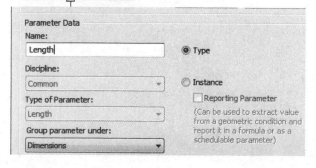

Under Name: type **Length**.
Set the group parameter to **Dimensions**.
Enable **Type**.
Press **OK**.

10.   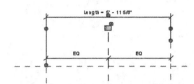   The dimension updates with the label.

11.   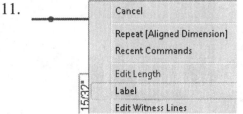   Select the overall vertical dimension.

Right click and select **Label**.

12.   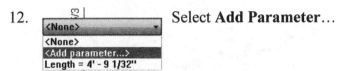   Select **Add Parameter**…

13.   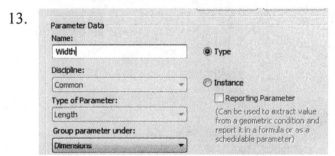   Under Name: type **Width**.
Set the group parameter to
Dimensions.
Enable **Type**.
Press **OK**.

14.   The dimension updates with the label.

15.   Select the **Family Types** tool on the Properties panel from the Modify ribbon.

16.   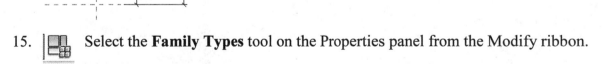   Select **New** under Family Types.

17.   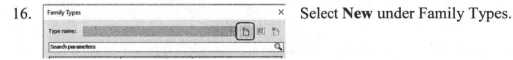   Enter **Small** for the Name.

Press **OK**.

18.

Dimensions		
Length	3' 0"	
Width	2' 6"	

Modify the values for the dimensions:
Length: **3' 0"**
Width: **2' 6"**

19.

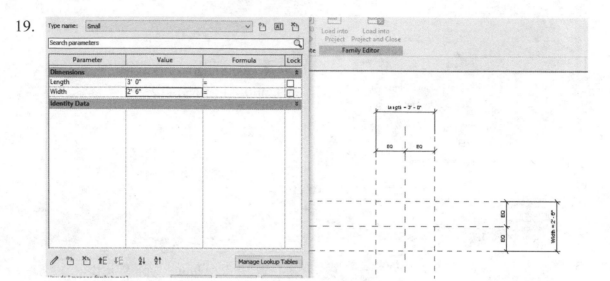

Move the dialog over so you can observe how the reference planes react.
Press **Apply**.
Note how the dimensions update.
This is called *flexing* the model.

20.

Type name:	Small	

Select **New** under Family Types.

21.

Name: Medium

Enter **Medium** for the Name.

Press **OK**.

22.

Type name: Medium	
Search parameters	
Parameter	Value
Dimensions	
Length	4' 0"
Width	3' 0"

Modify the values for the dimensions:
Length: **4' 0"**
Width: **3' 0"**

23. 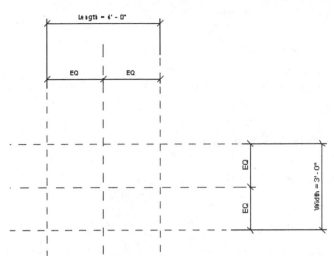 Press **Apply**.
Note how the dimensions update.

24. | Type name: | Medium | Select **New** under Family Types.

25. | Name: | Large | Enter **Large** for the Name.

Press **OK**.

26.
Type name:	Large
Search parameters	
Parameter	Value
Dimensions	
Length	5' 0"
Width	3' 6"

Modify the values for the dimensions:
Width: **3' 6"**
Length: **5' 0"**

27. 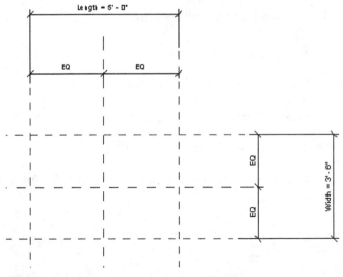 Press **Apply**.
Note how the dimensions update.

28.
| Family Types |
Name:	Large
	Large
	Medium
Paramet	Small

Press **OK** to close the dialog.

Left click on the drop-down arrow on the Name bar.
Switch between the different family names and press Apply to see the dimensions change.

29. 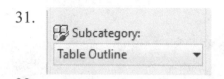 Activate the **Front** elevation.

| Zoom In Region |
| Zoom Out (2x) |
| Zoom To Fit |

Perform a **Zoom to Fit** to see the entire view.

30. Select the **Reference Plane** tool on the Datum panel from the Create ribbon.

31. On the ribbon:

Subcategory:

Table Outline

Set the Subcategory to **Table Outline**.

32. 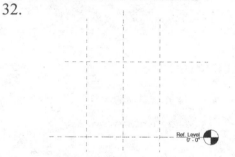 Draw a horizontal line above the Ref Level.

Ref. Level
0' - 0"

33. Select the reference plane you just drew.

Identity Data	
Name	Table Top
Subcategory	Table Outline

In the Properties palette:

Rename the reference plane **Table Top**.

34. Table Top   If you mouse over the reference plane, the name will appear.

If you select the reference plane, the name will be visible as well.

35. Select the **Reference Plane** tool on the Datum panel from the Architecture ribbon.

36. 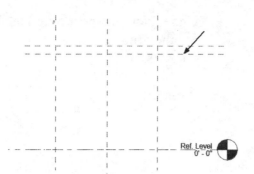 Draw a reference plane using the Table Outline subcategory below the Table Top Ref plane.

37. 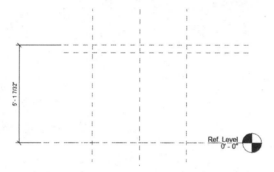 Select the new reference plane.

In the Properties pane:

Rename the reference plane **Table Bottom**.

38.  Place a dimension between the Ref. Level and Table Top ref. plane.

39. 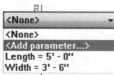 Place a dimension between the top ref. plane and the Table Bottom ref. plane.

40. Select the table height dimension, the dimension between the Ref. Level and Top ref. plane.

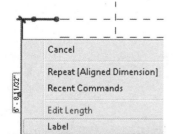 Right click and select **Label**.

Select *Add Parameter* from the **Label** drop-down on the Option bar.

41.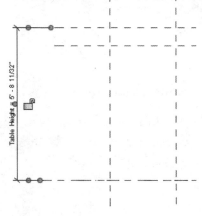

Under Name: type **Table Height**.
Set the group parameter to **Dimensions**.
Enable **Type**.
Press **OK**.

42. The dimension updates with the label.

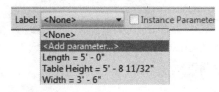

43. Select the table thickness dimension, the dimension between the Table Top ref. plane and the Table Bottom ref. plane below it.

Select *Add Parameter* from the **Label** drop-down on the Option bar.

44.

Under Name: type **Table Thickness**.
Set the group parameter to **Dimensions**.
Enable **Type**.
Press **OK**.

45.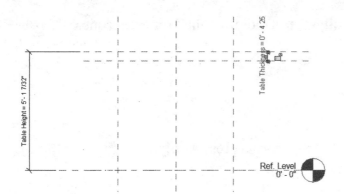

The dimension updates with the label.

46. Floor Plans
    Ref. Level
Ceiling Plans
    Ref. Level

Activate the **Ref. Level** under Floor Plans.

47.   Select the **Extrusion** tool on the Forms panel from the Create ribbon.
Extrusion

48.   Select **Work Plane → Set**.
Set

49. Specify a new Work Plane
    ● Name   Reference Plane : Table Bottom
    ○ Pick a plane
    ○ Pick a line and use the work plane it was sketched in

Enable **Name**.
Select the **Table Bottom** work plane from the drop-down list.
Press **OK**.

50.   Select the **Rectangle** tool from the Draw panel.
Draw

51.

Place the rectangle so it is aligned to the outer reference planes.

Select each lock so it is closed.

This constrains the rectangle to the reference planes.

3' - 6'

5' - 0"

52.  Select the **Family Types** tool from the Properties pane.

53. Name: Large / Large / Medium / Small   Select each type and press **Apply**.

Verify that the rectangle flexes properly with each size.
If it does not flex properly, check that the side has been locked to the correct reference line using the Align tool.

Press **OK** to close the dialog.

54. Materials and Finishes / Material <By Category> / Identity Data   Select the small button in the far right of the Material column.

This allows you to link the material to a parameter.

55. Family parameter: Material

Parameter type: Material

Existing family parameters of compatible type:

Search parameters

<none>

Select **New parameter**.

56. Parameter Data

Name:

Table Top

Discipline:

Common

Type of Parameter:

Material

Group parameter under:

Materials and Finishes

◉ Type

◎ Instance

☐ Reporting Parameter

(Can be used to extract value from a geometric condition and report it in a formula or as a schedulable parameter)

Type **Table Top** for the name.

Enable **Type**.
Set the Group Parameter under **Materials and Finishes**.

Press **OK**.

57. 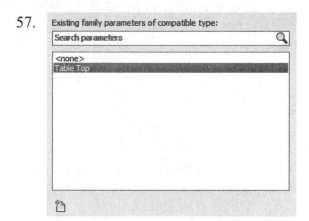 Table Top is now listed as a material parameter.

Press **OK**.

58. Select the **Family Types** tool.

59. 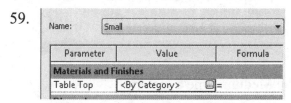 Select the **Small** Family type from the drop-down list.

Select the ... button next to Table Top to assign a material to the table top.

60. 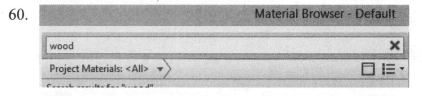 Type **wood** in the search field.

61. 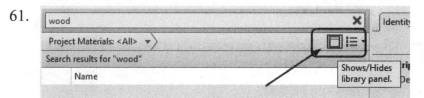 Select the icon that displays the library panel.

62. 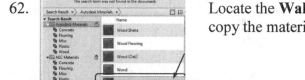 Locate the **Walnut** material and select the up arrow to copy the material to the active document.

63. 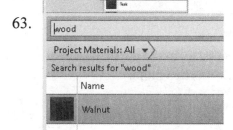 Highlight the **Walnut** material and press **OK**.

64.

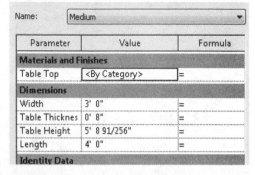

You now see the material listed in the Family Types dialog.

65.

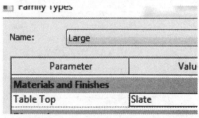

If you switch to a different type, you will see that the material is unassigned.

Switch to the **Medium** size.

Select the material column.

66.

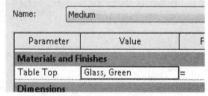

Assign the Medium type a material.

67.

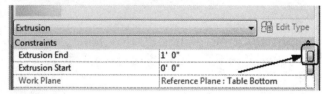

Assign the Large size a material.

Press **OK**.

68.

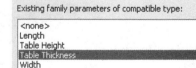

Click the button in the Extrusion End row.

69.

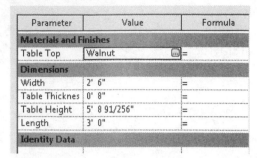

Select **Table Thickness** from the parameters list.

Press **OK**.

70.

Select the **Family Types** tool from the Properties pane.

71. Table Thickness has now been added as a parameter to the Family Types.

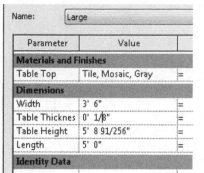

Change the value of the Table Thickness to 1/8″ for each type.

Press **Apply** before switching to the next type.

Press **OK**.

72. Select the **green check** under Mode to **Finish Extrusion**.

73. Elevations (Elevation 1)
    Back
    Front
    Left
    Right

    Activate the **Front** Elevation.

74. You see the table top which is equal to the table thickness.

75. Switch to a 3D view.

76.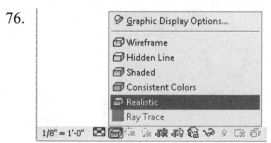

Set the Visual Style to **Realistic**.

77. Select the **Family Types** tool from the Properties pane.

78. Flex the model and see how the table top changes.

79. Activate the Ref. Level view.

80. Activate the Create ribbon.

Select the **Reference Plane** tool from the Datum panel.

81. Set the Subcategory to **Leg Locations**.

82. *You don't see the table because the table is above the reference level.*

Place four reference planes 4″ offset inside the table.

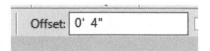

These planes will be used to locate the table legs.

83. Select the **Aligned Dimension** tool.

84. 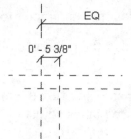 Place a dimension between the edge of the table and the inside reference plane.

85. 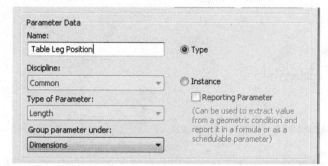 Add a parameter label to the dimension called **Table Leg Position**.

86. 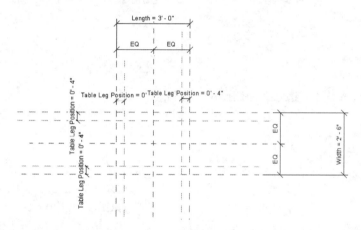 Add dimensions to constrain the reference planes to be offset four inches inside the table boundaries.

Make sure you place the dimensions between the reference planes and NOT between the table extrude edges and the reference planes. Do not lock any of the dimensions or reference planes or they won't shift when the table size changes.

87. Select each dimension and use the label pull-down on the Options bar to assign the Table Leg position parameter to the dimension.

88. Select the **Family and Types** tool.

89. 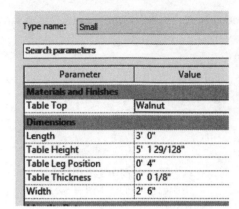 Modify the Table Leg Position parameter to **4″**.

Press **Apply** and verify that the reference planes flex.

Repeat for each family name and flex the model to verify that the reference planes adjust.

Press **OK**.

90. Select **Blend** from the Create ribbon.

Blend

91. Select the **Set Work Plane** tool.

Set

92. 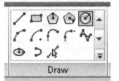 Select the **Ref. Level** for the sketch.

Press **OK**.

93. Select the **Circle** tool from the Draw panel.

94. 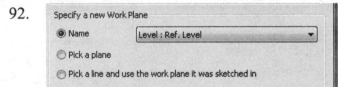 Draw a circle with a 1/2″ radius so the center is at the intersection of the two offset reference planes.

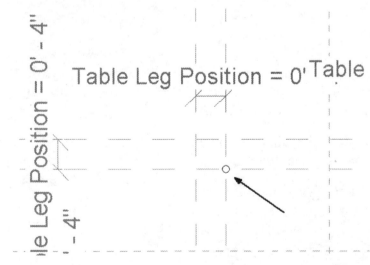

☑ Radius: 0′ 0 1/2″

You can set the radius in the Options bar.

95. Select **Edit Top** from the Mode panel on the ribbon to switch to the top sketch plane.

96. Select the **Circle** tool from the Draw panel.

97. Place a circle with a 1″ radius so the center is at the intersection of the two offset reference planes.

Modify | Measur
Radius: 0' 1"

You can set the radius in the Options bar.

98. On the Properties palette:

Assign a parameter for the material.

99. Select **New parameter**.

100. Type **Table Legs** for the name.

Press **OK**.

101. Highlight **Table Legs**.

Press **OK**.

102. Select the **Family Types** tool.

103. Table Legs is now a material to be assigned for each type.

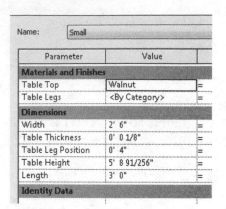

Select the **...** button in the Table Legs Value column.

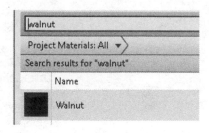

104. 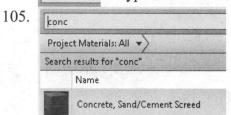 Type **concrete** in the search field.

105.  Locate a concrete material.

Select **Add material to document**.

106. 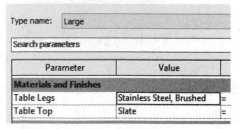 Type **pine** in the search field.

107. 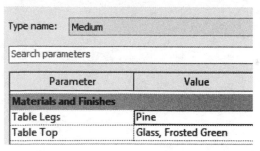 Locate the **pine** material.

Select **Add material to document.**

Close the dialog.

108. 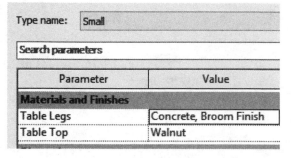 Each size of the table should be assigned a different material for the table top and the table legs.

Press **OK**.

Type name:	Large	
Search parameters		

Parameter	Value	
Materials and Finishes		
Table Legs	Stainless Steel, Brushed	=
Table Top	Slate	=

Type name:	Medium
Search parameters	

Parameter	Value
Materials and Finishes	
Table Legs	Pine
Table Top	Glass, Frosted Green

Type name:	Small
Search parameters	

Parameter	Value
Materials and Finishes	
Table Legs	Concrete, Broom Finish
Table Top	Walnut

109. 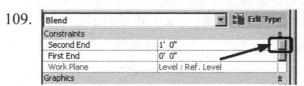 Select the button on the Second End row to associate the table legs length with a parameter.

110. Select **New parameter** at the bottom of the dialog.

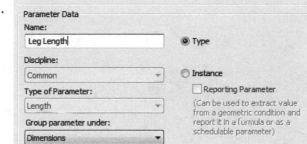

111. Enter **Leg Length** for the Name.

Enable **Type**.

Press **OK**.

112. 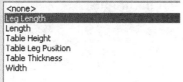 Highlight **Leg Length** in the parameter list.

Press **OK**.

113. Select **Family Types** from Properties pane on the Create ribbon.

114.

Dimensions		
Leg Length	5' 5 89/128"	= Table Height - Table Thickness
Length	3' 0"	=
Table Height	5' 5 105/128"	=
Table Leg Position	0' 4"	=
Table Thickness	0' 0 1/8"	=
Width	2' 6"	=

In the Formula column, enter **Table Height – Table Thickness**.

Press **OK**.

Note, you only have to enter the formula once and it works for all three sizes.

The variable names must match exactly! If you have a spelling error or use an abbreviation, you have to copy it.

115. 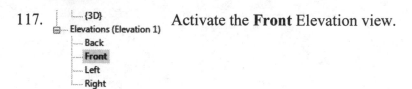 The value for Second End will update.

116. ✗ ✓ Mode   Select the **green check** on the Mode panel to finish the extrusion.

117. {3D}
     Elevations (Elevation 1)
        Back
        Front
        Left
        Right

     Activate the **Front** Elevation view.

118. You see a tapered leg.

119. Floor Plans
        Ref. Level

     Activate the **Ref. Level**.

120. **0' - 4"**   Select the leg so it highlights.

121. Select the **Copy** tool from the Modify panel.

122. 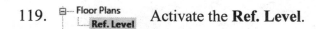 Enable **Multiple** on the Options bar.

123. Select the intersection where the center of the circle is the base point.

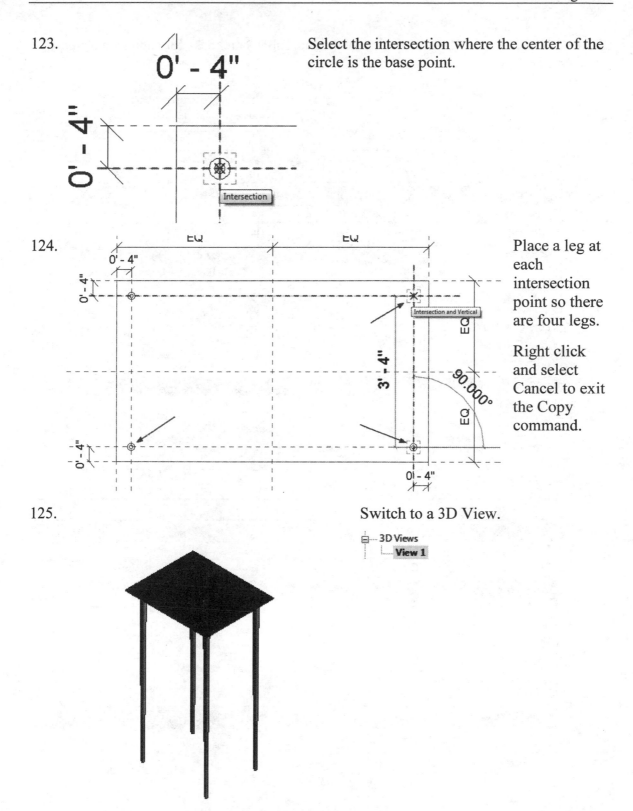

124. Place a leg at each intersection point so there are four legs.

Right click and select Cancel to exit the Copy command.

125. Switch to a 3D View.

126. Select **Family Types** from Properties pane on the Architecture ribbon.

Flex the model to check if the legs adjust position properly.

127. Change the value for the Table Height and Table Thickness for each size to see what happens to the model.

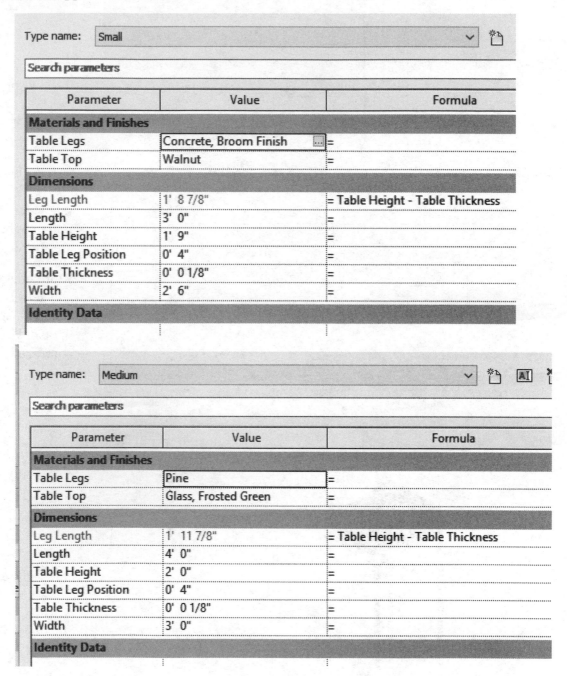

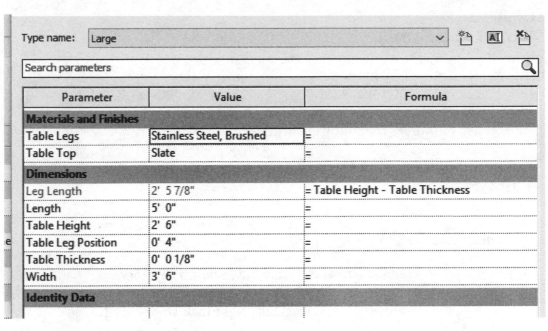

128. Save the family as *Coffee Table.rfa*.

129. Close the file.

Exercise 10-8
Modifying a Family

Estimated Time: 10 minutes
File: Office_2.rvt

This exercise reinforces the following skills:

❑   Standard Component Families
❑   Types

1. 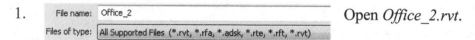 Open *Office_2.rvt*.

2. 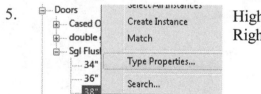 In the Browser, locate the *Families* category.
    Locate the *Doors* category.
    Under *Doors*, locate the **Sgl Flush** Door Family.

3.  Right click and select **New Type**.

4. 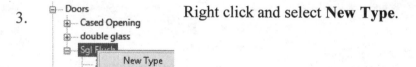 Name the new type: **38″ x 84″**.

5. 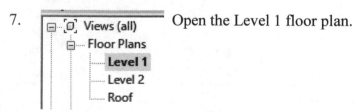 Highlight the **38″ x 84″** door type.
    Right click and select **Type Properties**.

6. | Dimensions | |
    |---|---|
    | Height | 7' 0" |
    | Width | 3' 2" |
    | Rough Width | |
    | Rough Height | |
    | Thickness | |

    Change the Width to **3′ 2″**.
    Press **OK**.

7. Open the Level 1 floor plan.

    - Views (all)
      - Floor Plans
        - **Level 1**
        - Level 2
        - Roof

8. 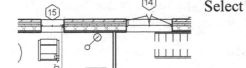 Select a door in the graphics window.

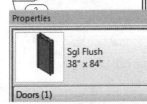

9. **Properties** In the Properties palette:

Select **38″ x 84″** from the drop-down list.

Sgl Flush
38″ x 84″

Note that the door updates.

Doors (1)

10. Close without saving.

Exercise 10-9
Adding a Shared Parameter to a View Label

Estimated Time: 60 minutes
File: basic_project.rvt

This exercise reinforces the following skills:

- ❑ View Labels
- ❑ Shared Parameters
- ❑ View Properties

1. Open *basic_project.rvt*.

2. Activate the Manage ribbon.

 Select **Shared Parameters**.

Shared
Parameters   *If you cannot locate the text file, browse for custom parameters.txt in the exercise files.*

3. Parameter group: Select the **General** parameter group from the drop-down.

General

4. Parameters Select **New** under Parameters.

New...

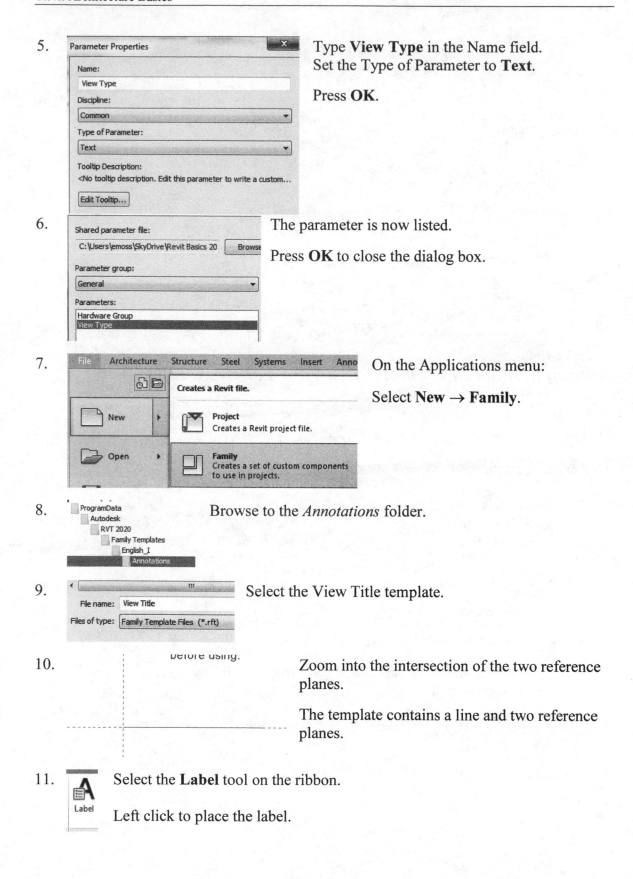

5.

Type **View Type** in the Name field.
Set the Type of Parameter to **Text**.

Press **OK**.

6.

The parameter is now listed.

Press **OK** to close the dialog box.

7.

On the Applications menu:

Select **New → Family**.

8.

Browse to the *Annotations* folder.

9.

Select the View Title template.

10.

Zoom into the intersection of the two reference planes.

The template contains a line and two reference planes.

11.

Select the **Label** tool on the ribbon.

Left click to place the label.

12. 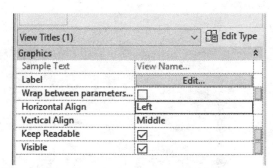   Highlight **View Name** from the list on the left pane.

Select add parameter to label (middle icon) to add the parameter to the right panel.

Press **OK**.

13. 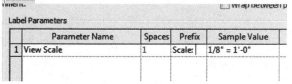   Locate the View Name above the line.

14. In the Properties panel:

Set the Horizontal Alignment for the label to **Left.**

You must have the label selected in order to see its properties.

15. Select the **Label** tool on the Create ribbon.

Left click to place the label.

16. 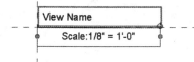   Select **View Scale** to add the parameter to the label.

Type **Scale:** in the prefix field.
Press **OK**.

17. 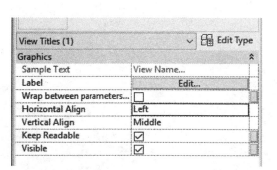   Place the Scale label below the view name.

18. In the Properties panel:

Set the Horizontal Alignment for the label to Left.

You must have the label selected in order to see its properties.

19. Select the **Label** tool on the Create ribbon.

Left click to place the label.

20. 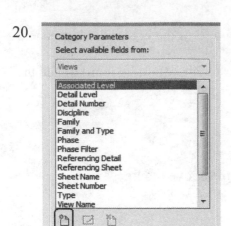 Select the **New Parameter** tool at the bottom left of the dialog.

21.  Press **Select**.

22. 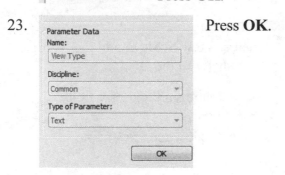 Select the **General** parameter group.

Select the **View Type**.

Press **OK**.

23. Press **OK**.

24. View Type is now listed as an available parameter.

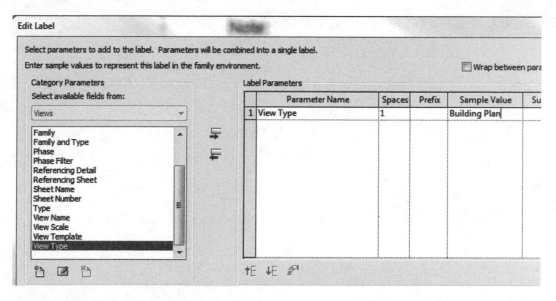

Add it to the Label Parameters box.
Type a Sample Value, like Building Plan.

Press **OK**.

25. 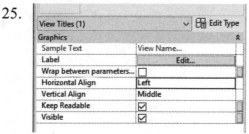 Set the Horizontal Align to **Left.**

26.

Building Plan

View Name

Scale: 1/8" = 1'-0"

This is what you should have so far.

Select the label for the View Type.

27. 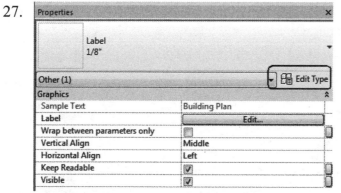 Press **Edit Type**.

28. Duplicate... Press **Duplicate.**

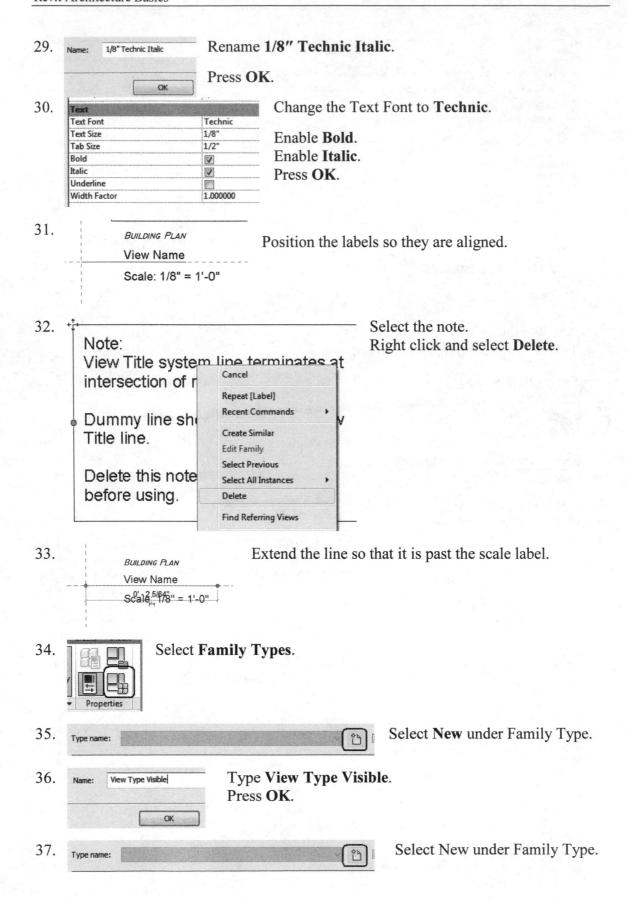

29. Name: 1/8" Technic Italic

Rename **1/8″ Technic Italic**.

Press **OK**.

30.

Text	
Text Font	Technic
Text Size	1/8"
Tab Size	1/2"
Bold	☑
Italic	☑
Underline	☐
Width Factor	1.000000

Change the Text Font to **Technic**.

Enable **Bold**.
Enable **Italic**.
Press **OK**.

31. BUILDING PLAN

View Name

Scale: 1/8" = 1'-0"

Position the labels so they are aligned.

32. Note:

View Title system line terminates at intersection of

Dummy line sh... Title line.

Delete this note before using.

Cancel
Repeat [Label]
Recent Commands ▶
Create Similar
Edit Family
Select Previous
Select All Instances ▶
Delete
Find Referring Views

Select the note.
Right click and select **Delete**.

33. BUILDING PLAN

View Name

0' 2 5/64"
Scale: 1/8" = 1'-0"

Extend the line so that it is past the scale label.

34. Properties

Select **Family Types**.

35. Type name:

Select **New** under Family Type.

36. Name: View Type Visible

OK

Type **View Type Visible**.
Press **OK**.

37. Type name:

Select New under Family Type.

38. Name: | View Type Hidden |

OK

Type **View Type Hidden**.
Press **OK**.

39.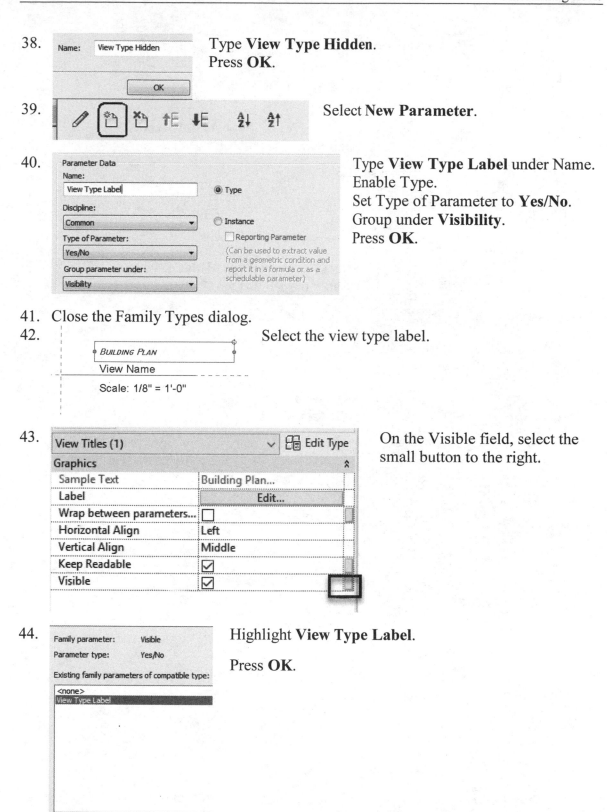

Select **New Parameter**.

40. Type **View Type Label** under Name.
Enable Type.
Set Type of Parameter to **Yes/No**.
Group under **Visibility**.
Press **OK**.

41. Close the Family Types dialog.

42. Select the view type label.

43. On the Visible field, select the small button to the right.

44. Highlight **View Type Label**.

Press **OK**.

45. Select **Family Types**.

46. 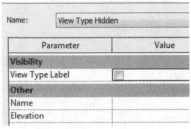 Select the **View Type Visible** type.
Enable the **View Type Label**.

47. 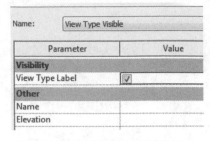 Select the **View Type Hidden** type.
Disable the **View Type Label**.

Press **OK**.

48. 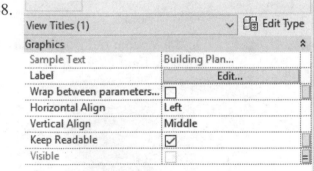 Note that in the Properties pane Visible is unchecked.

49. 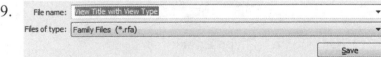 Save as *View Title with View Type.rfa.*
Press **Save**.

50.  Load into the *basic_project.rvt* and close.

51. 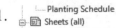 Activate Sheet: **A101-Site Plan**.

52. 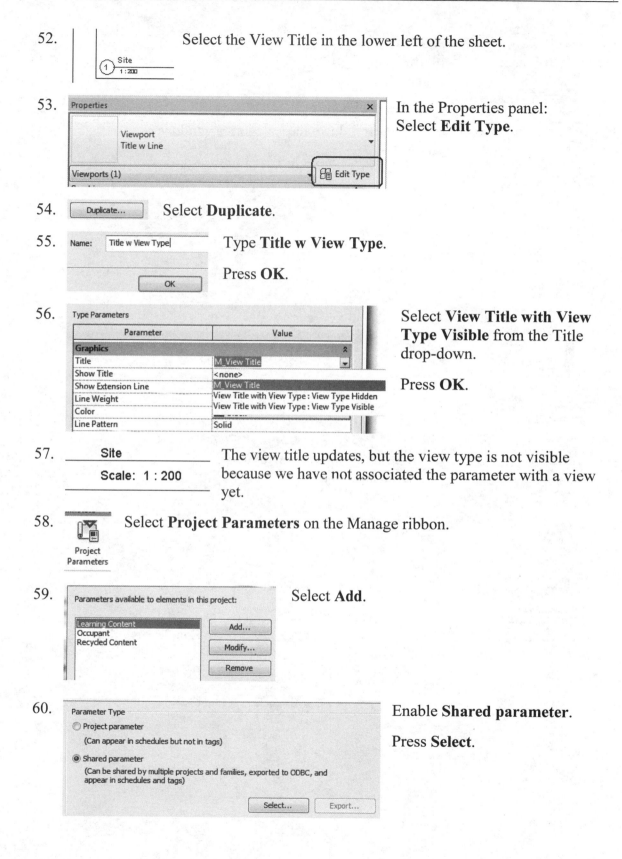 Select the View Title in the lower left of the sheet.

53. In the Properties panel:
Select **Edit Type**.

54. Select **Duplicate**.

55. Type **Title w View Type**.

Press **OK**.

56. Select **View Title with View Type Visible** from the Title drop-down.

Press **OK**.

57. The view title updates, but the view type is not visible because we have not associated the parameter with a view yet.

58. Select **Project Parameters** on the Manage ribbon.

59. Select **Add**.

60. Enable **Shared parameter**.

Press **Select**.

61. Select **General**.
     Highlight **View Type**.
     Press **OK**.

62. Scroll down the Categories window.

     Place a check next to **Views**.

63. Set the Group parameter under:
     Identity Data.

64. Enable **Instance**.

     Press **OK**.

65. Activate **Level 1** under Floor Plans.

66.

Sun Path	☐
Identity Data	
View Template	<None>
View Name	Level 1
Dependency	Independent
Title on Sheet	
Sheet Number	A102
Sheet Name	Plans
Referencing Sheet	A103
Referencing Detail	1
View Type	Building Plan

Scroll down to **Identity Data**. Type **Building Plan** in the View Type field.

67.

Floor Plans
   Level 1
   Level 2
   Site

Activate **Level 2** under Floor Plans.

68.

Sun Path	☐
Identity Data	
View Template	<None>
View Name	Level 2
Dependency	Independent
Title on Sheet	
Sheet Number	A102
Sheet Name	Plans
Referencing Sheet	A103
Referencing Detail	1
View Type	Building Plan
Extents	

Scroll down to **Identity Data**. Type **Building Plan** in the View Type field.

Note you can use the drop arrow to select a value.

69.

Floor Plans
   Level 1
   Level 2
   Site

Activate **Site** under Floor Plans.

70.

Identity Data	
View Template	<None>
View Name	Site
Dependency	Independent
Title on Sheet	
Sheet Number	A101
Sheet Name	Site Plan
Referencing Sheet	A103
Referencing Detail	1
View Type	Survey Plan

Scroll down to **Identity Data**. Type **Survey Plan** in the View Type field.

71.

   Planting Schedule
Sheets (all)
   A001 - Title Sheet
   A101 - Site Plan
   A102 - Plans
   A103 - Elevations/Sections
   A104 - Elev./Sec./Det.
   A105 - Elev./ Stair Sections
Families

Activate Sheet: **A101 - Site Plan**.

72.

SURVEY PLAN

Site
_____

Scale: 1 : 200
_____

The view title has updated.

If you don't see the view type, activate the view and verify that the View Type is set to Survey Plan.

73.

   A001 - Title Sheet
   A101 - Site Plan
   A102 - Plans
   A103 - Elevations/Sectio
   A104 - Elev./Sec./Det

Activate the **A102 - Plans** sheet.

74. **Viewport Title w Line**

    Viewport
    No Title
    Title w Line
    Title w View Type

Select the view title for each view and change to the **Title w View Type** using the Type Selector on the Properties pane.

75. BUILDING PLAN

    Level 2

    Scale: 1 : 100

The view title should update.

If you don't see the view type, activate the view and verify that the View Type is set to Building Plan.

Exercise 10-10
Managing Family Subcategories

Estimated Time: 60 minutes
File: Coffee Table.rfa

This exercise reinforces the following skills:

- ❏ Families
- ❏ OmniClass Number
- ❏ Family Subcategories
- ❏ Managing Family Visibility

1. Open the *coffee table.rfa*

2. Activate the Create ribbon.

   Select **Family Categories.**

3. 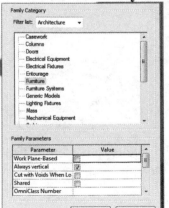 Highlight **Furniture**.

4. Select the Browse button next to OmniClass Number.

5. Locate the number for coffee tables.

Press **OK**.

The OmniClass Title will automatically fill in.

Press **OK**.

6. Select the table top.

Note in the Properties panel there is no subcategory assigned.

Subcategories can be used to manage level of detail (LOD).

7. Activate the **Manage** ribbon.
Select **Object Styles**.

8. Select **New** under Modify Subcategories.

9. Type **Table Top**.

Press **OK**.

10.

Modify Subcategories

New   Delete

Select **New** under Modify Subcategories.

11.

Name:

Table Legs

Subcategory of:

Furniture

OK

Type **Table Legs**.

12.

Category	Line Weight		Line Color	Line Pattern
	Projection	Cut		
Furniture	1		Black	
Hidden Lines	1		Black	Dash
Table Legs	1		Black	Hidden
Table Top	3		Blue	Solid

Change the Lineweight of the Table Top to **3**.

Change the Line pattern of the Table Legs to **Hidden**.

Change the Line Color of the Table Top to **Blue**.

13.

Category	Line Weight		Line Color	Line Pattern	Material
	Projection	Cut			
Furniture	1		Black		
Hidden Lines	1		Black	Dash	
Table Legs	1		Black	Hidden	Walnut
Table Top	3		Blue	Solid	Walnut

Assign the Walnut Material to both subcategories.

14.

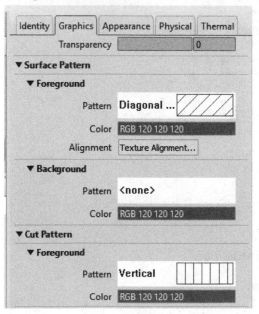

Modify the Walnut material to have a surface pattern and a cut pattern.

Press **OK**.

Close the Object Styles dialog.

15.

Material	Walnut
Identity Data	
Subcategory	Table Top
Solid/Void	Solid

Select the Table Top and assign the Table Top subcategory.

16.

Identity Data	
Subcategory	Table Legs
Solid/Void	Solid

Select the Table Legs and assign the Table Legs category.

17.

Views (all)
  Floor Plans
    Ref. Level

Switch to the **Ref Level** view.

18.

Graphic Display Options...

Wireframe
Hidden Line
Shaded
Consistent Colors
Realistic
Ray Trace

Switch to different displays and observe how the model appears.

Notice how the subcategories control the display of the two different element types.

19. Save the family as *Coffee Table 2.rfa*.

The following families are not cuttable and are always shown in projection in views:

- Balusters
- Detail Items
- Electrical Equipment
- Electrical Fixtures
- Entourage
- Furniture
- Furniture Systems
- Lighting Fixtures
- Mechanical Equipment
- Parking
- Planting
- Plumbing Fixtures
- Specialty Equipment

Families which are cuttable, where you can control the line style, color, and visibility in plan, RCP, section, and elevation views:

- Casework
- Columns
- Doors
- Site
- Structural Columns
- Structural Foundations
- Structural Framing
- Topography
- Walls
- Windows

Lesson 10 Quiz

True or False

1. Family subcategories control the visibility of family geometry.
2. A model family can be loaded using the Insert ribbon.
3. The Symbol tool lets you place 2D annotation symbols.
4. Dimensions in a family should be placed on sketches.
5. Shared parameters cannot be added to loadable families.
6. Shared parameters cannot be used in View Titles.
7. 2D and 3D geometry can be used to create families.
8. Geometry in families should be constrained to reference planes.
9. The ALIGN tool can be used to constrain geometry to a reference plane in the Family Editor.
10. Automatic dimensions are not displayed by default.

Multiple Choice

11. The following families are not cuttable (meaning the edges are always shown as solid):

    A. Furniture
    B. Walls
    C. Entourage
    D. Lighting Fixtures

12. The following families are cuttable (meaning line style, color and visibility can be controlled):

    A. Furniture
    B. Walls
    C. Entourage
    D. Doors

13. The intersection of the two default reference planes in a family template indicates:

    A. The elevation
    B. X marks the spot
    C. The insertion point
    D. The origin

14. A model family has the file extension:

    A. rvt
    B. rft
    C. rfa
    D. mdl

15. The graphical editing mode in Revit that allows users to create families is called:

    A. The Family Editor
    B. Family Types
    C. Family Properties
    D. Revit Project
    E. Draw

16. OMNI Class Numbers are assigned in this dialog:

    A. Type Properties
    B. Properties panel
    C. Family Types
    D. Family Categories

17. Use a wall-based template to create a model family if the component is to be placed:

    A. in or on a wall
    B. in or on a roof
    C. in or on a reference plane
    D. in a project

18. To insert an image into a title block family:

    A. Load a family
    B. Use the Insert ribbon
    C. Use the Image tool
    D. Use the Import tool

ANSWERS:
1) T; 2) T; 3) T; 4) F; 5) F; 6) F; 7) T; 8) T; 9) T; 10) T; 11) A, C, & D; 12) B & D; 13) C & D; 14) C; 15) A; 16) D; 17) A; 18) C

Revit Hot Keys

3F	Fly Mode	EH	Hide Element
3O	Object Mode	EL	Spot Elevation
3W	Walk	EOD	Override Graphics in View
32	2D Mode	EOG	Graphic Override by Element
AA	Adjust Analytical Model	EOG	Graphic Override by Element
AD	Attach Detail Group	EOH	Graphic Override by Element
AL	Align	EOT	Graphic Override by Element
AP	Add to Group	EP	Edit Part
AR	Array	ER	Editing Requests
AT	Air Terminal	ES	Electrical Settings
BM	Structural Beam	EU	Unhide Element
BR	Structural Brace	EW	Arc Wire
BS	Beam System	EX	Exclude
BX	Selection Box	FD	Flex Duct
CC	Copy	FG	Finish
CG	Cancel	FP	Flexible Pipe
CL	Structural Column	FR	Find/Replace Text
CM	Component	FS	Fabrication Settings
CN	Conduit	FT	Foundation Wall
CO	Copy	GD	Graphic Display Options
CP	Apply Coping	GL	Global Parameters
CS	Create Similar	GP	Group
CT	Cable Tray	GR	Grid
CV	Convert to Flex Duct	HC	Hide/Isolate Category
CX	Reveal Constraints Toggle	HH	Hide/Isolate Objects
DA	Duct Accessory	HI	Hide/Isolate Objects
DC	Check Duct Systems	HL	Hidden Line
DE	Delete	HR	Reset Temporary Hide/Isolate
DI	Aligned Dimension	HT	Help Tooltip
DF	Duct Fitting	IC	Isolate Category
DL	Detail Lines	KS	Keyboard Shortcuts
DM	Mirror - Draw Axis	LC	Lose Changes
DR	Door	LD	Loads
DT	Duct	LF	Lighting Fixture
EC	Check Electrical Circuits	LG	Link
EE	Electrical Equipment		
EG	Edit Group		

LI	Line
LL	Level
LO	Heating and Cooling Loads
LW	Linework
MA	Match Properties
MD	Modify
ME	Mechanical Equipment
MM	Mirror - Pick Axis
MP	Move to Project
MR	Multipoint Routing
MS	Mechanical Settings
MV	Move
NF	Conduit Fitting
OF	Offset
PA	Pipe Accessory
PB	Fabrication Part
PC	Check Pipe Systems
PI	Pipe
PF	Pipe Fitting
PN	Pin
PP or VP or Ctl+1	Properties
PP	Pin Position
PR	Properties
PS	Panel Schedules
PT	Paint
PX	Plumbing Fixture
R3	Define a new center of rotation
RA	Reset Analytical Model
RB	Restore Excluded Member
RC	Remove Coping
RD	Render in Cloud
RE	Resize\Scale
RG	Render Gallery
RH	Reveal Hidden Elements
RM	Room
RN	Reinforcement Numbers
RO	Rotate

RL	Reload Latest Worksets
RP	Reference Plane
RR	Render
RT	Room Tag
RW	Reload Latest Worksets
RY	Ray Trace
S	Split Walls and Lines
SA	Select All Instances in Entire Project
SB	Structural Floor
SC	Snap to Center
SD	Shading with Edges On
SE	Snap to Endpoint
SF	Split Face
SH	Snap to horizontal/vertical
SI	Snap to Intersection
SK	Sprinkler
SL	Split Element
SM	Snap to Midpoint
SN	Snap to Nearest
SO	Snaps OFF
SP	Snap to Perpendicular
SQ	Snap to Quadrants
SR	Snap to Remote Objects
SS	Turn Snap Override Off
ST	Snap to Tangent
SU	Sun Settings
SW	Snap to Workplane Grid
SX	Snap to Points
SZ	Snap to Close
TF	Cable Tray Fitting
TG	Tag by Category
TL	Thin Lines
TR	Trim/Extend
TW	Tab Window
TX	Text
UN	Project Units
UG	Ungroup
UP	Unpin

VH	Category Invisible
VI	View Invisible Categories
VG	Visibility/Graphics
VOG	Graphic Override by Category
VOH	Graphic Override by Category
VOT	Graphic Override by Category
VP	View Properties
VR	View Range
VU	Unhide Category
VV	Visibility/Graphics
WA	Wall
WC	Window Cascade
WF	Wire Frame
WN	Window
WT	Window Tile
ZA	Zoom to Fit
ZC	Previous Zoom
ZE	Zoom to Fit
ZF	Zoom to Fit
ZN	Zoom Next
ZO	Zoom Out (2X)
ZP	Zoom Previous
ZR	Zoom in region (window)
ZS	Zoom to Sheet Size (limits)
ZV	Zoom Out (2X)
ZX	Zoom to fit
ZZ	Zoom in region (window)
//	Divide Surface
Alt+F4	Close Revit
Alt+Backspace	Undo
Ctl+F4	Close Project file
Ctl+1	Properties
Ctl+	Activate Contextual Tab
Ctl+=	Subscript

Ctl+B	Bold
Ctl+C	Copy to Clipboard
Ctl+D	Toggle Home
Ctl+F	Project Browser Search
Ctl+I	Italic
Ctl+N	New Project
Ctl+O	Open a Project file
Ctl+P	Print
Ctl+Q	Close Text Editor
Ctl+S	Save a file
Ctl+U	Underline
Ctl+V	Paste from Clipboard
Ctl+W	Close window
Ctl+X	Cut to Clipboard
Ctl-Y	Redo
Ctl-Z	Undo
Ctl+Shift+=	Superscript
Ctl+Shift+A	All Caps
Ctl+Shift+Z	Undo
Ctl+Ins	Copy to Clipboard
F1	Revit Help
F5	Refresh Screen
F7	Spelling
F8	Dynamic View
F9	System Browser
F11	Status Bar

Notes:

About the Author

Autodesk
Certified Instructor

Elise Moss has worked for the past thirty years as a mechanical designer in Silicon Valley, primarily creating sheet metal designs. She has written articles for Autodesk's Toplines magazine, AUGI's PaperSpace, DigitalCAD.com and Tenlinks.com. She is President of Moss Designs, creating custom applications and designs for corporate clients. She has taught CAD classes at DeAnza College, Silicon Valley College, and for Autodesk resellers. She is currently teaching CAD at Laney College in Oakland. Autodesk has named her as a Faculty of Distinction for the curriculum she has developed for Autodesk products and she is a Certified Autodesk Instructor. She holds a baccalaureate degree in mechanical engineering from San Jose State.

She is married with three sons. Her older son, Benjamin, is an electrical engineer. Her middle son, Daniel, works with AutoCAD Architecture in the construction industry. Her youngest son is attending a local community college. Her husband, Ari, has a distinguished career in software development.

Elise is a third-generation engineer. Her father, Robert Moss, was a metallurgical engineer in the aerospace industry. Her grandfather, Solomon Kupperman, was a civil engineer for the City of Chicago.

She can be contacted via email at elise_moss@mossdesigns.com.

More information about the author and her work can be found on her website at www.mossdesigns.com.

Other books by Elise Moss
AutoCAD 2020 Fundamentals
Revit 2020 Certification Guide

Notes: